Founding Mothers and Others:

Women Educational Leaders During the Progressive Era

Edited by

Alan R. Sadovnik

and

Susan F. Semel

palgrave

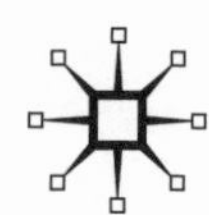

FOUNDING MOTHERS AND OTHERS

First published 2002 by PALGRAVE™
175 Fifth Avenue, New York, N.Y.10010 and
Houndmills, Basingstoke, Hampshire RG21 6XS.
Companies and representatives throughout the world

PALGRAVE is the new global publishing imprint of St. Martin's Press LLC Scholarly and Reference Division and Palgrave Publishers Ltd (formerly Macmillan Press Ltd).

ISBN 0–312–23297–7 hardback
ISBN 0–312–29502–2 paperback

Library of Congress Cataloging-in-Publication Data

Founding mothers and others : women educational leaders during the progressive era / edited by Alan R.
Sadovnik, Susan F. Semel.
p. cm.
Includes bibliographical references.
ISBN 0–312–23297–7 — ISBN 0–312–29502–2 (pbk.)
1. Women educators—United States—Biography. 2. Progressive education—United States—History—20th century. I. Sadovnik, Alan R. II. Semel, Susan F., 1941-

LA2311.F69 2002
370'.82'0973—dc21

[B]2001052311

A catalogue record for this book is available from the British Library.

Design by Letra Libre, Inc.

First edition: May 2002
10 9 8 7 6 5 4 3 2 1

Printed in the United States of America.

Contents

ACKNOWLEDGMENTS

This book began as a symposium on "Founding Mothers" at the American Educational Research Meetings in 1991. After this symposium, we turned our attention to many of the progressive schools founded by some of these women, which resulted in the publication of "*Schools of Tomorrow," Schools of Today: What Happened to Progressive Education* (Peter Lang, 1999). We continued, however, to work on female founders and other women educational leaders, culminating in this book.

During this period, we presented other parts of this work as a symposium at the American Educational Studies Association Meetings in 1996 and at the Joseph Cornwall Center for Metropolitan Studies at Rutgers University–Newark in 2001.

This book would not have been possible without the help of a number of individuals. At Palgrave Macmillan, Michael Flamini provided extraordinary editorial support and friendship. Amanda Johnson, Donna Cherry, and Sabahat Chaudhary all provided essential assistance in the production process.

Ian Steinberg provided exemplary technical assistance in the preparation of the final manuscript and index. The contributors provided timely commentary on the introduction and conclusion and in securing photographs, when available.

At Rutgers–Newark Bette Jenkins and LaChone McKenzie provided important and helpful secretarial support.

Most of all, we acknowledge the legacies of the remarkable women examined in this book.

CONTRIBUTORS

LOUISE ANDERSON ALLEN is Assistant Professor of Curriculum and Supervision at the University of North Carolina at Charlotte where she is also the graduate coordinator for the doctoral program in educational leadership. She was awarded the first postdoctoral fellowship offered by the Avery Research Center for African American History and Culture at the College of Charleston. Much of the research on Laura Bragg's involvement with the African American community, which is documented in Chapter 12 was supported by the postdoctoral fellowship. Her first book, *A Bluestocking in Charleston: The Life and Career of Laura Bragg,* was published in 2001. Her current research interests focus on southern women educational leaders of the Progressive Era and their curriculum reforms.

JACKIE M. BLOUNT is Associate Professor of Historical, Philosophical, and Comparative Studies in Education at Iowa State University (ISU). She has received the Thomas Urban Award for Education Research and the ISU Award for Early Achievement in Teaching. Her work has been published in such journals as the *Harvard Educational Review* and *The Review of Educational Research.* She has written *Destined to Rule the Schools: Women and the Superintendency, 1873–1995,* and is completing a book exploring the history of gender transgression, broadly defined, in public school employment. Currently she serves as Associate Dean of the ISU College of Education.

JODY HALL is Assistant Professor in the Department of Curriculum and Instruction at the University of Massachusetts at Boston. Her academic research interests are in the areas of history, psychology, and schooling. She is the author of "John Dewey and Pragmatism in the Primary School: a Thing of the Past?" in *Curriculum Studies* (1996) and "Psychology and Schooling: the Impact of Susan Isaacs and Jean Piaget on 1960s Science Education Reform" in *History of Education* (2000). Her practical research interest is elementary science education. She is the author of *Organizing Wonder: Making Inquiry Science Work in the Elementary School* (1998).

BLYTHE HINITZ is Professor of Elementary and Early Childhood Education and Coordinator of the P-3 Certification Program at the College of New Jersey. She currently holds a grant from Educational Equity Concepts and is participating in a group Fulbright Alumni Initiative Award. She is a founding member of the History Seminar of the National Association for the Education of Young Children. Blythe Hinitz is the author of *Teaching Social Studies to the Young Child: A Research and Resource Guide* (1992) and coauthor of *History of Early Childhood Education* (2000).

MARY HAUSER is Associate Professor of Education and Director of the Graduate Program in Education at Carroll College, Waukesha, Wisconsin. Her research interests are centered in the sociocultural context of education. She coedited *Intersections: Feminisms/Early Childhoods* (1998) and has published in several early childhood books and journals. She has presented research on Caroline Pratt in Readers Theatre format at AERA, ACEI, and at the Reconceptualizing Research in Early Childhood Education conferences and is currently working on a biography of Caroline Pratt.

CRAIG KRIDEL is Professor of Educational Foundations and Research at the University of South Carolina. His research interests focus on progressive education, educational biography, and documentary editing, and he is currently completing a history of the Eight Year Study and a biography of Harold Taylor. Kridel's publications include *Books of the Century, Writing Educational Biography, Teachers and Mentors,* and *The American Curriculum,* and he serves on the editorial board of the *History of Education Quarterly* and as columnist in the *Journal of Curriculum Theorizing.*

GAIL L. KROEPEL is Senior Professor of Computer Information Systems at DeVry Institute of Technology in Chicago and has been a computer consultant for Midwest businesses and corporations. Her scholarly writings have focused on the life and contributions of the progressive educator Flora J. Cooke and the philanthropist Anita McCormick Blaine; scholarly articles by Dr. Kroepel have been accepted for publication in the *American Educational History Journal* and the *Proceedings of the Midwest Philosophy of Education Society,* and she has presented papers at the History of Education Society, the Midwest History of Education Society, and the Midwest Philosophy of Education Society.

SUSAN MCINTOSH LLOYD taught history, music, and urban studies for 29 years at Phillips Academy in Andover, Massachusetts, and the same disciplines more briefly at both the elementary and college level. She served for seven years as a founding member of the National Board for Professional Teaching Standards. Since taking early retirement, she has been secretary of the Tinmouth, Vermont, School Board, chamber music coach and secretary-director for Rutland County's youth orchestra, and chair/editor for an international chamber music society. Most of her scholarship has focused on the history of American education. She has written two books, *A Singular School: Abbot Academy, 1828–1973,* University Press of New England, 1979, and *The Putney School: A Progressive Experiment,* Yale University Press, 1987, as well as numerous articles; and, with her Urban Studies students, she has written *Growing Up in Lawrence, Massachusetts,* I (1985) and II (1998).

JOSEPH W. NEWMAN is Professor and Chair of Educational Leadership and Foundations at the University of South Alabama. The author of *America's Teachers: An Introduction to Education,* 4th ed. (2002), he focuses his research on teachers, teacher organizations, and education in the South. Newman is active in the History of Education Society and Division F, History and Historiography, of the American Educational Research Association. Now serving as an officer of the American Educational Studies Association, he will become president in 2003.

KATHERINE C. REYNOLDS is Associate Professor of Higher Education Studies, and director of the Museum of Education, at the University of South Carolina. Her historical studies have resulted in several books, including *Park City: A History* (Weller Press), *Vi-*

sions and Vanities: John Andrew Rice of Black Mountain College (Louisiana State University Press), and *Carolina Voices: 200 Years of Student Experiences* (University of South Carolina Press). Her reviews and articles have appeared in numerous journals, including *Education and Culture, The Journal of Educational Thought, The North Carolina Literary Review, History of Education Quarterly,* and others.

KATE ROUSMANIERE is Associate Professor in the Department of Educational Leadership at Miami University, Oxford, Ohio. Her research interests center on the history and politics of teachers and methodological questions in the social history of education. Her publications include *City Teachers: Teaching and School Reform in Historical Perspective* (1997), *Silences and Images: The Social History of the Classroom* (coedited with Ian Grosvenor and Martin Lawn, 1999), and numerous publications on the history of teachers and historical research methods. She is currently writing a biography of Margaret Haley, the turn-of-the-century leader of America's first teachers' union.

ALAN R. SADOVNIK is Professor of Education and Sociology at Rutgers University, Newark, New Jersey, and Chair of its Department of Education. He is the author of *Equity and Excellence in Higher Education* (1995); coauthor of *Exploring Education: An Introduction to the Foundations of Education* (1994, 2001); editor of *Knowledge and Pedagogy: The Sociology of Basil Bernstein* (1995); and coeditor of *Exploring Society* (1987), *International Handbook of Educational Reform* (1992), *Implementing Educational Reform: Sociological Perspectives on Educational Reform* (1995), *"Schools of Tomorrow," Schools of Today: What Happened to Progressive Education* (1999), and *Sociology and Education: An Encyclopedia* (2001). He received the Willard Waller Award in 1993 from the American Sociological Association's Sociology of Education Section for the outstanding article published in the field, and American Educational Studies Association Critics Choice Awards in 1995 for *Knowledge and Pedagogy* and in 2000 for *"Schools of Tomorrow,"* He is coeditor, with Susan F. Semel, of the *History of Schools and Schooling* series at Peter Lang Publishing.

SUSAN F. SEMEL is Associate Professor of Education at the City College of New York. She is the author of *The Dalton School: The Transformation of a Progressive School* (1992); coauthor of *Exploring Education: An Introduction to the Foundations of Education* (1994, 2001); and coeditor of *International Handbook of Educational Reform* (1992); and *"Schools of Tomorrow," Schools of Today: What Happened to Progressive Education* (1999). She received American Educational Studies Association Critics Choice Awards in 1994 for *The Dalton School* and in 2000 for *"Schools of Tomorrow,"* She serves on the editorial board of *History of Education Quarterly.*

SAM STACK is Associate Professor of Social and Cultural Foundations of Education, Department of Advanced Educational Studies, West Virginia University, Morgantown. His academic interests include history of progressive education, educational theory, and educational biography. His publications have appeared in various monographs and journals, including *Journal of Thought, Vitae Scholasticae, Lock Haven International Review, Journal of Philosophy and Social Science, History of Education Quarterly,* and *Educational Studies.*

WAYNE J. URBAN is Regents' Professor and Acting Chair of the Department of Educational Policy Studies and Professor of History at Georgia State University, Atlanta. He is the author of *Gender, Race, and the National Education Association: Professionalism and Its Limitations* (2000), *More Than the Facts: The Research Division of the National Education Association: 1922–1997* (1998), *Black Scholar: Horace Mann Bond, 1904–1972* (1992), and *Why Teachers Organized* (1982). He is Secretary of the International Standing Conference for the History of Education and a past president of the History of Education Society and the American Educational Studies Association.

Charlotte Hawkins Brown (center, back row) with teachers at Palmer Memorial Institute (approx. 1918). Courtesy of North Carolina Office of Archives and History. Reprinted with Permission.

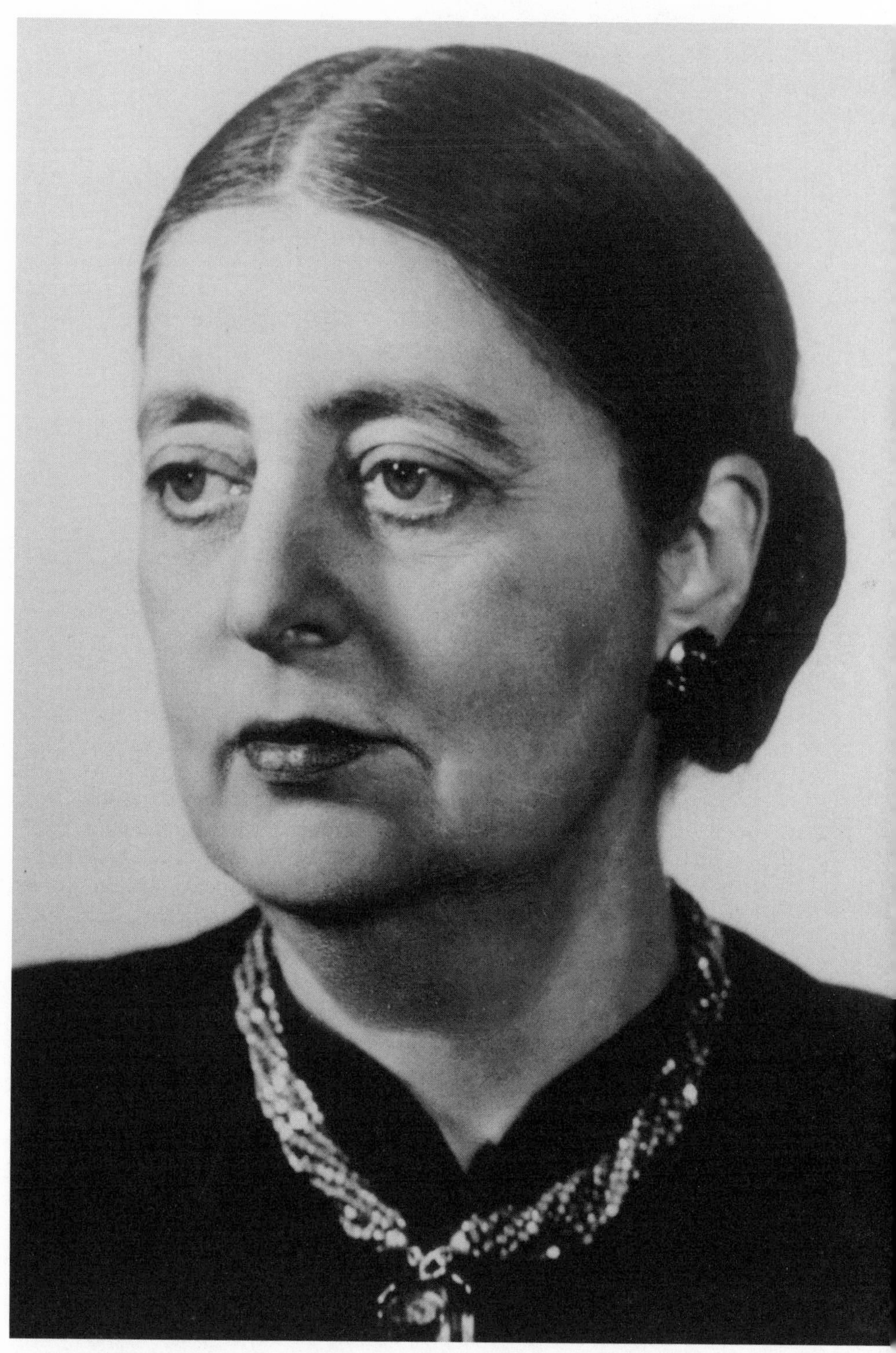

above: Margaret Naumburg, n.d. Courtesy of Thomas Frank, M.D., personal collection.
left: Marietta Johnson, n.d. Used by permission of the Marietta Johnson Museum, 440 Fairhope Ave, Fairhope, Ala. 36532.

left: Helen Parkhurst, n.d. Used by permission of the Dalton School, New York City.

above: Caroline Pratt, n.d. Used by permission of City and County School Archives.
right: Carmelita Hinton, n.d. Used by permission of the Putney School Alumni Office, Putney, Vt.

above: Flora Cooke. Used by permission of the Francis W. Parker Archives, Chicago, Ill.
right: Margaret A. Haley, n.p., ca. 1910–1915. Used by with permission from the Chicago Historical Society, Prints & Photographs Department, 1601 N. Clark Street, Chicago, Ill. 60614–6099.

Ella Flagg Young, (n.p.); (n.d), by Louis Betts (American). Used with permission from the Chicago Historical Society, Prints & Photographs Department, 1601 N. Clark Street, Chicago, Ill. 60614–6099.

Margaret Willis traveling in Egypt, c 1946, used by permission of the Museum of Education, University of South Carolina.

INTRODUCTION

ALAN R. SADOVNIK AND SUSAN F. SEMEL*

Interest in progressive education and feminist pedagogy has gained a significant following in current educational reform circles. Given this interest, the purpose of this book is to provide short educational biographies of a number of women educational leaders during the Progressive Era. These include both female founders of progressive schools in the early twentieth century and the schools that they founded, as well as a number of women leaders of educational organizations and movements, including public school districts, teacher unions and museums.

In the early part of the twentieth century, progressive reformers tended to concentrate their efforts in public education, applying scientific management techniques to the administration of schools, reforming curriculum, and creating secondary vocational schools.[1] However, as historian Lawrence A. Cremin indicated, a second trend developed during this period, as many progressive educators began to focus "on a select group of pedagogical innovative independent schools catering principally to middle class children."[2] A number of independent, progressive schools developed during this period.[3] The majority of them were founded by women during a time when "a great divide in the history of progressive education"[4] was occurring—one in which the thrust toward "social reformism was virtually eclipsed by the rhetoric of child-centered pedagogy."[5]

Educators commonly referred to these schools as "child-centered." They were often founded by female practitioners "spurred by the revolt against the harsh pedagogy of the existing schools and by the ferment of change and new thought of the first two decades of the twentieth century."[6] Although historians tend to group these child-centered schools together, each has a distinct philosophy and practice according to the particular vision of its founder. For example, the City and Country School, founded by Caroline Pratt, emphasized the idea of self-expression and growth through play—in particular, through play with wooden blocks.[7] Another school, Walden School, founded by Margaret Naumburg, who was heavily influenced by Freudian psychology, emphasized the notion of "individual transformation." Under the leadership of Naumburg's sister, Florence Cane, the school encouraged "children to paint exactly what they felt impelled to paint."[8] Other examples include the Bank Street School, founded by Lucy Sprague Mitchell,[9] the Dalton School, founded by Helen Parkhurst, which introduced the Dalton Plan,[10] and the Lincoln School, founded by Abraham Flexner, which became a laboratory school for Teachers College, Columbia University.[11] Outside of New York City, where the aforementioned four schools were located, other examples of progressive education sprang up. Among these were the Organic School in Fairhope, Alabama, founded by Marietta Johnson[12]; the Francis W. Parker School in Chicago, founded by one of the early pioneers of progressive pedagogy, Colonel Francis W. Parker;[13] the Putney School, a boarding school in Putney, Vermont, founded by Carmelita Hinton;[14] and the Shady Hill School in Cambridge, Massachusetts, founded by Edward Yeomans.[15]

Whereas these child-centered progressive schools were almost all independent private schools, public education was dominated by the social engineering strand of progressivism. From the transformation of the high school from an exclusively academic institution at the turn of the century to a host for life adjustment functions by the 1930s, to the social class- and race-based tracking systems that separated academic and vocational education, public progressive education from the 1930s to the 1960s stressed life adjustment rather than intellectual functions and often helped to reproduce rather than ameliorate social class, race, and gender inequalities. Progressive experiments in public schools did exist, for example, in Gary, Indiana, and Winnetka, Illinois.[16] Although public education during this period came to be dominated by administrative progressivism and male administrators, some women educators did have an influence in the public sector as principals, superintendents, and union leaders.

In the 1930s and early 1940s, a third strain of progressivism, social reconstructionism, had a limited effect on the school practice. Based on the work of Harold Rugg, George Counts, and Kenneth Benne, as well as the journal *Social Frontier,* a number of schools adopted a philosophy that espoused a radical reconstruction of the social order, especially with regard to inequalities.[17] Examples of these schools include Malumet, Hessian Hills, Arthurdale, and Downtown Community School.[18] Although they represented an important attempt to use schools to change society, they neither had a wide following nor a lasting influence.

Progressive education transcended a narrow focus on schools alone and extended its reach into communities and other educative institutions, including urban school systems, museums, teacher unions, and universities. This book, therefore, expands the traditional view of progressivism by broadening its focus to include education and educative institutions in the widest sense.

This book examines a number of women educational leaders during the Progressive Era—some of whom founded progressive schools, others who were leaders of schools, unions, museums and university programs—in an attempt to understand their particular visions of progressive education and the implications of their work for current educational reformers. One of the main objectives of this book is to document the contributions of women to the Progressive Education movement. Although much has been written about progressive education, there has been little codified in the literature on the contributions of women leaders to this movement.

Additionally, when appropriate, the chapters address particular administrative styles of these women and relate them to current thinking about feminist pedagogy and administrative styles. A branch of feminist theory suggests that there are significant differences between male and female administrative styles. Based in part on Gilligan's research, feminists argue that female pedagogy and leadership are more humane, less authoritarian, more democratic, and more concerned with caring and relationships than abstract goals. Moreover, their qualities are viewed as being at odds with the male-defined model of school administration that emerged in the early twentieth century.[19] Through an exploration of the schools and their female founders, the book illustrates whether or not these leaders' administrative styles and behaviors lend support to these feminist claims.

In *"Schools of Tomorrow," Schools of Today: What Happened to Progressive Education,* we argued that there are significant lessons to be learned from the history of early progressive schools for contemporary educational reform. This is no less the case for the lessons to be learned from women educational leaders during the Progressive Era. Over the past two decades a significant body of research on the importance of school leadership for educational reform has highlighted the critical role of school leadership.[20] The chapters in this book provide important case studies on educational leadership that provide lessons for contemporary school reformers, especially in the burgeoning charter school movement in which the cult of personality has proven so critical to charter school success or failure.[21]

ORGANIZATION OF THE BOOK

Although we did not ask contributors to follow a standard format, the focus of each chapter is on the women's lives and their educational careers (and in the case of founders, the school that they founded). Within this context, each chapter focuses on the following:

1. The important biographical aspects of the woman's career.
2. The relationship between her life and the times in which she lived.
3. Her educational philosophy and accomplishments, especially with respect to the school that she founded or movement of which she was a part. The influence of others (i.e., Dewey, Montessori, etc.).
4. Her lasting effects on education, her school, or movement. What happened to her school or movement after her departure or death?
5. Her leadership qualities and style and, if appropriate, whether or not this style supports feminist theories of leadership.

The book is organized into two parts. The first part examines the lives of female founders of schools and is organized chronologically with respect to the founding of the school. Part II chronicles the lives of other educational leaders. In Chapter One, Katherine C. Reynolds examines the life of Charlotte Hawkins Brown, founder of the Palmer Institute, a school for African American children, founded in Sedalia, North Carolina, in 1902. In Chapter Two, Joseph W. Newman looks at the life of Marietta Johnson, the founder of the Marietta Johnson School of Organic Education, founded in Fairhope, Alabama, between 1904 and 1907. In Chapter Three, Blythe Hinitz examines the life of Margaret Naumburg, founder of Walden School, founded in New York City in 1914. In Chapter Four, Mary Hauser examines the life of Caroline Pratt, founder of the City and Country School, founded in New York City also in 1914. In Chapter Five, Susan F. Semel examines the life of Helen Parkhurst, who founded the Dalton School, in New York City in 1919. In Chapter Six, Sam Stack chronicles the life of Elsie Ripley Clapp, who founded the Arthurdale Schools, in rural West Virginia in 1934. In Chapter Seven, Susan McIntosh Lloyd examines the life of Carmelita Chase Hinton, founder of the Putney School, in Putney, Vermont, in 1935.

The second part looks at other educational leaders. In Chapter Eight, Gail L. Kroepel examines the life of Flora J. Cooke, the first principal of the Francis W. Parker School in Chicago, founded in 1901 by Colonel Francis Parker. In Chapter Nine, Kate Rousmaniere chronicles the life of Margaret Haley, who was head of the Chicago Teachers' Federation starting at the turn of the century. In Chapter Ten, Jackie M. Blount examines the life of Ella Flagg Young, who became superintendent of the Chicago public schools in 1909, the first women elected to such an urban school superintendency. In Chapter Eleven, Louise Anderson Allen chronicles the life of Laura Bragg, founder of the Charleston Museum in South Carolina in 1909 and inventor of the Bragg box, which became an important feature of both museum and community outreach education. In Chapter Twelve, Wayne J. Urban examines the life of Charl Williams, who was elected the first Southern woman president of the National Education Association in 1921. In Chapter Thirteen, Craig Kridel examines the life of Margaret Willis, a teacher and educational leader at the Ohio State University School, from 1932 to 1967. In Chapter Fourteen, Jody Hall looks at the life of Susan Isaacs, Director of the Malting House School in Cambridge, England, from 1924 to 1927 and head of the Department of Child Development at the Institute of Education, University of London, from 1933 to 1943. She had a profound influence on the development of British progressive education and through Lillian Weber and Deborah Meier on U.S. progressive education from the 1960s to the present.

Chapter Fifteen provides an analysis of some the issues that emerge through an examination of the lives and works of these women.

NOTES

* Parts of this introduction are adapted from the introduction to *"Schools of Tomorrow," Schools of Today: What Happened to Progressive Education* (New York: Peter Lang, 1999).

1. David Tyack and Elisabeth Hansot, *Managers of Virtue: Public School Leadership in America, 1920–1980* (New York: Basic Books, 1981).
2. Lawrence A. Cremin, *American Education: The Metropolitan Experience* (New York: Harper and Row), p. 229.
3. For a detailed discussion of these schools and historical case studies, see Semel and Sadovnik, eds., *"Schools of Tomorrow," Schools of Today: What Happened to Progressive Education.*
4. Lawrence A. Cremin, *The Transformation of the School: Progressivism in American Education, 1876–1975* (New York: Random House, 1961), p. 179.
5. Ibid., p. 181.
6. O.F. Kraushaar, *American Nonpublic Schools: Patterns of Diversity* (Baltimore and London: Johns Hopkins University Press, 1972), p. 81.
7. For a full discussion of Caroline Pratt's philosophy, see her book *Experimental Practice in the City and Country School* (New York: E.P. Dutton and Co., 1924).
8. Cremin, *The Transformation of the School,* p. 213.
9. Joyce Antler, *Lucy Sprague Mitchell: The Making of a Modern Woman* (New Haven: Yale University Press, 1987).
10. Susan F. Semel, *The Dalton School: The Transformation of a Progressive School* (New York: Peter Lang, 1992).
11. Cremin, *The Transformation of the School,* pp. 280–286.
12. Joseph W. Newman, "Experimental School, Experimental Community: The Marietta Johnson School of Organic Education in Fairhope, Alabama." In Semel and Sadovnik, *"Schools of Tomorrow," Schools of Today,* pp. 67–102.

13. Marie Stone, ed. *Between Home and Community: Chronicle of the Francis W. Parker School* (Chicago: Francis W. Parker School, 1976) idem, *The Progressive Legacy: Chicago's Francis W. Parker School (1901–2001)* (New York: Peter Lang, 2001).
14. Susan Lloyd, *The Putney School: A Progressive Experiment* (New Haven: Yale University Press, 1987).
15. Edward Yeomans, *The Shady Hill School: The First Fifty Years* (Cambridge, Mass.: Windflower Press, 1979).
16. For a discussion of the Gary schools see Ronald Cohen, *Children of the Mill: Schooling and Society in Gary, Indiana, 1906–1960* (Bloomington: Indiana University Press, 1990), and Ronald Cohen and Raymond Mohl, *The Paradox of Progressive Education: The Gary Plan and Urban Schooling* (Port Washington, N.Y.: Kennidat Press, 1979); for a discussion of the Winnetka schools, see Carleton W. Washburne and Sidney P. Marland, *Winnetka: The History and Significance of an Educational Experiment* (Englewood Cliffs, N.J.: Prentice Hall, 1963).
17. See Michael James, *Social Reconstruction Through Education* (Norwood, N.J.: Ablex, 1995).
18. Susan F. Semel and Alan R. Sadovnik, "Lessons from the Past: Individualism and Community in Three Progressive Schools," *Peabody Journal of Education* (Summer 1995): 56–84.
19. See Carol Gilligan, *In a Different Voice* (Cambridge, Mass.: Harvard University Press, 1982); Nell Noddings, *Caring: A Feminist Approach to Ethics & Moral Education* (Berkeley: University of California Press, 1984) and *The Challenge to Care in School: An Alternative Approach to Education* (New York: Teachers College Press, 1992). For more recent applications and a discussion of webs of inclusion, see Judy B. Rosener, "Ways Women Lead," *Harvard Business Review* (November–December, 1990): 119–125, and Joyce Fletcher, *Disappearing Acts: Gender, Power, and Relational Practice at Work* (Cambridge, Mass.: MIT Press, 1999) idem, *Inside Women's Power: Learning from Leaders* (Wellesley, Mass.: Wellesley Centers for Women Research Report, 2001), for a full discussion of this research.
20. For a summary of this research, see National Center for Educational Statistics, *School Quality: An Indicators Report* (Washington, D.C.: U.S. Department of Education, 2000).
21. Bruce Fuller, *Inside Charter Schools* (Cambridge, Mass.: Harvard University Press, 2000).

CHAPTER ONE

CHARLOTTE HAWKINS BROWN AND THE PALMER INSTITUTE

KATHERINE C. REYNOLDS

The educational work of Charlotte Hawkins Brown (1883–1961) was an ill fit for progressive referents like "child centered" or "classroom community." On the surface, the Alice Freeman Palmer Memorial Institute (Palmer Institute), the school Brown founded for African Americans in rural Sedalia, North Carolina, appeared rule bound in method and controlling in philosophy. Dress codes, schedules, and student work programs loomed large at the all-grades private boarding school. However, when progressive education is viewed in terms of social development in the context of regional and economic realities, Brown's efforts can be located within reform endeavors embraced by Progressive Era educators. Her experiences and contributions, as well as those of a number of other African Americans seeking schooling for blacks in an overwhelmingly hostile environment, reflect Kliebard's notion that progressive educators could be found in "reform subgroups" of similar ideologies, rather than in a large movement under a single coherent umbrella.[1]

When Brown opened the Palmer Institute in 1902, provisions for the education of southern black citizens were nearly nonexistent. The South was a place of grinding poverty and rural isolation, showing only sparse signs of recovery from the economic devastation of the Civil War and Reconstruction. Illiteracy among the region's citizens over 10 years old was 12 percent among whites and 48 percent among blacks. Schools for African American students were very separate and not at all equal, typically open only a few months each year and extending only through the elementary grades. Among all southern schools at the turn of the century, attendance per child averaged only 72 days a year; and an estimated one-third of the South's school age children did not enroll. Teacher pay, whether measured daily or annually, was shockingly low throughout the region, and lower still among African American teachers who on the average earned one-third to one-fourth the comparable wages of their white counterparts.[2] The situation was exacerbated by child labor and a high proportion of citizens under 20 years of age.

As summarized by Joseph Kett, regional differences confronted by progressive-minded reformers meant that "southern Progressives had to swim in a different ocean, to confront a radically different society than the one northern Progressives lived in."[3] Without doubt, the undercurrent that shaped the swells and tides of that ocean was racism. Progressive educational reform in the mode of Dewey and Kilpatrick was rare in the South and was, as C. Vann Woodward noted, "progressive for whites only."[4] In some instances, funding in public schools for blacks was reduced when increases were necessary to support reform implementation in white schools. Such situations prompted William A. Link to observe that "school reform—in more obvious ways than other varieties of reform—exposed a fatal flaw in Southern Progressivism."[5]

EARLY INFLUENCES

Charlotte Hawkins Brown was a fortunate exception among southern blacks who were her contemporaries and in comparison to the students she would eventually teach. Her family, guided by a strong matriarch, Caroline (called Carrie) Hawkins, was somewhat above the poverty line. When Charlotte was not quite six years old, they joined the surging black migration north, settling in Boston, where Charlotte attended high school and normal school.

Carrie Hawkins' ingenuity and determination put her daughter on a path that was marked by religious fervor, social correctness, and educational appreciation. Born to a fair-skinned, half-white plantation slave, Carrie lived in the elegant Raleigh home of her mother's white sister, Jane Hawkins, after emancipation. Aunt Jane enjoyed attempting to imbue her with social graces and academic abilities, but the relationship ended when niece Carrie became pregnant out of wedlock at age 16. Carrie gave birth to her daughter Charlotte (called Lottie) in her own mother's house in Henderson, North Carolina, where Charlotte Hawkins Brown spent her earliest years.[6]

Although the extended Hawkins family lived in the South at a time replete with institutionalized racism, Jim Crow laws, employment discrimination, and frequent lynchings, Carrie Hawkins, her mother, and siblings managed to establish a substantial five-room house that was home to various members of the family, as necessary. With the family's penchant for hard work and social acceptability, the spotlessly clean home included a pedal organ in the front hall and copies of fine works of art throughout. Nevertheless, the laws that forced racial segregation in the post-Reconstruction Era meant limited opportunities and low wages for hard-working southern blacks who sought economic security. Inevitably, many heard of better opportunities in the North, and those determined to seek social and economic betterment moved.

A number of members of the Hawkins family moved to Boston in 1888, including Charlotte Hawkins Brown, her mother, grandmother, brother, and several aunts. They soon settled in Cambridge, and Carrie Hawkins married another native of North Carolina, Nelson Willis, a day laborer who became a stepfather to her children during a union that lasted 35 years.[7]

Carrie Hawkins Willis arranged the family's Cambridge household around three areas of strong commitment: education, religion, and social acceptability. The foundation for all three was a strong work ethic, with the Willis family taking in everything from boarders to laundry and with young Charlotte's contribution of ironing and baby-sitting. Charlotte attended Allston Grammar School and Cambridge English High School. Most of her fellow students in the school's rigorous general, commercial, and scientific courses were white, and the principal of Cambridge English High School was an honors graduate of Brown University. In later years, Charlotte was quick to recall, "No teacher or student ever made me uncomfortable in any way. I knew nothing of segregation."[8]

Charlotte Brown's acquisition of social graces and moral values continued with voice and piano lessons and regular attendance at Cambridge's all black Union Baptist Church. She organized the church kindergarten and, admired for her oratorical ability, often served as a youth speaker. Dress, manners, and an artfully decorated home soon became as important to Brown as they were to her mother. At age 14, Brown was in the audience at another Baptist church when she heard Booker T. Washington lecture on "the Negro in the South." Washington left an impression on her, marked by selfless humanitarianism and the ability to suspend animosity in the face of racial bigotry. He also informed his audience of the great need for well-educated blacks in the North to come back to their southern roots and help those of their own race.[9]

During her senior year at Cambridge English High School, Charlotte tended the baby of a friend of one of her teachers' for several hours each afternoon. While pushing the stroller and reading *Virgil* for a school assignment at the same time, she was stopped by a woman who commented on the cute baby and asked her where she went to school. Soon after, Brown's high school principal told her that Alice Freeman Palmer, the revered Wellesley College president, had asked him to identify the brown-skinned girl who wheeled a baby carriage near the school. That chance meeting with its scant conversation eventually would have great consequences for Brown's future.

Although Charlotte Hawkins Brown longed to attend Radcliffe College, her mother encouraged her to use the fine education she had already acquired to go right into teaching. Eventually, mother and daughter compromised on a two-year normal school education for Charlotte. Investigating normal schools in the area, Brown found that Alice Freeman Palmer, as a member of the Massachusetts Board of Education, was influential at the state normal school in Salem, Massachusetts. When Brown wrote to Palmer to inquire about the school, she reminded her of their chance meeting. Palmer replied almost immediately and, after checking on Brown's high school record, offered her financial assistance for her normal school education.

After a year as a commuter student at Salem Normal School, Brown was approached by a field secretary for the American Missionary Association (AMA) and asked to consider immediate work in one of the many AMA schools that had sprung up in the rural South since emancipation. Already deeply moved by the idea of returning to the South to contribute to the progress of the people she called her own, Brown was thrilled to find that she was needed at a small school supported by the AMA in McLeansville (later Sedalia), North Carolina, just east of Greensboro. She and her family had visited their North Carolina relatives regularly, and she understood the depths of educational need among rural blacks of the area. In 1901, Brown accepted an AMA appointment as a teacher at Bethany Institute, a church and school whose enrollment reached more than 100 pupils during the few months it was in session, before the children were needed by their families for sharecropping work.[10]

A SCHOOL IN THE BACK WOODS

Joining one other young teacher and the school's principal, Manuel Baldwin, Brown plunged quickly into her preparatory work with the upper grade students who someday planned to be teachers. Classes were held in the dilapidated Bethany Church that stood in a clearing in a wooded area crossed with dirt roads, which was also home to a small general store, blacksmith shop, and a number of nearly collapsing residences. Although Brown had hoped to use the more progressive techniques and curriculum she had experienced in Cambridge, she found that she first needed to apply a great deal of remedial work. Her persistence throughout the first year, however, gained her good friends among the parents and others in the community who helped her add a boarding facility by renovating the abandoned blacksmith building.

By the time the financially strapped AMA announced a year later that it would no longer support Bethany Institute, students and parents had encouraged Brown to try to fund and establish a school on her own. Principal Baldwin bowed out of the Bethany Church and school, although he continued as pastor of two other nearby AMA churches. Coupled with the sudden death of Alice Freeman Palmer in 1902, this turn of events sparked Brown's determination to continue to run Bethany as a private school and eventually to found the Palmer Institute, an all-grades boarding school, in Sedalia, North Carolina.

Funding issues came first; curricular decisions would have to wait. With her New England contacts, Brown was able to raise just enough money to start her school during a summer visit home to Cambridge. Her earliest approach was to gather many small donations by speaking at black church services and passing the plate. But she netted only enough in her summer solicitations to get through the 1902–1903 academic year. Later, she needed to elicit funds from wealthy white benefactors whom she met through contacts from her school years in Cambridge. Additional benefactors were found among wealthy New Yorkers who owned hunting lodges located near the school, where they came for sport and vacation. However, even after Brown's nearly full-time work soliciting likely donors, sufficient moneys to construct the first building of the new school were not available until 1905. George Herbert Palmer gave permission to use his wife's name for the school, although he never responded enthusiastically to Brown's urgings concerning his own involvement. The Palmer Institute received its official Charter of Incorporation at the start of the 1907 school year and completed that year with 90 boys and girls in attendance and a staff of 5.[11]

Clearly, Brown's initial objective in founding a school was simple: to provide educational opportunities to poor and isolated African American children who otherwise would have no avenue toward personal betterment or economic security. The route to those opportunities was detailed further in Brown's determination that the aim of education was "to produce sound character, mental discipline, and an elevated spirit."[12] The curriculum supporting that objective became a complex amalgam of Brown's personal experiences and beliefs, preferences of potential donors, and the immediate needs of children and their families. Brown built into the educational experience ample amounts of education in the liberal arts (e.g., foreign languages, mathematics, literature, and music), industrial and agricultural subjects, spiritual commitment, personal morality, and etiquette.

Opportunity—generally related to funding—determined the extent to which one area of endeavor might be emphasized over another at any given time. For example, the corporate charter filed with the State of North Carolina stated the aim of the Palmer Institute as, "education of the colored race and . . . at such institution to teach said colored race improved methods and agriculture and industrial pursuits generally."[13] However, Palmer Institute recruitment literature labeled the school, "a little bit of New England in North Carolina . . . [t]he only college preparatory school of its type in America for Negro youth."[14]

Brown displayed little of the romantic philosophy about children that had prompted early progressive educators to embrace the learning possibilities of child-centered creative activities and playfulness. Like many southern citizens interested in African American social and economic uplift following the Civil War, she may have been appropriately suspicious of progressive educators' support of manual training and of learning without reliance on books, possibly viewing these approaches as more likely to slow progress toward racial betterment. As William Reese has noted, "In the South following Reconstruction, white racists cited [Johann] Pestalozzi approvingly, saying schools which emphasized books had no place in the education of African Americans."[15]

Like many working-class whites, Brown viewed education in a context of time and place that was far different from the environment surrounding middle-class suburban public schools or northern private experimental schools. In Brown's context, child centeredness meant providing children with the basic tools for survival, and perhaps even the good life, in a society more hostile than democratic in its legalized racial discrimination and inequality of opportunity. Compared to what was generally available for southern black children who picked cotton and tended family hog pens, almost any consistent and long-term education that opened new opportunities could be labeled as progressive reform.

ACCOMPLISHMENTS AND COMPROMISES

Among the early offerings at Palmer Institute were classes in music, sewing and laundering, literature, languages, mathematics, agriculture, and carpentry. Teacher training was soon added for the upper grades, enabling the school to tap into some county funds. Brown also quickly added community outreach to her list of activities, first by enlisting the support and advice of Virginia Randolph, who was well known for her success in starting School Improvement Leagues supported by the Negro Rural School Fund. Part of the improvement agenda was school and community physical beautification, and part was student work in service to school and community. Randolph urged Brown to emphasize carpentry projects that would keep school buildings in good condition and to teach students to "sew, make shuck mats, baskets, darn, and anything that will help them in their homes."[16] Community outreach efforts, as well as Palmer Institute programs such as commencements and student musical productions, soon made Palmer Institute and Bethany Church the two mainstays of Sedalia social life.

At times, Charlotte Hawkins Brown may have seemed accommodating to Northerners, who liked the idea of industrial training for a black work force that could contribute to industry without threatening the elite white power bases. When she solicited funds from the wealthy whites, Hawkins herself emphasized that ideal, as well as the concept of training the female students to be excellent domestic workers in the kitchens and laundry rooms of the white captains of industry.[17] Later, she compared Palmer Institute to a New England preparatory academy, and she took great pride in noting that of the seven students who graduated from the senior class of 1909, two became school principals, one a physician, one a pastor, and one a teacher.[18] Accommodation was difficult to distinguish from manipulation.

Brown's northern benefactors were, for whatever motives, generous in their support. They included wealthy New York businessmen, O. W. Bright and Charles Guthrie, whose wives had alerted them to Brown's cause. They enlisted others and were soon joined by the most dedicated long-term contributor, Boston financier, Galen L. Stone. Valuable local support came from Greensboro, North Carolina, banker, Charles Bray, and from Charles D. McIver, president of the State Normal and Industrial School for Girls in Greensboro.

The young school founder traveled frequently to New York and Boston and was an eloquent and energetic force for her school. On one of her trips, she met Harvard student, Edward S. Brown, who was boarding at her mother's house in Cambridge. They married in 1911, and Brown moved to Sedalia with his new wife for a time. However, they were never able to reconcile the complex issues of the unanticipated role reversal and rural life at "her" school. They separated after a year and divorced in 1916. Charlotte Hawkins Brown married again in 1923—to a Palmer Institute teacher, 15 years her junior, John W. Moses. That union ended in less than a year, when Moses quickly proved to have a wandering eye for women.[19]

Most major contributors also served on the school's board of trustees, where they were positioned to advise and guide in directions that required a great deal of patience and compromise from Brown. New York donor, Frances Guthrie, for example, sent Brown a 26-page document with suggestions for the school, including many insisting that black students were inferior learners—"nearly all very ignorant people," who were not like Brown herself. She urged Brown to stick to simple Bible stories and instruction in cleanliness and modesty. "Essential reading and writing skills," Guthrie added, "could be achieved concurrently through topics such as 'How I Clean a Room.'"[20]

Other wealthy northern women who supported the school felt free to advise Brown on her own conduct. Contributor Helen Kimball wrote from Boston, "I want to suggest your taking time to fold your letters evenly, and if you thought beforehand how to condense, it would save time writing."[21]

Brown mustered a great deal of patience with her benefactors to launch Palmer Institute. Undoubtedly, she sensed that the education she could offer there would make an important difference in young lives without other immediate options for betterment. She was determined to instill in her students equal measures of religion, discipline, practical know-how, and social skills, as well as liberal learning that might be found in the best academies for white children. Students were bound by strict rules of deportment, dress codes, limited hours for socializing between the sexes, and required religious observances. As one observer noted, "Everything is done 'exactly right' at Palmer with no allowance being made for the fact that Palmer Memorial is a rural school. When an academic atmosphere is transplanted to a rural section, as has been done at Palmer, there is an imposing majesty surrounding the whole scheme of things. At Sedalia, fine art parallels practical art in most pleasing proportions."[22]

By 1922, Palmer Institute became the only rural high school in Guilford County to earn accreditation. The nine-month high school year included courses in English, math, physics, chemistry, French, Latin, history, and civics. Domestic science, agricultural, and industrial subjects rounded out the offerings. Students also selected from a full range of physical education and extracurricular activities.

Independent and self-assured, Brown often walked a path of passive resistance by listening to her advisors but doing what she herself thought was best. Her ongoing determination to expand the school, buy more acreage, and build more buildings often conflicted with one or more board members, who thought that she should have money in hand before obligating it. Helen Kimball expressed the sentiment shared by a number of contributors: "I was surprised to hear that you were engaged in building. I have said that the work ought not to be enlarged until the money is in hand, and I shall not be ready to help pay off any debts incurred. This is the fault of all the southern schools; they want to get in too fast. . . ."[23]

Brown, nevertheless, continued to build, stretch her credit, and search for funds. Even with contributions from individual donors, support from the Rosenwald and Slater funds, and some county and federal funding, Palmer Institute never experienced financial stability. Exasperated that Brown continued to incur debts without any certainty of paying them off, Galen Stone responded to her pleas for support in 1921 with a refusal that noted, "You are the victim of the unfortunate sudden increase in costs of doing all work, such as that of erecting your building. You are also unfortunate in having planned to spend money and continued to spend it before you had raised it, having counted on your ability to get it when the emergency might arise and having failed in that ability."[24]

Still, Brown continued to build and to expand. By 1924, the school housed more than 200 students, mostly boarders, on 300 acres with four large buildings. Although Palmer included elementary schooling, its recruitment literature labeled it, "an accredited high school offering special instruction in fine and practical arts."[25] Later, Brown added a junior college. Only by convincing the AMA to take over the school in 1927—leaving operating control in her own hands—did Brown manage to keep it afloat. Six years later, however, the relationship proved a test of wills—Brown's and the AMA's—too unpleasant to continue, and Brown took back control.

INDEPENDENT THOUGHT AND REFORM-MINDEDNESS

The same audacity and stubborn determination that enabled Brown to launch and shape a school that demonstrated the significant possibilities of African American education were characteristics that did not always endear her to those who expected deference and agreement from a black woman. Her quick mind, impeccable manner and dress, and impressive speaking ability were off-putting to paternalistic or controlling white men and women. Thus, many relationships she had with trustees and others interested in the school ended on a sour note.

However, Brown used these experiences to better understand the complexities of race relations and to spur her own involvement in activities aimed at racial uplift. In 1912, she became the president of the North Carolina Federation of Negro Women's Clubs, an office that she held for 20 years. In 1919, she published her first book, *Mammy: An Appeal to the Heart of the South,* a novel of unveiled emotional invitation. Recounting the trials of a former slave as the unappreciated but blindly devoted servant to an uncaring southern white family, the story delved into the personal consequences of racial prejudice. The volume sold well, especially among southern women, but many of Palmer Institute's benefactors decided that Brown had gone too far.[26]

Brown generally reserved her activism on behalf of African Americans for women and through women's organizations. She took an active role in the National Council of Colored Women and the International Council for Women of the Darker Races, and she served on the advisory board of the National Urban League. She also networked tirelessly with other African American women educators, such as Mary McLeod Bethune, Anna Julia Cooper, and Nannie Helen Burroughs. She frequently attended interracial women's meetings and spoke to white women's clubs, putting herself on the leading edge of the push for racial integration in all areas, but especially in the schools. She insisted that the races needed to become aware of one another, and school was where that should start. In this way, "the un-American spirit of superiority of race" would be replaced with "a common standard of superiority by character, intelligence, and natural ability."[27]

Summarizing her position, Brown insisted, "Negro woman wants everything—education, power, influence—in fact everything that the white woman wants but her white husband."[28]

By the 1930s, with the encouragement of the very liberal benefactor Galen Stone and with the knowledge of the democratic ideals undergirding progressive education, Brown managed to move the curriculum at Palmer away from accommodation with white views of the potential and place of African Americans. She was concerned that her students were becoming "more versed in books than in habits of living."[29] This concern appeared to fit well with progressive education themes of the time, especially the ideal of active learning rooted in the life interests of learners. However, Brown's response to her quandary was progressive only in terms of her own concepts for racial uplift. She decided that proper social and cultural training, was the answer. Echoing W. E. B. DuBois, she determined further that cultural uplift was particularly essential for the most select group of African American youths who would become the leaders of all—"leaders whose very appearance will inspire youth to seek ideals of truth, beauty, and goodness . . . (with) appreciation of all that is fine and beautiful."[30]

Predictably, Brown's second book was a preachy, somewhat nagging, how-to volume, titled *The Correct Thing to Do, to Say, to Wear.* Self-published in 1940, the slim volume was commercially published a year later to favorable acceptance by parents and teachers. Chapters included "How to Behave," "Poise," "Earmarks of a Lady," and "Earmarks of a Gentleman." Most of the earmarks found their way into the rules and regulations of Palmer Institute, where female students wore skirts and male students wore sport jackets and where strict rules governed the use of radios, gum chewing, and mixing across gender boundaries. Compulsory chapel and a campus work program also remained as crucial elements of the cocurriculum.

Palmer Institute Junior College started as a pilot program in 1928 and quickly launched the school's commitment to culture and refinement, although Brown's stated purpose was a two-year normal school and institute of post secondary art and music training. Theater groups and music groups became the cement that brought a sense of community to the Palmer curriculum, with students embracing active learning in set design, stage construction, costume making, lighting, and performing. Student learning undoubtedly increased when shows took to the road in North Carolina and beyond. One theater troupe performed at Boston's Symphony Hall in 1928; and Palmer's touring musical group, the Sedalia Singers, traveled up and down the East Coast, eventually performing at the White House.[31]

Still, Brown kept a tight rein on the curriculum and the cocurriculum, building an image of a fine finishing school during Palmer's third decade of operation, with "cultural training and musical training which is rarely equaled by any of the Negro schools in the South."[32] Thus, as public schools for African American students began to slightly improve and compete, Brown was able to recast Palmer as an elite cultural oasis, with the "numbers among its students representatives from some of the most outstanding families in America."[33]

IN RETROSPECT

This final repositioning of Palmer Institute represented the philosophy that Charlotte Hawkins Brown had adopted early in her career as an educator—an opportunistic mix of what would assist students to confront their environments successfully and what would assist the school to survive. Brown was a street-savvy promoter of what she believed in, and she understood short-term compromises and long-term payoffs. When she retired from Palmer Institute in 1952, the school had educated hundreds of black youths who would have otherwise had little or no opportunity to advance past the most elementary levels. Although Brown died in 1961, Palmer continued on with its second president, Wilhemina Crosson, until her retirement in 1966. After several other leaders and various attempts to rejuvenate a faltering private institution amid improved public schooling conditions for African Americans, Palmer Institute closed in 1971.

Brown and Palmer Institute demonstrate the need for broader interpretations concerning educational reforms of the Progressive Era. The South, with its very few black youngsters in formal educational settings early in the twentieth century, was a location where social reform struggled to gain a foothold. Educational reforms contributed to stumbling progress but concentrated on most basic needs concerning facilities, literacy, adequate school-year lengths, and teacher availability and pay—especially for the black citizenry. Rural isolation and racial inequality shaped a Progressive Era South that, with only a few scattered pockets of exception, was decades behind the educational progress witnessed in other regions. Progressive curriculum and pedagogy were, from that perspective, far down the list of concerns for educational reform in the South.

Certainly, however, Brown could be classified as an educational reformer who took issues of curriculum and pedagogy very seriously. Well traveled and for two decades (1912–1932) the president of the North Carolina Teachers' Association, she undoubtedly knew of progressive ideas and experiments committed to student self-expression and interest. She understood, as well, that Palmer Institute was far from the cutting edge, with its rigid curricular requirements and rule-bound discipline. Instead, Palmer represented a complex mix of Brown's own style, the aims and desires of available benefactors, and the immediate needs of black students confronting a "democracy" that institutionalized racial injustice and enabled political and economic inequality. If democracy in school life was preparatory for life in a democratic society, educators of blacks and whites needed to take into account the two very different democracies. From limited and mostly menial employment opportunities to segregation, political disenfranchisement, and lynching, the black citizen of the South had concerns directed more toward survival than the good life. With some boundaries prescribed by financial concerns and her own preferences, Brown did strive to prepare her students for life in a democratic society. She succeeded in uplifting the abilities and visions of hundreds of students who ultimately managed active and contributing lives well beyond what their circumstances would have predicted.

Just as Brown and a number of other southern educators redefined Progressive Era school reform to fit conditions unique to their region, Brown also needed to recast female leadership approaches to succeed with her benefactors, her staff, and her students. Her natural inclination toward stubborn persistence softened to passive resistance among her financial backers and hardened to unbending discipline among her students. Although her long-term goals for the black youngsters of Palmer Institute fit well with feminist notions about humane largesse and social goodness, her more immediate practices demonstrated qualities that were both strategic and authoritarian. Students generally recalled Brown as a role model and rule overseer first and a motherly friend only rarely. However, Brown played the role of mother for years to several nieces and nephews who lived with her at various times, notably niece Maria Hawkins Cole, who became the wife of musician Nat King Cole.[34]

Brown was well revered in her own time, earning honorary doctoral degrees at Wilberforce, Lincoln, and Howard universities and becoming a speaker who was sought after among forces striving for improved race relations. Her concerns for racial uplift, rather than pedagogical or curricular enhancements, were at the heart of her educational agenda. The school that she built and the educational agenda that she brought to it also reflected a particularly southern brand of progressivism, summarized by George Brown Tindall as "traditionalistic, individualistic and set in a socially conservative milieu."[35] Therefore, Charlotte Hawkins Brown's contributions as an educator can only be assessed in terms of the regional and racial divides that tore through the American landscape during her lifetime.

NOTES

1. Herbert M. Kliebard, *The Struggle for the American Curriculum, 1893–1958,* 2nd ed. (New York: Routledge, 1995), p. 243.
2. Robert A. Margo, *Race and Schooling in the South, 1880–1950: An Economic History* (Chicago: University of Chicago Press, 1990). See also James D. Anderson, *Education of Blacks in the South, 1860–1935* (Chapel Hill: University of North Carolina Press, 1988), and Russell Sage Foundation, *A Comparative Study of Public School Systems in the Forty-Eight States* (New York: Russell Sage Foundation, 1912).
3. Joseph F. Kett, "Women and the Progressive Impulse in Southern Education," in *The Web of Southern Social Relations: Women, Family and Education* (Athens: The University of Georgia Press, 1985), p. 171.
4. C. Vann Woodward, *Origins of the New South, 1877–1913* (Baton Rouge: Louisiana State University Press, 1951), p. 369.
5. William Link, *The Rebuilding of Old Commonwealths* (Boston: St. Martin's Press, 1996), p. 110.
6. Information concerning the early life of Charlotte Hawkins Brown is derived from Tera Hunter, "The Correct Thing: Charlotte Brown and The Palmer Institute," in *Southern Exposure,* 11, no. 5 (September–October 1983): 37–43; Cecie Jenkins, "The Twig Bender," no date, typescript microfilm in Charlotte Hawkins Brown file, North Carolina State Division of Archives and History Raleigh, N.C.; Constance H. Marteena, *The Lengthening Shadow of a Woman* (Hicksville, N.Y.: Exposition Press, 1977); Sandra Smith and Earle H. West, "Charlotte Hawkins Brown," in *Journal of Negro Education,* 51, no. 3 (1982): 191–206; and Charles W. Wadelington and Richard F. Knapp, *Charlotte Hawkins Brown and Palmer Memorial Institute* (Chapel Hill: University of North Carolina Press, 1999).
7. Jenkins, "Twig Bender"; Wadelington and Knapp, *Charlotte Hawkins Brown.*
8. Charlotte Hawkins Brown, "Some Incidents in the Life of Charlotte Hawkins Brown Growing Out of Racial Situations at the Request of Ralph Bunch," 1937, typescript, p. 2, Charlotte Eugenia (Hawkins) Brown Papers, 1900–1961, Arthur and Elizabeth Schlesinger Library, Radcliffe College, Cambridge, Mass. Microfilm in North Carolina State Division of Archives and History Raleigh, N.C. (hereafter cited as Brown Papers).
9. Charlotte Hawkins Brown, "Booker T. Washington's Philosophy of Education and Business," typescript of a speech delivered August 29, 1944, at Negro Men's League, Birmingham, Ala., Brown Papers; Wadelington and Knapp, *Charlotte Hawkins Brown.*
10. See Cecie Jenkins, "Early Life." First draft of "Twig Bender of Sedalia," (1945) in Holgate Library, Bennett College, Greensboro, N.C.
11. *Palmer Memorial Institute Charter of Incorporation,* November 23, 1907, Brown Papers.
12. Charlotte Hawkins Brown, June 2, 1944. Speech delivered at Howard University, Washington, D.C., upon receiving an honorary Ed.D. degree.

13. Ibid.
14. Palmer Memorial Institute, "Some Interesting Facts About Our School," recruitment brochure (Sedalia, NC, 1924), p. 3, Brown Papers.
15. William J. Reese, "The Origins of Progressive Education," in *History of Education Quarterly,* 41 (Spring 2001): 1–24.
16. Virginia Randolph to Brown, July 27, 1909. Brown Papers.
17. Hunter, "The Correct Thing."
18. Smith and West, "Charlotte Hawkins Brown," pp. 191–206.
19. Wadelington and Knapp, *Charlotte Hawkins Brown,* pp. 68–73.
20. Frances A. Guthrie to Brown, no date, 1907. Brown Papers.
21. Helen F. Kimball to Brown, June 5, 1905. Brown Papers.
22. Gordon B. Hancock, *The Calm at Sedalia* (Norfolk, Va: The Guide Publishing, no date) p. 2.
23. Kimball to Brown, August 17, 1907. Brown Papers.
24. Galen L. Stone to Brown, May 9, 1921. Brown Papers.
25. Palmer Memorial Institute bulletins, 1824, 1931–1932. Brown Papers.
26. Charlotte Hawkins Brown, *Mammy: An Appeal to the Heart of the South* (Boston: Pilgrim Press, 1919). See also Henry L. Gates, Jr., ed., *African-American Women Writers, 1910–1940* (New York: G.K. Hall, 1995).
27. Brown to Max Loeb, October 1933. Brown Papers.
28. "The Importance of Overcoming Discrimination," 1930 typescript in Brown Papers. Quoted in Smith and West, "Charlotte Hawkins Brown," p. 202.
29. "Pronouncement of the Ideals for the New Palmer," 1930 typescript in Brown Papers. Quoted in Smith and West, "Charlotte Hawkins Brown," p. 197.
30. Ibid.
31. Wadelington and Knapp, *Charlotte Hawkins Brown,* p. 144.
32. Palmer Memorial Institute Bulletin, 1935–1936. Brown Papers.
33. Ibid.
34. Marteena, *Lengthening Shadow;* Wadelington and Knapp, *Charlotte Hawkins Brown,* pp. 195–196.
35. George Brown Tindall, *The Emergence of the New South, 1913–1945* (Baton Rouge: Louisiana State University Press, 1967), p. 7.

CHAPTER TWO

MARIETTA JOHNSON AND THE ORGANIC SCHOOL

JOSEPH W. NEWMAN

The founder of the school that historian Lawrence Cremin described as "easily the most child-centered of the early experimental schools"[1] deserves a special place among the "founding mothers and others" discussed in this book. A member of the same generation as John Dewey and Jane Addams, Marietta Johnson caught the same spirit of social reform and put it to work at the School of Organic Education, which she established in 1907 in the utopian community of Fairhope, Alabama. In the early twentieth century, as women emerged as leaders in educational reform across the United States, Johnson became one of the first to gain national, even international recognition.[2]

Like some of her counterparts discussed in this book, Johnson was, in Joyce Antler's apt phrase, a "quiet feminist." The biographies of such women, Antler argues, exemplify feminism as an "individual's struggle for autonomy." In this chapter I pay close attention to Johnson's personal life as just such a struggle. Johnson also developed what Antler calls a "conscious, political strategy for altering the social order," albeit not along explicitly feminist lines, and I follow this aspect of Johnson's career, too.[3] Along the way I consider whether Johnson's personal style embodied some of the traits that feminist scholars in the tradition of Carol Gilligan have identified as characteristic of women leaders.[4]

A TRADITIONAL TEACHER AND HER CONVERSION EXPERIENCE

Born Marietta Louise Pierce on October 8, 1864, near St. Paul, Minnesota, she grew up in a close-knit farming family that included a twin sister and six other siblings. Her parents, Clarence and Rhoda Martin Pierce, were devout members of the Christian Church and raised their children in the faith. The family's economic circumstances seem to have been comfortable, and the Johnsons encouraged all their children to go to school and make something of themselves. Marietta's twin sister chose nursing. A brother graduated from the University of Minnesota and became the university's alumni director. We know little else about the family.[5]

After her father died while Marietta was still young, her mother helped support the family by opening a school for neighborhood children in the Pierce home. Watching and assisting her mother at work probably influenced Marietta to announce regularly, from age ten on, "I am going to be a teacher when I grow up."[6] After attending public schools in St. Paul, she graduated in 1885 from the State Normal School at St. Cloud and began her career as a teacher in rural Minnesota.[7]

Enthusiastic, charismatic, and ambitious, Marietta Pierce moved quickly up the occupational ladder. Certified for elementary schools, she managed to teach every elementary grade as well as several high school subjects during her first five years of teaching. In 1890 she returned to St. Paul to become a "training" or "critic" teacher in the city normal school. She took a similar position at Moorhead State Normal School in 1892, signing on at Mankato State Normal School three years later.[8]

Discovering a talent for supervising adults and young adults, Pierce genuinely enjoyed critic teaching. In later years, training teachers would become an important part of her agenda for educational reform. From her vantage point inside state normal schools, she came to understand how the public education bureaucracy worked. Then, she felt comfortable in the system, with all its rules, regulations, and standards, but that feeling would soon change.[9]

Meanwhile, Pierce distinguished herself as an academic traditionalist. She was especially proud of her skills in reading instruction. In a semiautobiographical, book-length manuscript she drafted during the 1930s, she looked back on her early career and painted an ironic portrait of the achievement-driven teacher she had been: "It was a great joy to me when six-year-old children in the first grade could read through four first readers in three months!! . . . Of course, it was high pressure, but they could do it! The student teachers were thrilled with these results and no doubt went out into teaching positions determined to reach these standards. The parents were pleased. I was a success! And we all truly believed that the children loved to do it, and that it was good for them!"[10]

After devoting years to perfecting teaching techniques she considered tried and true, Johnson (who married in 1897) felt her faith in traditional education shake to its foundation when she "underwent a conversion experience." The religious imagery is appropriate: Johnson herself used it to explain the intensity of her ordeal. According to Lawrence Cremin, Caroline Pratt of the City and Country School in New York City and several other founders of progressive schools experienced much the same thing.[11] Johnson's conversion began in 1901, around the time of her thirty-seventh birthday, when her superintendent at Mankato State "thrust a book" into the hand of the critic teacher, saying, "Unless education takes this direction, there is no incentive for a young man to enter the profession." Passing lightly over the gender message in the superintendent's remark—"I thought there was every incentive for everybody to enter the profession"—she took the volume home and read it.[12]

The book was *The Development of the Child* (1898) by Nathan Oppenheim, a pediatrician at Mt. Sinai Hospital in New York City. "The world has a wrong idea of its children," Oppenheim insisted. Parents and teachers who think of children as "adults in small" and of childhood as a time for mastering adult behavior are causing children harm—in some cases irreparable harm. According to Oppenheim, children are "absolutely different from adults, not only in size, but also in every element which goes to make up the final state of maturity." Constantly changing, children need a "special treatment and environment" to guide and encourage their development.[13]

As Johnson read on, she began to question virtually everything she had learned about teaching; indeed, she felt appalled at what she had been teaching teachers. She convinced herself she had been a "child destroyer" whose efforts violated the "order of development of the nervous system. I realized that my enthusiasm was destructive, and the more efficient I was, the more I injured the pupils!"[14] For the rest of her life, Johnson would refer to Oppenheim's work as a "scale dropper." It opened her eyes and cleared her vision.

Johnson proceeded to repent her educational sins. She took Oppenheim's work as her "educational Bible" and embarked on a personal reading program that led her from the latest studies of child physiology and psychology to earlier works by Jean Jacques Rousseau and Freidrich Froebel and back again into the works of contemporary educators.[15]

One of her strongest influences was John Dewey. The Dewey she first met on her intellectual quest was the University of Chicago professor who had just written *The School and Society* (1899), the young scholar who was still working actively with children as director of the university's experimental Laboratory School.[16]

Even more important in initially reshaping Johnson's thought was C. Hanford Henderson, former headmaster of New York's Pratt Institute, a Brooklyn trade school of art and design. From Henderson's *Education and the Larger Life* (1902) she appropriated the concept of "organic" education: treating each child as a complete organism—a "whole" child, as the progressive slogan would soon have it—paying balanced attention to body, mind, and spirit.[17]

Johnson turned these books over and over in her mind, revisiting them for inspiration and support for the rest of her life. A friend who took in Johnson and her family when they first moved to Fairhope in 1902 recalled that Marietta would ask her to put the Johnsons' infant son to bed after dinner every night so she could stay up "'til all hours of the morning, studying Dewey, Oppenheim, Henderson."[18] Well before the move to Fairhope, she began "experimenting" with her own child and a few others from the neighborhood. Johnson confessed that as the idea of education keyed to child development took shape in her mind, it "took possession of me and I could not rest until I had started a school."[19]

BALANCING MARRIAGE, MOTHERHOOD, AND CAREER

Marietta Pierce was 12 years into a successful career and happy with her work at Mankato State when she married John Franklin Johnson in 1897. Also a native of St. Paul, Frank was a carpenter and cabinetmaker who enjoyed the outdoors and dreamed of becoming a rancher. The couple was quite compatible and their relationship remarkably egalitarian by the standards of the day. But Frank considered himself the main breadwinner of the family, and he moved the Johnsons several times in search of better places to pursue his dream. Soon after marrying they relocated to cattle ranches in Montana and North Dakota. Their first child, Clifford Ernest, was born in North Dakota in the spring of 1901. Her new roles as wife and mother made Marietta reevaluate her long-held career plans: Would she ever return to teaching?[20]

The answer turned out to be yes. Ranching was hard work, the Johnsons discovered over the course of four long years. Frank's eyesight was failing, and he developed lung problems. Marietta found her new situation taxing, mentally as well as physically. Even though she had grown up on a farm, ranching and housekeeping were not her favorite work. Teaching, the career she had chosen at age ten, had already brought her success and satisfaction.

Being a mother proved to be particularly challenging, which is surprising given how easily she worked with children as a teacher. Younger children, in particular, treated Johnson with immense respect, trying so hard to please her people came away wondering what sort of magic she commanded.[21] But teaching and mothering were different, she found. Of all the children she worked with when she began her educational experiment, "my own child was the hardest one to manage—because I was his mother." Clifford Ernest was not sufficiently in "awe" of her, she believed. Whereas other children readily accepted her guidance, even when it "[ran] directly counter to their wishes," her son would "scream it down."[22]

Later, in her work at the Organic School, Johnson never related to children as a "cuddly" or "coddling" teacher, nor did she believe teachers should try to take the place of mothers or fathers. Approachable, friendly, and sincerely interested in people, Johnson nevertheless inspired respect, a quality that became central to her leadership style.[23]

A NEW BEGINNING IN FAIRHOPE

A few months after Clifford Ernest was born in 1901, the Johnson family moved back to Minnesota, Marietta returned to her duties as a critic in the St. Paul normal school, and her educational conversion began. As she read, reoriented her thinking, and put her new insights into practice with her son and other children, Johnson looked forward to opening a school.[24]

Frank was still determined to succeed at ranching—or farming—and some Minnesota friends told the Johnsons about Fairhope, an unusual community located on the Alabama Gulf Coast, where the mild winters might improve their health. After visiting several times, the Johnsons moved to Fairhope over the Christmas season of 1902. They felt at home immediately because, like many others drawn to the small village (population 100) in its early years, they were neither native Southerners nor conventional in their politics.[25]

The Johnsons were socialists in the upper midwestern tradition, believers in the Christian Socialism that fueled the Social Gospel movement during the late nineteenth and early twentieth centuries. The Johnsons' strong commitment to social and economic reform helps explain why Marietta was so receptive to the ideas on educational reform she first encountered in her mid-thirties.[26]

Fairhope was a utopian community, a single-tax colony founded in 1894 on the ideas of Henry George. Of the millions who read *Progress and Poverty,* George's late nineteenth-century bestseller,[27] a few disciples took his words seriously enough to put them into practice. This Deep South colony, which its founders believed had a "Fair Hope" of succeeding, was the first and largest single-tax experiment in the nation.[28]

The original colonists, most of them from the Midwest, resolved to make their new home a model of *cooperative individualism.* Under the terms of the community experiment, the colony owned the land, with individuals and families holding 99-year leases on their plots. The annual rent paid to the colony constituted the "single tax" on land, which generated funds for such public amenities as parks, a beach, and a library. The colony also owned the utilities. These arrangements, George's followers believed, would help control the gap between rich and poor by preventing the wealthy from monopolizing the land and its resources. George himself, though, doubted an experimental single-tax colony could succeed.

From the very start, Fairhopers won a well-deserved reputation as an intellectual group united in their belief that the economic system of the United States was flawed. Beyond that central conviction, the early residents agreed to disagree. With *cooperation* at one end of their ideological axis and *individualism* at the other, spirited debates broke out in the many organizations they formed, such societies as the Progressive League, Socialist Club, and School of Philosophy. The socialists who came to dominate public affairs in the early 1900s, with their emphasis on collectivity, often found themselves at odds with the colony's single-tax founders, ardent defenders of individualism, and other values central to private enterprise. In Fairhope, women and men alike voted and participated in town-meeting politics. The climate became particularly charged after the village incorporated in 1908 and a socialist defeated a single taxer in the first race for mayor.[29]

Marietta Johnson bridged the two camps better than almost anyone else. She viewed business, especially big business, with suspicion—indeed, there was hardly a trace of the profit motive in her—but she admired Henry George and incorporated the principles of single taxation into her teaching. Although there was never a doubt where she stood on an issue, Johnson rarely appeared to be grinding an axe or, in the parlance of the day, "riding a hobby." By making her positions seem reasonable to people and gaining their confidence, Johnson was able to build consensus. These qualities served her well throughout her career.

Visitors came to Fairhope in ever-larger numbers as the little community cultivated its reputation as a place to enjoy nice winters, natural beauty, off-center politics, and literate conversation. To artists, writers, and intellectuals, along with well-to-do people who simply wanted to soak up the atmosphere, Fairhope seemed a marvelous place to get away.

That said, Fairhope's main attraction was Marietta Johnson's school. The first genuine celebrity to visit was Upton Sinclair, who wintered in Fairhope in 1909 while his son David attended the newly established School of Organic Education for one term. Sinclair had published his first best-seller, *The Jungle,* in 1906, a socialist critique of "wage slavery" in the Chicago meat packing industry. Apparently expecting to find more in Fairhope than he actually did, he captured the spirit of the early community with a few words in his *Autobiography* (1962): "Here were two or three hundred assorted reformers who had organized their affairs according to the gospel of Henry George. They were trying to eke out a living from poor soil and felt certain they were setting an example for the rest of the world."[30]

ANOTHER TRANSITION

Marietta Johnson wanted to add an educational dimension to the Fairhope experiment. With Alabama still struggling to build a viable state school system during the early 1900s, it was easy for newcomers to make unfavorable comparisons between the midwestern and northern schools they had left behind and the schools they found in rural Baldwin County, where the colony was located. Beyond raising additional local funds and trying to attract better teachers, though, the colonists had no special ideas on how to improve schooling. Nor did Henry George have much to say on the subject. Johnson, however, did.[31]

Her chance to try her ideas out in a school setting finally came in January 1903, less than a month after her move to Fairhope, when the 38-year-old woman took charge of the public elementary school. After more than a year of reading, introspection, and experimentation at home, Johnson was eager for the chance. In what became a pilot project for her later work at the Organic School, she added gardening and manual arts to the curriculum, invited adults to school to make music and tell stories, and organized a normal course. At the start of the 1903–1904 school year, Johnson announced ambitious plans to expand the normal course, develop a high school program, and recruit out-of-state students.

These plans were cut short by Frank's decision to move the family to Mississippi early in 1904. Upton Sinclair was right: The countryside around Fairhope was not prime farmland. Frank prevailed on Marietta to allow him one more chance to follow his dream, and the family bought a pecan farm in Lauderdale County, Mississippi, more than 150 miles northwest of Fairhope. For more than three years they tried to make a go of it.

The Mississippi sojourn, traumatic in several ways, marked a transition in both their domestic relationship and their careers. The Johnson farmhouse burned in 1905, Frank's eyesight steadily worsened, and Marietta discovered how much she missed living and teaching in Fairhope. She stayed in touch with friends by writing letters and visiting as often as she could. In the spring of 1905, at the age of 40, she gave birth to a second child, Franklin. Longing to take her experiment back into a school setting, she contented herself with teaching her two sons and several other children, including some who came from Fairhope to live with her.

She returned to Fairhope in the summer of 1906 to conduct a teachers' institute at the request of her close friends, Lydia and Samuel Comings, who had operated a private school before moving to Fairhope and were especially interested in physical culture and manual arts. When the Johnsons finally decided to give up their pecan farm during the summer of 1907, the Comingses persuaded them to move back to Fairhope, where with the Comings's backing, Marietta opened a kindergarten that soon grew into the School of Organic Education.

From 1907 on, Marietta's career took precedence over Frank's. Although he would serve on the Colony Council and win election as mayor of Fairhope in 1912 on the Socialist ticket, his primary work outside the home became teaching manual arts in the School of Organic Education, where he used his skills as a carpenter and cabinetmaker. He also kept busy within the Johnson household. Gentle and unassuming, Frank played a supportive role as "Marietta Johnson's husband," taking over most of the family's domestic responsibilities, including much of the child raising, so his wife could put in long hours at school and continue to do school work, late into the night, at home.[32]

PUBLIC ELATION, PRIVATE GRIEF

Overjoyed to be back in Fairhope, Johnson was finally able to open her school in November 1907. It began modestly enough as a kindergarten in the small cottage where she and her family lived. The Comingses provided a monthly subsidy of $25, from which the teacher spent $15 for rent and $10 for salary and supplies. Six kindergartners, two of them the Johnsons' own sons, enrolled the first day and were quickly joined by a handful of older elementary students. As a harbinger of the prominence the school would soon achieve, Hanford Henderson came to Fairhope for the opening of the school, probably at the Comings's invitation. In one of Johnson's first public speeches on organic education, she proudly addressed the Progressive League alongside the educator whose "organic" concept had helped turn her thinking around.[33]

Early in 1908 the Colony Council appropriated an additional $25 per month for the school, enabling the Johnson family to live a little less frugally, Frank to join the work as a shop teacher, and Marietta to operate the school as a quasi-public institution that required no tuition of local children. A young woman who wanted to be a teacher apprenticed herself as an assistant. With more than 30 students enrolled, Johnson's ongoing educational experiment was entering a new phase. She had every reason to be optimistic.[34]

Privately, though, Johnson was grieving. Just after school began, her two-and-a-half-year-old son Franklin died from a fall. Carrying right on with her public work, Johnson held her sorrow inside. For the rest of her life she would refuse to discuss Franklin's death, even with those closest to her. In her conversations, speeches, and writings, she would make singular references to "my son"—Clifford Ernest—even when discussing events that occurred during Franklin's lifetime. Compounding her grief, her friend and benefactor Samuel Comings died of a stroke on Christmas Eve of 1907. When Johnson reopened her school a few days later, outwardly she appeared to be back to normal. She handled herself in the same way, years later, when her husband Frank died in the summer of 1919: controlling her emotions and moving ahead with her public work. Johnson's closest friends understood how deeply but how privately she grieved.[35]

THE FLUSH OF SUCCESS

Most people who met Johnson during the early years of the Organic School never saw her private side. Instead, they came away struck by her dedication, sincerity, and persuasiveness. Some who disagreed with her socialist politics or raised their eyebrows at what seemed a lack of structure and standards in her school still had to confess their admiration for what she was doing.

Joseph Fels, founder of the Fels Naphtha Soap Company and financial backer of single-tax causes throughout the nation, visited Fairhope in 1908 and saw the Organic School at the end of its first academic year. Although he appreciated Johnson's commitment to single taxation, Fels got to know her well enough to debate some of their differences—particularly her socialism. It must have been something to see: a traditionally educated millionaire standing face to face with a teacher who believed competitive schools were breeding grounds for insincere businessmen.[36]

Johnson won Fels over. He left so impressed with organic education and Johnson's "great goodwill to all peoples" he made one of the largest donations that was ever given to the school during Johnson's lifetime. Beginning with an initial gift of $1,000, Fels contributed $10,000 more over the next few years—a fantastic sum considering the monthly stipends of $25 and $50 that supported the school during its first year.[37]

The Fels donation enabled Johnson to move the school in 1909 to a new location one block off Fairhope's main street, to a ten-acre site the colony provided rent and utility free. Johnson, Lydia Comings, and four other Fairhope women signed papers incorporating the school that same year, with Comings serving as business manager and president of the board. The Bell Building, a former public school renamed for the bell that rang out at regular intervals across the village, served as the focal point of the new campus. A total of ten buildings, including the School Home, a money-generating dormitory for boarding students, dotted the campus by the early 1920s.[38]

The plain but impressive new facilities made it easier for Johnson to recruit well-to-do children, first from the Northeast and Midwest and later from the West as well, students whose parents were enamored of what would soon be called "progressive education." Some of these students, like Upton Sinclair's son, enrolled for just the winter while their parents vacationed in the South. Resort communities were thriving along the Eastern Shore of Mobile Bay, and Fairhope, a town of over 500 residents by 1910 and more than 800 by 1920, had a special appeal. The two younger sisters of Margaret Mead, for instance, spent time at the Organic School during the early 1920s. Tuition- and board-paying students made up approximately one-third of the enrollment and subsidized the attendance of local children, who attended tuition free.[39]

Johnson and her school enjoyed the flush of success, but with most of the Fels money invested in the physical plant, she had to turn her attention to fundraising. She often spent her late-night hours writing letters that generated a steady stream of invitations to speak and conduct teachers' institutes. Johnson went out on the lecture circuit. Beginning in 1910, she spent every summer away from Fairhope, soon adding trips throughout the school year.[40] As the Organic School grew, she hired other teachers and became more of a director, which was then her title.

Johnson's administrative style fits the profile that feminist scholars have developed of women leaders. She emphasized human relationships and built a strong sense of community within the experimental school, interwoven as it was with the life of the experimental community. Many graduates recall that students who lived in Fairhope usually walked home for long, leisurely lunches—and, for younger children, naps—but when the big bell rang to signal the end of the school day, nobody left. Students and teachers took their time finishing what they were doing. Johnson sometimes had to "shoo" them away several hours later.[41] Traditional lines of separation faded as school and community blended seamlessly. This aspect of education at the Organic School, remarkable even in a progressive institution, became one of the fondest memories of its graduates. And Johnson was at the center of it all.

What made the school work so well, even when its director was away raising money, was that students and teachers alike usually tried to act in a given situation as they thought she would want them to. Johnson had to reconcile her rejection of "external, competitive" standards with her own powerful influence. Her goal was for students to cultivate "inner, human" standards, the feeling of "inner necessity" that developed in the "absence of external demands."[42] Holding the same goal for teachers, Johnson gave them room to try their own innovations. In her school, people worked to satisfy themselves, not to please others—or so she intended. But she had to walk a fine line indeed when students and teachers confessed that part of their motivation for working hard was pleasing her.[43]

Johnson herself rarely passed up a chance to teach. Often she would return to Mobile from a trip by train, catch the bay boat to Fairhope, and all but run to get back into the classroom with the children.[44]

As exhausting as her travels were, Johnson enjoyed them. Her husband's willingness to stay at home and take care of their son afforded her a freedom available to few married women, even those with careers in the public sphere. When Frank died, Clifford Ernest was almost 18—old enough to take care of himself. Marietta seemed to thrive on meeting people, and one productive contact seemed to lead to another.

SOCIETY FRIENDS, THE *NEW YORK TIMES*, AND PROFESSOR DEWEY

A chance encounter on a train with W. J. Hoggson of New York City and Greenwich, Connecticut, turned out to be one of the biggest breaks of her career. Hoggson, whose sincerity made Johnson "believe for the first time in my life that it is possible for a businessman to be a Christian," bought her a train ticket and covered some of her other expenses after she discovered she had left her money at home. Listening attentively to her educational ideas and asking good questions, he invited her to speak at his home in Greenwich. There she met a group of socially prominent women who were interested in educational reform.[45]

Across the nation in the late nineteenth and early twentieth centuries, civic-minded women organized "school improvement associations" to reform urban public schools. The educational vision of the Greenwich women Johnson met was more private and *sub*urban, but they believed she could help privileged children such as theirs while she was developing a model for school reform on a large scale. She became especially close to Mrs. Charles D. Lanier, daughter-in-law of the poet Sidney Lanier. Johnson's new friends adopted her and formed an organization to support her work.[46]

With Mrs. Lanier as president and Mr. Hoggson as treasurer, the Fairhope League (renamed the Fairhope Educational Foundation in 1920) became Johnson's northeastern—and soon national—sponsor. Her well-connected Greenwich friends booked as many lectures and other engagements as she could handle. For 20 years Johnson conducted an annual Fairhope Summer School in Greenwich, a hands-on demonstration program for teachers, parents, and others interested in organic education. Until 1927, she also served as the salaried director, mostly in absentia, of the private Edgewood School founded by Lanier, a kind of "Organic School North." Johnson used the money these activites generated to keep the Fairhope school afloat, spending little beyond living expenses on herself and her family. The greatest benefits of her metropolitan New York connections came in 1913, when her supporters arranged an interview with The *New York Times* and used it to get John Dewey's attention and persuade him to visit Fairhope.[47]

The full-page article that ran in the Sunday, March 16, 1913, edition of the *Times* was at least as important to Johnson as it would be to an educator today—a career-making breakthrough. The *Times* reporter wrote a brief introduction and filled the rest of the page with Johnson's own words. She spoke confidently, fluently, and anecdotally, using well-rehearsed lines from her speeches to explain how she had lost faith in traditional education and resolved to find an alternative: "The test of the school, as I have often said before and cannot say too often, should not be what the child does for the school, but what the school does for the child. . . . Any system through which one child flourishes while another unjustly languishes is most imperfect and breeds discontent with the social system which tolerates it." Going into detail about her work at the Organic School, Johnson explained that "the ability of children to develop when they are measurably left alone is nothing less than marvelous." And so Johnson, already skilled at winning over audiences ranging from one to several hundred, explained her ideas to the much larger readership of The *New York Times,* Professor Dewey included.[48]

The publicity value was enormous. The *Times* followed up with another article in July 1913 and published at least three more stories on Johnson over the course of her career. Favorable articles appeared in *Scientific American* and *Literary Digest* in 1914. The *New Republic* ran a glowing report in 1915.[49]

During the summer of 1913, Johnson's Greenwich friends invited Dewey to Fairhope to see the Organic School. Then at Columbia University and in an extraordinarily productive phase of his career—*Democracy in Education,* his magnum opus, would appear in 1916—Dewey accepted the invitation provided he could visit over the Christmas season, the only spot available on his calendar.[50] Johnson knew the stakes were high, calling Dewey's visit "the most critical experience of my life!"[51] After she explained the significance to her students, they voted to hold school during the holidays, and boarders made arrangements to stay over.

Dewey was conducting research with his daughter Evelyn for a book on experimental schools. Although Evelyn made all the other site visits for their project, Dewey himself went on this one. He brought along his 14-year-old son Sabino, who attended the Organic School for a week and liked it so much he wanted to stay. "All the children are crazy about the school," Sabino told his father.[52] Dewey informed the many readers of *Schools of To-Morrow,* as the book was titled when it was published in 1915, that he was one of many "students and experts" who had "made pilgrimages" to see the Organic School.[53]

Johnson was conducting an educational experiment as part of a community experiment, she explained to the professor. Fittingly, he titled his chapter on the school "An Experiment in Education as Natural Development," pronouncing it a "decided success."[54] He was especially impressed with the manual arts program, the health and enthusiasm of the students, and the lack of pressure in the school. And Fairhopers were impressed with him. Local mythology still includes a "Dewey in Fairhope" story that has the bespectacled, white-haired philosopher playing Santa Claus and bouncing students on his knee.

The *New York Times* articles, favorable reviews in other periodicals, and John Dewey's blessing catapulted the Organic School into the front ranks of experimental schools. Invitations for Johnson to speak and conduct institutes came flooding in. Fundraising became easier. She was able to attract talented teachers and affluent boarding students from around the country, with total enrollment varying between 100 and 220. Marietta Johnson had instant name recognition in experimental education circles. The future looked bright indeed.

INSIDE THE ORGANIC SCHOOL

Dewey said the Organic School was the best of its type he had ever seen. While other teachers and school directors talked about putting the needs and interests of children first, Johnson actually did.

She modified the single-year grade groupings used in traditional schools, trying to make them correspond more closely to the stages of child development as she understood them. She placed 4- and 5-year-olds together in the kindergarten, 6- and 7-year-olds in the "first life" class, 8- and 9-year-olds in "second life," 10- and 11-year-olds in "third life," and 12- and 13-year-olds in "fourth life." For the last four years, students attended high school.

Mindful of Nathan Oppenheim's warnings against forcing and rushing children, the once-eager reading teacher now steered them away from books until age eight or nine, convinced children could learn more through direct experience with the environment. In the face of strong parental pressure not to wait quite so long, Johnson could recite a litany of damage caused by "excessive or too early use of books," from eyestrain and cramped posture to unclear thinking, unsocial attitudes, and nervous breakdowns.[55] Although she had the gift of making anecdotal evidence sound credible when talking to parents of prospective students or trying to sway a crowd, this aspect of her pedagogy was too radical ever to gain wide acceptance.

Usually lost in the wash of controversy was the larger point Johnson was trying to make: Reading books was neither the best way for a young child to learn nor the only way for anyone else to learn. Dewey and other visitors to the school were impressed that so much education took place outdoors, where students worked math problems on the walls of a deep gully or studied geography in the woods, fields, streams, and bay. "Our shop," Johnson noted, sometimes with fond references to her husband Frank, the Comingses, and Hanford Henderson, is the "most important place on the campus."[56] Every student spent time in the shop every day, learning the manual arts while working with wood, metal, and ceramics. Students and teacher joined hands in daily folk dancing, too, which became one of the hallmarks of the school.[57]

Such features attracted so much attention it was easy for visitors to overlook a curriculum that became more demanding and rigorous as students gained maturity. Once in high school, every student took four years of literature, history, math, and science and two years of Latin and French.

Before high school, though, students did no homework and took no tests, and no student of any age ever received grades or a report card. Johnson refused to compare one student to another, rejecting external standards that lent themselves to quantification and measurement in favor of the internal standard of simply doing one's best. To critics who charged that Organic was a "do-as-you-please" school—a common caricature of child-centered schools—Johnson countered that "children have no basis for judgment [and] do not know what is good for them." Teachers and parents, she insisted, were responsible for taking charge and directing young people away from "unwholesome" activities.[58]

"Mrs. Johnson is trying an experiment under conditions which hold in public schools," Dewey stated, "and she believes her methods are feasible for any public school system. . . . Any child is welcome."[59] Trying to build an operating model of democracy in education, Johnson worked especially hard to create an egalitarian climate at the Organic School. She provided equal opportunities for females and males. She made no distinctions based on social class. She recruited "backward" (disabled) children who could not attend public schools.

Johnson held enlightened views on race for her day, going so far as to denounce racial prejudice in her speeches and writings. Some of Fairhope's nonsouthern founders shared her views, but they feared they would jeopardize the entire community experiment if they treated African Americans equally. Johnson reluctantly yielded to community pressure and did not admit black students. With the silent acceptance that often surrounded matters of race, even John Dewey did not see fit to comment on the situation.[60]

JOHNSON AT HER PEAK IN THE 1920S

As Johnson traveled throughout the nation, she met some of the other "founding mothers" whose lives and careers are discussed in this book. She taught Margaret Naumberg at the second Fairhope Summer School in Greenwich in 1914, which seems to have played a part in Naumberg's conversion experience. As willing to learn as she was ready to teach, Johnson enjoyed interacting with the founder-peers she met on the national circuit, most of whom ran elite private schools. She may have forged close friendships at this level, as some of her few extant personal letters suggest.[61] On the other hand, she seems to have felt apart from this group. Not only did she have a less autocratic leadership style than some of her peers, her sense of mission was also different. Johnson operated the Organic School as a quasi-public institution, never giving up the hope that her experiment would provide a model for reforming public education.

Moving into her fifties as she and her school hit their stride, Johnson began thinking seriously about a national association to promote organic education. For several years she had been organizing organic education societies in some of the cities where she spoke, and now she was ready to take a bigger step. In 1918 she shared her plans with

Stanwood Cobb, then a disillustioned instructor at the U.S. Naval Academy and soon-to-be founder of the Chevy Chase Country Day School in Maryland. Cobb had heard about Johnson from Gertrude Stevens Ayres, founder of the Washington School in the District of Columbia, who told him she had patterned her school after the Organic School in Fairhope. Johnson supplied Cobb with the names of several other women in Washington who were supporters of organic education.[62]

These details of the relationship between Johnson and Cobb are suggestive of the extent of Johnson's influence in Washington as well as in other cities. She clearly had a strong following, especially among women interested in private, child-centered education. Her more serious followers founded at least nine "satellite" schools directly on the Fairhope model, with Johnson's active participation, and her ideas seem to have been the inspiration behind several times as many other schools.[63]

But Cobb was skeptical of Johnson's plans for a national organic education society, worried that a group devoted to promoting only one approach would fail. When Johnson came back with a proposal for a more comprehensive society that would publicize a wider range of educational experiments, Cobb became more interested. He took the initiative in organizing the Progressive Education Association (PEA) in 1919 in Washington, D.C.[64]

A founding member and one of the five speakers at the first meeting, Johnson stayed with the organization for the rest of her life. The seven principles the PEA adopted to guide its work represented a synthesis of her ideas and those of Eugene Randolph Smith, headmaster of Baltimore's Park School. But while acknowledging the value of different approaches to child-centered education, Johnson still tried to use the PEA to promote her own. The contacts she made in the organization led to her involvement in the New Education Fellowship, the PEA's European counterpart, and her appearance at conferences in Cambridge, England (1922); Heidelberg, Germany (1925); Locarno, Switzerland (1927); and Dublin, Ireland (1933).[65]

If Johnson became an international star, her heart remained in Fairhope, with her original experiment. As her reputation spread, she attracted teachers who were willing to work for very little money in order to work with her. Then in a position to be selective, she hired teachers whose academic credentials were generally as strong as their admiration for her. Johnson took special pride in assembling a faculty whose members held degrees from major state universities, the Ivy League, even Oxford and Cambridge. Some of these teachers came south to work at the Organic School and spent the rest of their lives in Fairhope; others served apprenticeships and moved to other progressive schools; a few went on to found their own schools based on the organic model.[66]

The example par excellence of the caliber of teacher Johnson attracted was Charles Rabold. A student of the noted English folklorist Cecil Sharp, Rabold left a position in the music department at Yale to teach at the Organic School. During the 1920s he turned the already well-established dance program (which Sharp himself had helped Johnson develop) into one of the defining features of the school. With a confident, outgoing personality that complemented Johnson's, Rabold could help even the most awkward students and teachers learn to enjoy English folk dancing—country, Morris, and sword. Rabold became Johnson's friend, advisor, and assistant director. This relationship became her closest after her husband's death.[67]

As Fairhope's population neared 1,500 at the end of the 1920s, Johnson redoubled her efforts to stay in touch with her neighbors. Despite her heavy travel schedule, she remained a visible presence in Fairhope—not only at the Organic School but on village streets, in colony and town meetings, and in the pages of the *Fairhope Courier.* Never too busy to stop and talk, she often complimented people on something personal: a hat, a dress, an accomplishment. Some of her former students felt comfortable enough to call her "Aunt Mettie" or "Ma Johnnie." An adult course on organic education she first conducted in 1921 turned into an annual winter event that sometimes featured Clarence Darrow or other prominent speakers. By the end of the decade the winter course was pulling in people from throughout the nation—pilgrims who, like Dewey, wanted to learn about the educational experiment, the community experiment, or as Johnson always emphasized, the connection between the two. Not every resident of Fairhope was a true believer in either single taxation or organic education, of course, but most locals felt proud Johnson was putting their small town on the intellectual map.[68]

DEPRESSION AND DECLINE

In what should have been the crowning touch to a successful decade, Johnson published her first book, *Youth in a World of Men,* in 1929. At the urging of her Greenwich friends, particularly Mrs. Lanier, to whom she dedicated the book, Johnson tried to put down on paper the essence of her speeches and demonstrations. But the results of her efforts as a writer were as mixed as the reviews the book received. She had won a well-deserved reputation as a practitioner—a doer—yet *Youth in a World of Men* struck most readers as detached from experience, even naive. Instead of offering insight into the Fairhope experiment, Johnson touted organic education as a panacea, promising more than she or it could possibly deliver. The self-confidence and persuasiveness that carried her message so well in person fell flat in print.[69]

Then the Great Depression hit, and the Organic School felt the impact almost immediately. The number of boarding students dropped off sharply, decimating the operating funds from this vital source. The financial crisis reminded the school's supporters of how precarious its situation had always been, even in better days. In 1924, for instance, Johnson and other members of the corporation had been forced to mortgage the school to keep it open. Much of this debt remained outstanding when the stock market crashed in October 1929. As the decade of the 1930s began, the school's economic position slipped from marginal to dangerous.[70]

A personal tragedy foreshadowed the trying years ahead. In January 1930 Charles Rabold was killed in an airplane crash in California. Rabold was the faculty member whose charisma and other qualities best matched Johnson's, and she was planning to turn the school over to him when she retired. Rabold's death, coming late in her life and at a difficult time for the school, was a loss she never got over. In contrast to the public resilience she showed after the earlier tragedies in her life, this time her grief was visible.[71]

As the depression wore on, Johnson had to work at maintaining her characteristic optimism. She did manage to keep up her busy schedule of lectures and demonstrations, for they continued to generate money for the school. But while invitations to speak regularly crossed her desk, Johnson had to face the fact that her name was losing some of its magic. As social reconstructionists seized leadership of the PEA, Johnson and other child-centered educators lost out and moved to the margins of the Progressive Education movement. By the time she turned 70 in 1934, she was being dismissed as past her prime and written off as a "play schooler" despite her unwavering commitment to single taxation and other social reforms. No publisher would touch her manuscript for a second book, which had to wait 40 years to be published as *Thirty Years with an Idea*

(1974). Some of her New York and Greenwich friends remained loyal, but the collapse of the Fairhope Educational Foundation during the 1930s dried up yet another source of revenue for the school.[72]

Even more seriously, Fairhope was changing. By the early 1930s, socialist influence in community life had all but disappeared, and Johnson found her longstanding political views out of step with those of town leaders. Prominent and successful single taxers, those whose families had gotten in on the community's ground floor, cloaked themselves in individualism and free enterprise, carefully distancing themselves from fundamental criticism of the economic order. With the town's population approaching 3,000 as the decade of the 1930s drew to a close, more and more residents were viewing Fairhope as just a nice place to live.

Desperate to save the school, Johnson had long since cut teacher salaries and reduced other expenses. What else could she do? Every summer it appeared the school would not reopen in the fall, but Johnson refused to give up, turning to donations from friends, supporters, and her personal savings to keep the school alive. Students and parents noticed a decline in the quality of instruction as key faculty members, trying to remain loyal but unable to survive on their salaries, reluctantly said goodbye. Johnson also became more lenient in accepting tuition- and board-paying students with learning or behavioral problems.[73]

At last Marietta Johnson's struggle ended. In poor health, she died on December 23, 1938, 25 years after John Dewey's visit.

THE ORGANIC SCHOOL AFTER MARIETTA JOHNSON

The school's struggle continues to the present day. Over the past 60 years, it has sometimes drifted so far from Johnson's educational ideas, she almost certainly would have disowned it—a perfect example of what happens to many institutions after losing a founder or key leader.

Instead of a quasi-public school with a clear mission, the Organic School has become a strictly private institution in search of a population to serve, marketing itself at different times to Fairhope's local aristocracy, to the parents of child misfits, to segregation-minded whites, to almost anyone who could afford the tuition, and to those who "just want something different" for their children.

Fairhope, too, has changed over the years. A stronghold of conservative Republican politics since the early 1960s, Fairhope is now a pleasant community of 11,000 rated by *Money Magazine* as one of the best places in the nation to retire. Soaring property values have brought great wealth to some of the descendants of its single-tax founders. Roll over, Henry George.

Although Johnson would be disappointed that the Organic School is no longer the heart of the community—and never will be again, barring drastic changes in one or both—better days may lie ahead for the school. Its last two directors have tried to recapture the founder's vision, and they seem to have done a remarkably good job. At the start of the 2001–2002 school year, enrollment was 75 and rising. The students are more racially and ethnically diverse than ever. The high school, closed in 1983, is reopening grade by grade. And much to the disbelief—and occasional delight—of parents who inquire about enrolling their children, teachers give neither grades nor report cards. Today's Organic School is once again a model of what teaching and learning can be when they are not driven by standardized tests and other external standards.

CONCLUSION

A student who had just completed Marietta Johnson's organic winter school in 1926 summed up his experience with these words: "Into every phrase of [the school's] expression enters Mrs. Johnson, whose radiant spirit has so permeated the entire atmosphere that we have long ceased to wonder whether the glow be hers or our own. . . . Granted rare charm of personality, her extraordinary genius seems to have been won through a determined application of commonsense and selfless devotion to a vital purpose."[74] Flowery and excessive by today's standards, these words nevertheless capture Johnson's personal style and the powerful effect she had on many people she met.

Johnson ran the Organic School—indeed, ran her life—on charisma rather than authority. She could control situations without seeming to do so. She could make people believe that her ideas, some of which were genuinely radical, were actually theirs. An Indianapolis newspaper reporter covering a series of Johnson's lectures in 1922 tried to explain her gift: "As one hears her expound her theories, he keeps thinking all the time—exactly so! That's what I've always thought! And—naturally—what good sense she has."[75]

From a feminist perspective, Johnson's life reads as a quest for autonomy. Clearly preferring teaching to mothering, she neglected her private life to advance her public work. Her husband Frank understood; her son Clifford Ernest did not, and he became increasingly alienated from his mother as he grew older. Close friends in Fairhope and elsewhere sustained Johnson as she pressed ahead.

One of her strongest qualities, and one that enabled her to command all the more respect, was her total dedication to her work. No one ever accused Johnson of being a self-promoter; on the contrary, she made countless personal sacrifices to promote organic education. Over thousand of hours at the Organic School, over thousands of miles and hundreds of speeches on the road, Johnson devoted the second half of her life to her educational experiment, hoping it would do nothing less than reform public education in the United States—and eventually change the world.

In a recent dissertation on Johnson and her school, Janet McGrath assesses the impossibly long odds Johnson faced, even as she hoped to accomplish so much: "As a radically child-centered educator, she was utopian; she imagined an ideal egalitarian and non-competitive world, a world which could never exist in an era which worshiped science, efficiency and rationalism."[76]

In the history of the Organic School I wrote for *"Schools of Tomorrow," Schools of Today* (1999), I argued that Johnson's experimental school flourished in its early years because—and probably *only* because—it was an integral part of an experimental community. As long as Fairhope remained a community with a point to prove, it sustained the school, and vice versa. As the community changed, the school lost its critical foundation.[77]

Now, at the end of this biographical chapter, I want to give even more credit to this rare woman and all she accomplished. But I still have to conclude that, even under the favorable circumstances that prevailed for a short time, all the charisma and dedication Marietta Johnson brought to bear, and all the respect she commanded were not enough to win the larger battle she courageously chose to fight.

NOTES

1. Lawrence A. Cremin, *The Transformation of the School: Progressivism in American Education, 1876–1957* (New York: Vintage Books, 1964 [1961]), p. 152.

2. For a history of the school from Johnson's lifetime to the present, see Joseph W. Newman, "Experimental School, Experimental Community: The Marietta Johnson School of Organic Education in Fairhope, Alabama," *"Schools of Tomorrow," Schools of Today: What Happened to Progressive Education,* in Susan F. Semel and Alan R. Sadovnik, eds. (New York: Peter Lang, 1999), chap. 3.
3. Joyce Antler, "Feminism as Life Process: The Life and Career of Lucy Sprague Mitchell," *Feminist Studies* 7 (Spring 1981): 134. See also idem, *Lucy Sprague Mitchell: The Making of a Modern Woman* (New Haven, Conn.: Yale University Press, 1987).
4. Carol Gilligan, *In a Different Voice: Psychological Theory and Women's Development* (Cambridge, Mass.: Harvard University Press, 1982)
5. Sources of biographical information on Johnson include Paul M. Gaston, *Women of Fair Hope* (Athens, Ga.: University of Georgia Press, 1984), chap. 2; Robert H. Beck "Marietta Johnson: Progressive Education and Christian Socialism," *Vitae Scholasticae* 6 (Fall 1987): 115–159; and Johnson's own semiautobiographical account, *Thirty Years with an Idea* (University, Ala.: University of Alabama Press, 1974). The Marietta Johnson Museum, 10 South School Street, Fairhope, Ala. 36532, has reissued *Thirty Years with an Idea* and *Youth in a World of Men,* Johnson's other book (see note 69), as a combined volume titled *Organic Education: Teaching without Failure* (Montgomery, Ala.: Communication Graphics, 1996).
6. Johnson, *Thirty Years with an Idea,* p. 1.
7. Beck, "Marietta Johnson," p. 117; Gaston, *Women of Fair Hope,* p. 67.
8. Ibid.
9. Johnson, *Thirty Years with an Idea,* chap. 1.
10. Ibid., p. 2.
11. Cremin, *Transformation of the School,* pp. 147, 203.
12. Johnson, *Thirty Years with an Idea,* p. 6.
13. Nathan Oppenheim, *The Development of the Child* (New York: Macmillan, 1898), pp. 7–9; 11.
14. Johnson, *Thirty Years with an Idea,* p. 8.
15. Ibid., pp. 8, 12–13.
16. John Dewey, *The School and Society* (Chicago: University of Chicago Press, 1899).
17. C. Hanford Henderson, *Education and the Larger Life* (New York: Houghton Mifflin, 1902).
18. Quoted in Gaston, *Women of Fair Hope,* pp. 68–69.
19. Johnson, *Thirty Years with an Idea,* p. 12.
20. Gaston, *Women of Fair Hope,* p. 68; Beck, "Marietta Johnson," pp. 118, 135.
21. Interview with Dorothy Cain, Fairhope, 31 October, 1994. Cain is a graduate of the Organic School and cofounder of the Marietta Johnson Museum with her late husband, Kenneth Cain, who was Johnson's foster son.
22. Quoted in an interview conducted by Davis Edwards, "Mrs. Marietta L. Johnson of Fairhope, Ala., Discusses a System of Developing the Latent Power of Children and Points Out Weaknesses of Prevailing Methods of Teaching," The *New York Times* (16 March 1913).
23. Interview with Dorothy Cain, 31 October, 1994; interview with Mary Lois Adshead, former director of the Marietta Johnson Museum, 13 April, 1999.
24. Gaston, *Women of Fair Hope,* p. 69.
25. Beck, pp. 118–119.
26. Ibid., pp. 122–133.
27. Henry George, *Progress and Poverty: An Inquiry into the Causes of Industrial Depressions, and of Increase of Want with Increase of Wealth* (New York: W. J. Lovell, 1879).
28. Helpful sources on Fairhope's history include Paul M. Gaston, *Man and Mission: E. B. Gaston and the Origins of the Fairhope Single-Tax Colony* (Montgomery, Ala.: Black Belt

Press, 1993); Paul E. Alyea and Blanche R. Alyea, *Fairhope, 1894–1954* (Birmingham, Ala.: University of Alabama Press, 1956); and Larry Allums, *Fairhope, 1894–1994: A Pictorial History* (Virginia Beach, Va.: Donning Company Publishers, 1994). Gaston, a native of Fairhope and graduate of the Organic School, is working on a new history of the community.

29. Makato Ogura, "'Organic' Chemistry: The Utopian Visions of Fairhope and the School of Organic Education" (senior essay, Department of History, Yale University, April 1993), pp. 12–16; Laura Elizabeth Smith, "A Woman and Her Idea: Marietta Johnson and the School of Organic Education" (B.A. honor's thesis, Department of History, Harvard University, March 1991), pp. 5, 36.
30. Upton Sinclair, *The Jungle* (New York: Doubleday, 1906); idem, *Autobiography* (New York: Harcourt, Brace & World, 1962), p. 162.
31. The discussion in this section is based on Gaston, *Women of Fair Hope,* pp. 71–80.
32. Interview with Dorothy Cain, 31 October, 1994; Gaston, *Women of Fair Hope,* pp. 89–90.
33. Johnson, *Thirty Years with an Idea,* pp. 14–16; Gaston, *Women of Fair Hope,* p. 77. Samuel H. Comings was the author of *Industrial and Vocational Education, Universal and Self-Sustaining* (Chicago: C. H. Kerr, 1905?). Henderson wrote the introduction for the second edition of Comings's book. Henderson and Comings had similar educational interests and may have become acquainted while Henderson was headmaster at Pratt Institute.
34. Gaston, *Women of Fair Hope,* p. 78.
35. Interview with Dorothy Cain, 31 October, 1994; Gaston, *Women of Fair Hope,* pp. 77–78, 89.
36. Johnson, *Thirty Years with an Idea,* pp. 34–35; Alyea and Alyea, *Fairhope,* pp. 156–157; Beck, "Marietta Johnson," pp. 124–125.
37. Ibid.
38. "Welcome to the Marietta Johnson Museum," undated information sheet, Marietta Johnson Museum, Fairhope, Alabama; L. J. Newcomb Comings, "An Intimate History of the Early Days of the School of Organic Education," pp. 3–4, undated typescript, Marietta Johnson Museum. The museum is located in the restored Bell Building on the school's old campus.
39. Gaston, *Women of Fair Hope,* pp. 78, 81, 100–101; Margaret Mead, *Backberry Winter: My Earlier Years* (New York: Washington Square Press, 1972), pp. 68–69.
40. Johnson, *Thirty Years with an Idea,* chap. 3; Gaston, *Women of Fair Hope,* p. 82.
41. See, e.g., Charol Shakeshaft, *Women in Educational Administration* (Thousand Oaks, Calif.: Corwin, 1989), and Jackie M. Blount, *Destined to Rule the Schools: Women and the Superintendency, 1873–1995* (Albany: State University of New York Press, 1998). The museum has a large collection of videotaped interviews with students, teachers, parents, and others involved with the Organic School from its founding to the present. Dorothy Cain, who conducted most of the interviews, emphasized the close connection between school and community during our talk on 31 October, 1994.
42. Johnson, *Thirty Years with an Idea,* pp. 64–65, 84–87.
43. Gaston, *Women of Fair Hope,* p. 94.
44. Interview with Mary Lois Adshead, 13 April, 1999.
45. Johnson, *Thirty Years with an Idea,* pp. 37–39.
46. Ibid.; Cremin, *Transformation of the School,* chap. 3.
47. Johnson, *Thirty Years with an Idea,* pp. 39–40; George Allen Brown, "Memoir: Marietta Johnson and the School of Organic Education," a preface to Johnson, *Thirty Years with an Idea,* p. xiv.
48. Quoted in Davis Edwards, "Mrs. Marietta L. Johnson," (see note 22).
49. "New Ideas about Educating Children Are Being Demonstrated in a Greenwich School," The *New York Times* (27 July, 1913). Other stories appeared in the *Times* on 26 April,

1925; 27 March, 1932; and 2 April, 1933. See also Sidonie Matzner Gruenberg, "An Experiment in Organic Education," *Scientific American* Supplement 2028 (14 November 1914); "Fitting the School to the Needs of the Child," *Literary Digest* (5 December, 1914): 1118–1119; and the editorial in the *New Republic* (21 August, 1915): 64.

50. John Dewey, *Democracy in Education: An Introduction to the Philosophy of Education* (New York: Macmillan, 1916); Gaston, *Women of Fair Hope,* p. 85.
51. Johnson, *Thirty Years with an Idea,* pp. 40–41.
52. Quoted in Mary Lois Adshead, "Marietta Johnson: Visionary," *Alabama Heritage* 58 (Fall 2000): 31.
53. John Dewey and Evelyn Dewey, *Schools of To-Morrow* (New York: E. P. Dutton, 1915), Preface, p. 17; Gaston, *Women of Fair Hope,* p. 85.
54. Dewey and Dewey, *Schools of To-Morrow,* p. 39.
55. Johnson, *Thirty Years with an Idea,* pp. 61–62, 65;
56. Ibid. *Thirty Years with an Idea,* p. 105.
57. For a good summary of the school's activities, see the pamphlet *Daily Program of the School of Organic Education, Fairhope, Alabama* (Greenwich, Conn.: The Fairhope League, 1915?), originally a chapter in the second edition of S. H. Comings's *Industrial and Vocational Education.*
58. Johnson, *Thirty Years with an Idea,* pp. 64–65, 84–87, 95–96.
59. Dewey and Dewey, *Schools of To-Morrow,* p. 23.
60. Gaston, *Women of Fair Hope,* pp. 7–9.
61. Cremin, *Transformation of the School,* p. 211. Some of her letters and other personal papers are in the Marietta Johnson Museum and the archives of the School of Organic Education in Fairhope.
62. Patricia Albjerg Graham, *Progressive Education: From Arcady to Academe. A History of the Progressive Education Association, 1919–1955* (New York: Teachers College Press, 1967), pp. 17–20; Beck, "Marietta Johnson," pp. 151–152.
63. Gaston, *Women of Fair Hope,* pp. 107, 132.
64. Graham, *Progressive Education,* pp. 17–20; Beck, "Marietta Johnson," pp. 151–152.
65. Ibid.; Cremin, *Transformation of the School,* p. 243.
66. Gaston, *Women of Fair Hope,* pp. 101–103.
67. Ibid., pp. 99, 103; interview with Dorothy Cain, 31 October, 1994.
68. Gaston, *Women of Fair Hope,* pp. 104–106; interview with Mary Lois Adshead, 13 April, 1999; interview with Dorothy Cain, 31 October, 1994.
69. Marietta Johnson, *Youth in a World of Men* (New York: John Day, 1929); interview with Dorothy Cain, 31 October, 1994; Gaston, *Women of Fair Hope,* pp. 108–109.
70. "Organic School Building History," undated typescript in folder "Historical Documents of Organic School Campus," Marietta Johnson Museum.
71. "Mr. Rabold and Mrs. Bottstein Victims in California Plane Crash," *Fairhope Courier* (23 January, 1930), p. 1; "Then and Now," *Cinagro* (1947), pp. 7–8, Marietta Johnson Museum; interview with Dorothy Cain, 31 October, 1994. The museum has a collection of *Cinagros*, the school's yearbook, from the 1920s through the 1980s.
72. Gaston, *Women of Fair Hope,* pp. 109–117. Johnson's *Thirty Years with an Idea* is much more specific than *Youth in a World of Men* about her work at the Organic School.
73. Interview with Dorothy Cain, 31 October, 1994; videotaped interview with former students Claire Totten Gray, Joyce Totten Bishop, and Madeline Gibbs Scott, 30 December, 1991, Marietta Johnson Museum.
74. "Farewell to the Organic Winter School," *Fairhope Courier* (11 March 1926).
75. Kate Milner Rabb, "A Hoosier Listening Post," *Indianapolis Star* (1 April 1922).
76. Janet Ruie McGrath, "A School for Utopia: Marietta Johnson and the Organic Idea" (Ph.D. diss., Louisiana State University, Baton Rouge, 1996), p. 397.
77. Newman, "Experimental School, Experimental Community," p. 101.

CHAPTER THREE

MARGARET NAUMBURG AND THE WALDEN SCHOOL

BLYTHE HINITZ

Margaret Naumburg was the charismatic and forceful founder of Walden School in New York City. She went on to become an acknowledged leader in the "new school" movement, and later, the founder of art therapy in the United States. As art educator Judith Rubin says in her memorial tribute, "had she done nothing more than to found the Walden School (1914), a place where freedom and discipline in all of the arts were fostered and were considered central to normal children's healthy development . . . *Dayenu!* ('It would have been enough!')."[1] This chapter examines Margaret Naumburg's life and times, her educational philosophy and its implementation in the Children's School (Walden School), and her leadership qualities and style.

CHILDHOOD AND YOUNG ADULTHOOD

Margaret Naumburg was born on May 14, 1890, in New York City, the third daughter of Max and Therese (Kahnweiler) Naumburg. Max Naumburg came from Bavaria and Therese, from North Carolina. Margaret's upbringing followed the family's strict Germanic-Jewish tradition. One gains a sense of her childhood from the writings of her sister Florence, who kept a list of "Things My Mother Does to Me That I Won't Do to My Children."[2] Florence Cane also wrote that "youthful efforts at creation were either laughed at, or brushed aside, or even swept out with the rubbish as merely interfering with the order and cleanliness of a well-kept house." A companion from her later years wrote, "I shall never know with certainty the reasons for Margaret's advice to me not to bother with relationships, although I felt the echo of past pain in her words."[3]

Naumburg attended private kindergarten and grammar schools, which included three years at the Horace Mann School, she spent one year at Public School (P.S.) 87 in New York City, and prepared for college at the Sachs School. After matriculating at Vassar College in 1908, she remained there for only one year. She transferred to Barnard College, where she majored in philosophy and economics, completing her baccalaureate degree in 1912. During her college career, she studied with John Dewey and was elected president of the Socialist Club, beginning her lifelong interest in social reform. She also read Maria Montessori's book, *The Montessori Method,* which was in vogue at the time. Although she did not have a career path in mind when she graduated, she did not want any connection with education.[4]

In the summer of 1912, she went to Europe to study economic reform with Sidney and Beatrice Webb at the London School of Economics. She was concluding her studies and losing interest in the topic when she learned that Dr. Maria Montessori was beginning her first training class for American and English teachers. As "a would-be reformer without a cause," she traveled to Italy to study with her. In spite of some philosophical and personality differences with the *Dottoressa,* she received her diploma in 1913.

In *The Child and the World,* Naumburg stated that she did not go into education, as many do, because it was the obviously available field for women after college. According to Cremin, her own experiences with schooling had been so unsatisfactory that education was the last thing she wished to touch.[5] Naumburg wrote: "Only when a fundamental change in our present system became a practical possibility to me did I find myself committed to education. From the very beginning, I saw that there might be ways of modifying orthodox education, either to enter the system and work from within, or to make a fresh start, outside of all accepted institutions, and construct a plan with new foundations."[6]

Naumburg tried both approaches. She founded a private Montessori class that met for half a day and, at the same time, a half-day Montessori kindergarten in the New York City Public School system, with strikingly different results.

SETTING THE STAGE

To place Naumburg's work in context, let us briefly examine what was happening in New York City neighborhoods and in social movements in several countries. During the second decade of the twentieth century, the New York City neighborhoods of Greenwich Village and the West Side of Manhattan attracted "bohemian" and "intellectual" residents. These two groups of people provided many of the children for a new type of private school that subsequently would be called "progressive."[7] The founders of these schools were the philosophical descendants of Pestalozzi, Froebel, and Herbart and were knowledgeable about John Dewey's work. They were women of the suffragette generation, who directed their energies toward righting pedagogical wrongs instead of political ones.[8] In the draft notes for her unpublished autobiography, former Walden School Director Margaret Pollitzer (Hoben)[9] wrote: "Experimental education originated as a social protest. The pioneer experimental schools were nearly all started by educators who were social reformers at heart. They did not begin with the school. They had looked at life first and found it wanting. . . . They saw the mounting number of neuroses and maladjusted personalities. . . . The impetus toward a new education came primarily from the desire for a better society and protest against the sterility and authoritarianism of the contemporary public schools."[10]

Through the changes that they instituted in small schools in the United States and in Europe, these women influenced avant garde thought in education. Margaret Naumburg and Caroline Pratt, the founder of the City and Country School, were a part of this educational transformation. At one time they talked of beginning a school together. Ultimately their points of difference were too many, and their strong personalities did not permit this idea to be put into practice. However, they did communicate with each other often and sometimes socialized together amid the creative artists and writers of Greenwich Village and other New York City "intellectuals."[11]

Very early in her career Naumburg felt herself powerfully drawn to this revolutionary movement in education. She saw that in place of "a process of stuffing, education became a method of laying hold on the fundamental realities of life."[12] Curiosity, interest, imagination, individuality, and self-expression were deemed important. "[Naumburg] was sure that there would be nothing of value to communicate, and no action worth taking, unless the inner resources of feeling and sensibility were rich. Somehow the progressive education movement . . . with the exception of Caroline Pratt's Play School, revealed itself . . . as deficient in considered appreciation of these inner resources, the 'subjective inner life.'"[13]

EDUCATIONAL BEGINNINGS

After Naumburg returned to the United States from Italy, Lillian Wald offered her the opportunity to open a Montessori class at the Henry Street Settlement with a friend. During the year 1913–1914, she "decided that, while Montessori had great insight in looking to the senses for essential educational redirection, the didactic material was unsuited to do anything unusual or creative."[14] Naumburg developed a more eclectic program and continued to modify the use of the materials in the ensuing years. However, in an article published at the time of Montessori's first visit to the United States, Naumburg, in a feminist mode, praised her as a pioneer woman doctor and educator, and a champion of working women worthy of emulation.[15]

To continue her professional development, Naumburg spent the summer of 1914 studying "organic education" with Marietta Johnson in Fairhope, Alabama. In a later piece of writing, she described her school's "agreement with progressive schools in general . . . as a positive acceptance of the physiological, psychological, and sociological data revealed by modern science."[16] During the time period of 1914 to 1916, Naumburg returned to Columbia University for postgraduate work, resuming her study with John Dewey.

Naumburg had discovered Freudian teaching when it was introduced to U.S. readers in 1910. According to her family, she later was more attracted by the work of Carl Gustav Jung, a Swiss psychiatrist and psychologist. She was psychoanalyzed by Dr. Beatrice Hinkle, a Jungian practitioner, between 1914 and 1917. At that time, undergoing analysis demanded unusual courage and imagination because it was not a widely accepted practice. When she founded the Children's School, the teaching staff was asked to undergo psychoanalysis, so that they would be able to conduct a school where children could add to their emotional, as well as intellectual, power. Florence Cane, the art director of the school, and many of the other staff members, complied with the request.[17] Later Naumburg underwent a second analysis by a Freudian, Dr. A. A. Brill, whose work she had read a few years earlier. He was the parent of a Walden School student.

MARGARET NAUMBURG AND WALDO FRANK

Margaret Naumburg met Waldo Frank some time between 1913 and 1914. They moved in the same artistic/literary, socialist, and Jewish circles. According to his journals, Frank recalled nothing about Margaret Naumburg before the winter of his senior year at college. He later wrote that he met her in Paris, thinking that it was the first time. Their relationship may have continued at a musical evening hosted by their mutual friend, Claire Raphael.[18]

Waldo and Margaret carried on a chronicled courtship during which she wrote to him just about every day from February of 1914 until their marriage in 1916. The relationship was kept secret from Margaret's parents until she and Waldo were officially engaged to be married. However, her sister Florence Naumburg Cane, knew about and supported the relationship.[19]

The New York City Board of Education Montessori Kindergarten

From April 19, 1915, through February 1, 1916, Naumburg tried teaching a public school kindergarten class at P.S. 4 in the Bronx. She faced many obstacles while attempting to introduce a half-day Montessori kindergarten program into the New York City public schools. Her well-documented pleas for a room with heat in winter, for new materials, and for curriculum change went largely unheeded by the Froebelian supervisors and the Board of Education administrators. She used funds gathered from friends, her coteacher (Claire Raphael), and those from her own pocket to equip her classroom with the necessary furnishings, supplies, and materials. Correspondence with Waldo Frank during the spring of 1915 mentions her "satisfactory" discussions with Fannibelle Curtis, director of kindergartens for the New York City Board of Education, who supported the Montessori kindergarten experiment, and other people. When she resigned, Ms. Naumburg contacted every major newspaper in New York City. The archival clippings file, replete with headlines, documents the indignities to which the children and their teachers had been subjected.[20] The Naumburg correspondence file provides interesting insights into the inner workings of this urban public school system during the period 1915 to 1916.[21]

The year 1916 was significant in Margaret Naumburg's life. She made several important professional and personal decisions. The private class had a growing enrollment, and Naumburg decided it was time to move her school to larger quarters. She traveled to the Midwest to visit the home of the "Gary Plan," and the Chicago progressive schools. She convinced Waldo Frank that they should marry, after Claire Raphael eloped with Arthur Ries. Naumburg and Frank wed in December 1916. Frank wrote in his *Memoirs* about his "deliberate sabotage of the marriage vow" with Naumburg. He said, "We had wanted to live openly together because we loved each other. She was an educator of whom 'respectability' was expected; therefore, we had to be married. But it was understood between us that we were not really married."[22]

A Look at an Innovative Program

In May of 1916 Margaret Naumburg visited the Gary, Indiana, schools to research a series of articles for the *New York Evening Mail.* The New York City Board of Education had begun an experiment with Gary Plan schools, and Naumburg was asked to assess the workings of the school system in which the Gary Plan began. She met with William Wirt, the Gary superintendent of schools, visited many of the school sites, and met with Dr. Bachman, an investigator of the General Education Board of the Rockefeller Foundation. In a letter to Waldo Frank, she described her difficulty in separating the theory of the Gary schools from the practice and in ascertaining how the things that were inspirational were sustained.[23]

Her newspaper articles described the swimming pool, animals, gardens, creative work in the drawing and music studios, dramatics, and science laboratories. Interviews with Mr. Wirt discussed problems of the Gary Plan, the in-service training of teachers, the teacher of the future, and "socializing the teachers in Gary." The last article in the series depicted a revision of previous attendance practices. Every teacher in the school district was given the responsibility of keeping in touch with a group of 20 to 25 families. The teachers were responsible for home visits and for those families' communication with the schools. The noncompulsory "Saturday schools," in which no formal teaching was done, came closest to the ideas expressed in Naumburg's own later writing.[24]

On this trip Naumburg also visited the Francis W. Parker School in Chicago. She called it "the best private school I've ever seen" and wrote to Waldo, "They have some dramatic work that's fine." She also went to supper at Hull House and attended a "radical party" for Clarence Darrow. In a letter dated May 31, 1916, Naumburg wrote about meeting "Carl Sandburg, a newly arrived poet, brought out by Holt." She also described a discussion of Jung's book, which was new to the group in Chicago.[25]

BREAKING THE TIES WITH WALDO

The Franks lived and traveled together as husband and wife from 1916 through early 1923. Their only child, Thomas, was born on May 12, 1922. After the birth of his son, Waldo "wanted now to be truly married to my wife. But it was too late; she had suffered too much. . . ." During the decade they were together, they had a stormy relationship. He had several affairs with other women, one of which was used as the rationale for the divorce.[26] Waldo married three more times and had several more children. Margaret and Thomas went to Reno via "a Pullman train."[27] Naumburg was required to reside in Reno for six months prior to obtaining a divorce, during which time, she read and painted. Margaret had met and developed an intense relationship with African American author Gene Toomer in the year before her divorce. Toomer joined her briefly in Reno but then left for France. The Franks were divorced in 1924. Upon returning to New York City, Margaret enrolled her son as a student in Walden School. He later attended the Litchfield School, a boarding school in Connecticut. Thomas maintained his relationship with his father, spending vacations and summers with Waldo and his new family. Naumburg's relationship with Toomer ended in 1926.[28]

WIDENING CIRCLES OF ARTISTIC AND OCCULT INFLUENCES

Throughout her life, Naumburg was interested in the Orient, Egypt, and the occult. In late 1924, when she returned to New York City from Reno, she joined a group that followed Georges Ianovitch Gurdjieff, a writer, dance teacher, philosopher, and Russian emigré guru who lived in France. "The Work," led in New York by Alfred R. Orage a London editor, included self-observation, Gurdjieff's sacred dances (yoga-like exercises), and the study of mysticism.[29] After approximately three years, Naumburg shifted her attention to Pojodag House, another occult group. Her younger brother Robert, her sister Florence, and Florence's husband Melville Cane were also involved in this combination of astrology, Egyptian mythology and Christian religion. From the early 1930s through 1940, Naumburg was heavily involved with the medium Eileen Garrett.

This interest in parapsychology paralleled Naumburg's career shift from arts education and educational psychology to art psychotherapy.[30]

THE LAST PHASE OF MARGARET NAUMBURG'S LIFE

Between 1970 and 1974 Naumburg reviewed her life's work and produced a handwritten set of notes, including some reflections on Walden School. In one of her few references to teacher education, she wrote,

> A fundamental revolution in educational methods for the elementary school grades can only be brought about by a revolution in the training of teachers at all age levels. The learning of teachers to teach about specific areas of subject matter at each grade level, whether in public or private schools, has to be greatly modified. Teachers must first be re-educated to *know themselves* as people, to uncover their unconscious motivation. Then they can be prepared to understand and know their young pupils or students—beginning with nursery school and on through college. As yet there are surprisingly few voices to be heard among educational psychologists, who openly advocate a fundamental revolution in the education of young children, adolescents, and college and university students by their teachers.[31]

Dr. Thomas Frank described his mother's last years for a meeting of art therapists as follows:

> In the mid-1970s, my mother's powers began to wane, although she was still teaching at New York University after she had reached the age of eighty. She often recounted, with amusement and pride, that the only way she had avoided being forced to retire earlier was by lying about her age. By 1975 she had left New York City and moved to Boston in order to be close to her immediate family, consisting of her youngest granddaughter, myself, and my wife. During the final period of her life in Boston, . . . [she] became quite frustrated by memory loss, which precluded her continuing the active work schedule she had been so accustomed to. So when she died at the age of 93, . . . it was clearly after her long and productive life, involving the fields of Progressive Education and Art Therapy, had already come to its own natural closure.[32]

Margaret Naumburg died on February 26, 1983, in Brookline, Massachusetts.

THE ORIGIN OF THE CHILDREN'S SCHOOL

As a result of the success of her first Montessori kindergarten, and with the support of friends and colleagues, Naumburg began her own modified Montessori class with Claire H. Raphael in October, 1914. They rented a room at the Leete School on East 60th Street in New York City and advertised their program.[33] Naumburg was the curriculum developer and teacher, and Raphael was the music specialist and assistant teacher. In 1914, the prevailing theories of child care and education were beginning to be questioned. The Children's School opened during this time of educational ferment. As recorded in the school catalogs, the majority of documents date the school's founding from the 1914 inception of the psychoanalytic theory–based nursery school.[34]

Walden School began with a single class of preschool children, who were two, three, and four years of age. As these children grew, the school organization and program grew in proportion to their increasing developmental needs. The program was based on the threefold nature of children, as organisms needing a unified and balanced development of physical, emotional, and intellectual powers. According to Pollitzer, Walden was "one of the first two preschools to regard early childhood years as significant educationally."[35]

Dalcroze Eurhythmics, F. Matthias Alexander's Physical Coordination, Alys Bentley's Correlated Movements and Music, and Dr. Yorke Trotter's Rhythmic Method of Teaching Music were included in the nursery school curriculum. Naumburg studied all of the above-mentioned systems. She published an article about Dalcroze Eurhythmics, in which she explained the adaptation of a system originally developed to assist musicians in more accurate feeling for rhythm, for all children and adults. She connected the "idea of bringing out the inner sense of rhythm that all people have hidden within them," to such practical applications as mill hands working with a specific set of looms in a factory, and "active enjoyment of music" by everyone.[36]

When Naumburg decided to expand her program from a single class into a school, she began to formulate curricula based on her studies of these movement systems, the psychoanalytical theories of Freud and Jung, and an examination of the curricula and methods of the major progressive educators of her time. According to Paul Rosenfeld, Naumburg had developed a working knowledge base about "every sort of child, . . . the harmed as well as the whole, the troubled as well as the bright, the handicapped as well as the privileged. . . ."[37]

Margaret's relationship with Waldo enabled her to move in the inner literary and artistic "circles" in New York and, later, across the country. This benefited her school because she was able to meet, develop professional relationships with, and employ such creative personalities as Lewis Mumford (who taught English) and Ernest Bloch (who taught music). She also utilized and benefited from the expertise of family, friends and acquaintances in matters of publicity and finances.

Detailed descriptions of the Walden School philosophy, program, curriculum, and methodology may be found in later sections of this chapter.

THE GROWTH OF WALDEN SCHOOL

In 1916, Naumburg decided that the Children's School needed larger quarters and the school was moved into two brownstone houses at 34 West 68th Street. A third house was added in 1923 and a fourth in 1925. The school remained at this location until 1933. When the older elementary school children protested against the school name (around 1922) it was changed to Walden School, after Henry David Thoreau's *On Walden Pond.* The name honored the democratic tradition of the New England Transcendentalists. Naumburg said, "This is a current which asserts the rights of the individual against pressures for massive conformity, and the rights of minorities against the oppressions of majorities. In the area of social law it asserts the primacy of the law of conscience. In the area of socio-moral decisions, it questions systemizations and asserts the integrity of individual vision."[38]

The school attempted to be inclusive by offering scholarships to some children who were unable to pay the tuition; however, it was always necessary to have a majority of paying pupils for the school to remain open. In the lower school of 1929, an attempt was made "to have a representative group" of children from "different kinds of homes—workers' homes on the east side, professional people's homes, and the elaborate homes of the very wealthy in a large city."[39]

THE PHILOSOPHY AND THEORETICAL BASES OF WALDEN SCHOOL

The idea that the first five years of a child's life are most significant, most plastic and, therefore, most creative educationally, became the cornerstone of the school philosophy. A spontaneous community environment was developed for children aged two to four, planned on the basis of their needs. As these children moved on to higher groups, new two- to four-year-olds replaced them in the nursery and lower school. The curricular problems changed from year to year as the children gained in experiences; the basic ideas, however, were the same throughout the school. This structure made the nursery a unified, integral part of this free school.[40]

The Walden School staff believed that analytic psychology, the study of the unconscious, provided more satisfying solutions to many problems of child development than other theoretical formulations. They agreed with the psychoanalytic theorists who argued that all maladjustments in later life could be traced back to the influence of the first five years of life.[41] The observation and study of children's play was an important part of their methodology. They demonstrated, through their record keeping in an "environment which was planned to meet the needs of the children and was freely molded by them," that young children have "a much deeper, more complex, and more significant psychology than educators had supposed."[42]

The Walden School nursery contributed to research in the field of analytical psychology by keeping records of the emotional development of the children and disseminating them through professional publications. Staff study of and engagement in psychoanalysis deeply affected their handling of sexual problems. In frequent conferences with parents, they emphasized the need for answering the questions of small children about birth and their bodily structures in a natural, simple way, at the level of the understanding of the child.[43] The staff also believed in "habit formation." They felt that during the nursery years the child could be reached directly and taught anything. Therefore, this was the time to effect habit-formation. Habit clinics were held for both parents and children. Author and biographer Paul Rosenfeld complimented Naumburg on a relationship with parents that enabled her to make a case "for freedom of the personality." He emphasized that "her chief care is to preserve in each child its own identity."[44] Rosenfeld stated that the school was free of dogmas, placing Naumburg philosophically between the academicians and those who take a laissez faire attitude—resigned not to influence the child in any way.[45]

SOCIALIZATION OF THE INDIVIDUAL

Naumburg discussed social development in several published works. She said that two- to four-year-old children who had good emotional adjustment did not need the suggestions of adults to become socialized. She wrote that, "the moving center of socialization in a school lies not within the will of the teacher but within the wishes and impulses of the children. These we have sought to direct according to our aim."[46] She and her staff found that study of the individual interests and creative powers of each child was necessary, for adults to provide adequate outlets for expression at every age. Then the child's social development was enhanced, and each child was able to function within the social life of the school group.

This was an area of major disagreement between Dewey and Naumburg, as evidenced in several publications.[47] In 1928, Naumburg described her examination of unpublished teachers notebooks from the Laboratory School at the University of Chicago, shown to her by Dewey. She was "surprised to find how regularly all the schemes of making and doing were set beforehand."[48] Based on her own research, Laurel Tanner agrees, stating: "Certain conditions had to be met for children's activities to lead to learning. Dewey's curriculum was a structured curriculum. Teachers began with the concepts that they wanted children to learn and planned activities accordingly. In no sense was learning incidental—if by that is meant accidental. In too many efforts to follow Dewey, the difference has been misunderstood."[49] Naumburg was of the opinion that Dewey's "influence went deeper in transforming the curriculum than in forwarding the independent development and activity of the children themselves."[50]

SOCIALIZATION INTO THE GROUP

Naumburg stated that a new technique of education was needed for the "new schools." The psychological approach should devote the same care and attention advocated to an individual child to the life of a group of children passing through "its phases of physical, emotional, and intellectual adaptation." She reminded her readers that two groups can never be identical because they are composed of individuals who come together at a given time and place. Naumburg felt that the methodology used by teachers who facilitated group socialization was more complex than that of the "project method," as exemplified by the work of Dewey and Kilpatrick.[51]

In her writing, Naumburg described the process by which a child was socialized into a group. Each child entered Walden nursery school as an impulsive, "separate little world," who wanted to satisfy its own immediate needs. As each "self-seeking entity" encountered other children in sudden undesigned contact, their need for each other in their play scheme became, for the time being, greater than the compulsion to grab blocks and build alone. Thus, the need and desire for playful cooperation created the beginning of an organic group. Naumburg stated that if the teacher consciously plans the conditions of group life, and "knows just how to hold back or redirect the energies of the youngsters, it is possible for children as young as two or three" years old to engage in cooperative activity.[52] Pollitzer later wrote that the Walden staff had "learned a great deal by leaving the children alone, within the limits set by safety, and watching them work out their relations in their own plans." Through changes in the environment, and analysis of teaching methods and the school organization, the staff and the director collaboratively developed an educational philosophy that balanced and correlated the life of each child with the life of the entire social group.[53]

THE ORGANIZATION OF WALDEN SCHOOL

The Roles and Responsibilities of the Teaching Staff

Walden School teachers subscribed to the philosophical and psychological tenets of the founder. Teachers were an integral part of the development of the school philosophy and curriculum, along with the students. The school was organized so that every class from nursery school through high school had a teacher with whom the boys and girls in the group developed a close personal relationship. The teacher was enjoined to avoid an overemotional attitude or an overpositive or dynamically influential one. The ideal was to create "an informal, happy contact" between teacher and child. Instead of becoming emotionally involved with the child, the "new mother at school" demonstrated affection for the child and won his or her confidence. The teacher gradually was able to help the child do things for his or her own sake rather than for their effect on an adult. The teacher showed an equal interest in and affection for all members of the group. Pollitzer cites this relationship with the teacher as "an intermediate step between the home and the creating of new relations in life outside."[54]

In addition to the teachers with group responsibilities, trained creative specialists in art and science were introduced, as the children grew older. At the elementary school level, the special teachers tended to visit the group classrooms. For example, Mrs. Cane visited the teachers and students in their classrooms to monitor the children's progress and to offer any necessary individual assistance. On the other hand, the older children went to the laboratories, shops, and studios for particular studies.[55]

Class Size, Program and Time Schedule

Class size was limited to 15 or 16 students, at all levels. Based on the fact that no 16 children have the same degree of ability or speed in any given subject, every group was divided into two or three sections, according to their capabilities in each subject. This permitted briefer, more concentrated periods of work that allowed children to move ahead more quickly in their studies. The small groups had time periods of different lengths available to them. The schedule provided for alternation between periods of intense concentration under the constant supervision of a teacher, and periods wherein the child, having mastered sufficient technique, was allowed to work independently. Free periods in the program meant that the child was free to choose a school subject on which he or she preferred to spend *more than* the required minimal time.[56] The schedule also allowed the teacher time for individual instruction, when needed. Subgroup work did not supplant the use of full group meetings in any class.

For the first few years, the school had a half-day program for two-and-a-half to seven-year-olds. This was due to the lack of space for all of the activities that a full-day program for young children requires. After the move to 68th Street, Naumburg's plan for an "all day school" was implemented. Her rationale was that a longer day allowed every phase of the children's lives to be carried on in a world of their own, where they could work, play, sleep, and eat together. They would learn to share community interests, without neglecting their personal growth, through study and creation. The all-day plan "made possible a more significant and balanced arrangement of the city child's day, by substituting the control of sympathetic and interested teachers for the misguided efforts of maids and governnesses [*sic*]." It allowed for "wider use of outdoor life in special trips and play schemes."[57] Nitscheke and Goldsmith later reported that through the full-day program, each child gained a sense of responsibility toward himself and toward the school community.[58]

The full-day schedule was not compulsory for the nursery level group, so half of the children left at noon and the others remained until 4:30 P.M. Like Montessori, the Walden School staff thought of the daily routines in terms of self-activity on the part of the child. The program of the day was based solely on health routine. All other activities were chosen entirely by the children. Weather permitting, children spent the first hour indoors and the remainder of their activity time (morning and afternoon) outdoors.

Physical Environment

Naumburg and her staff attempted to create a homelike and intimate, rather than an institutional, atmosphere in the facility. "Gay print curtains, vari-colored tables and chairs, odd bits of pottery, Indian baskets made each room distinctive in color scheme and arrangement. And the children of every group then went ahead, added, improved, and changed the rooms, with paintings, wood carvings, pottery, and so on, of their own devising."[59]

After the move to the 68th Street buildings, the nursery groups were on one floor, with easy access to their own rooftop playground. This allowed the children to move easily between indoor and outdoor venues at will. "As the children of the school matured, a more systematic cooperation about the care and decoration of the school became possible."[60] The older children decorated their own classrooms and dining areas, as well as those of the nursery and lower school children.

Each of the groups below high school was allotted two classrooms so that louder and more active group movements would not interrupt quiet work and individual play. For the youngest children, one of the rooms was used chiefly for quiet work, such as painting, modeling with clay, stringing beads, playing house, etc. The other room was used for more lively activity, such as ball playing, riding kiddie cars, working with blocks, or taking trips via boat or auto. In the latter room, there was a workbench, equipped with light hammers and specially chosen little saws, a vise, nails, and a box of wood.[61] Teachers in the lower and elementary schools worked with a subgroup in one classroom. The balance of the class worked in the adjoining classroom—preparing a play, planning a trip, writing stories, or finishing required work, on their own initiative.[62]

All of the staff members regarded the inclusion of equipment and apparatus for physical development as important, because gross and fine motor skills form a foundation for cognitive learning. Small children acquire their contact with life largely through hands-on experimentation with materials and testing their own physical powers, therefore, simple, child-size materials and equipment were within easy reach of the students in each classroom. The environment was organized so that the teacher remained in the background as much as possible. The literature thus demonstrates that the mandate for hands-on constructivist experiences in nursery schools began not in the 1950s, but in the 1920s, or even earlier.[63]

Many kinds of apparatus leading to the development of large motor skills and muscular control were included in the setting, because they develop a general feeling of independence. The outdoor roof play area had various kinds of swings, a rope ladder, a slide, and a sandbox with "non-rustable" toys/kitchen utensils. It was the scene of much dramatic play.[64]

The Curriculum of Walden School

The curriculum evolved from the needs of the children. Naumburg wrote that its original aim was to develop the threefold nature of each child, individually. Later directors highlighted the school's policy of limited adult guidance and learning by personal experience rather than through external controls.[65]

Naumburg expected each student to do a minimum of required work in all of the regular school subjects whatever his or her chief interests or abilities might be. Rosenfeld wrote of the flexibility of the school, stating that any system or course of instruction was there tentatively—a scheme to be tried out against any new fact encountered by Naumburg and modified at the behest of experience. He observed that programs of instruction might be drawn up carefully each fall, but they remained perfectly flexible and subject to improvisations.[66]

Rosenfeld pointed out that the children drew pictures and wrote words about "the material presented to them in the form of a lecture or in the course of a sightseeing visit . . . about history and open-air markets alike." Math came from keeping the market books for the school; geography, from a visit to the market and seeing the variety of "foodstuffs" available. The children saw and spoke with people of many ethnic groups, which brought up discussions of immigration. This led to an anthropology course for the 12-year-olds, taught by Dr. A. A. Goldenweiser of the New School for Social Research, beginning with the study of the Iroquois Indians and continuing through primitive cultures, taboos, superstition, religion, morality and inheritance of acquired characteristics. Pupils formed self-constituted groups for playlets, discussion of a topic, and "the expedition of any commonly assumed business." In the science laboratory, seven- and eight-year-olds melted glass tubing, made ink, and experimented with magnets, batteries, and a steam engine.[67] Naumburg wrote of the school's general approach to subject matter in this way: "Progressive schools generally feel the need of a more unified and more significant correlation of subject matter. They agree to attempt a revision of subject matter by inclusion of material more stimulating and more closely related to the lives of the children; [and] they try to introduce better textbooks and special equipment." They are "anxious to improve the methods and approach of teacher to child. They involve the children in the selection of the year's work."[68]

Curriculum ideas and theories at Walden School changed over time. During the first few years, the school waited patiently for the child to "evince an appetite for plaiting rafia [*sic*] or drawing with crayons on paper." The child was surrounded by materials that he or she was physically capable of manipulating and left to select among them. When the child began to choose, the teachers could assist when instruction was called for.[69] As higher level classes were added, the curriculum and the available materials expanded.

Language and Literacy

Language acquisition is an important part of every child's development. Like Jean Piaget, Margaret Pollitzer realized that when a child acquires language, "adults are apt to assume an understanding on the child's part of the concepts behind the words. This is far from the rule. The child is apt to become lost in a haze of auditory and visual sensations, words and experiences, which are beyond his range, and which therefore carry no real meaning for him."[70]

The students of the Children's School had books and writing materials, as well as the construction materials mentioned above. Many of the children learned to write, and then to read, because they wanted to. The method of teaching each child was adjusted to that child's particular ability. Some learned to write first by touching the letters; some by sounding, and others by seeing them. Lucille Deming underscored Naumburg's initial reluctance to teach reading in the lower school. She wrote, "Miss Naumburg has no particular desire to have the children acquire 'the tools' so early, but home surroundings have doubtless stimulated their interest in reading and writing. The children seem to enjoy the learning for its own sake, as an interesting activity."[71]

In 1925, all restrictions on the teaching of reading to the six- and seven-year-olds were removed. Prior to that time, teachers had "waited for an interest to arise even if it should be deferred until the seven, eight, or nine year level. Experiences indicated that psychological difficulties arise when children do not begin to read until so late. This is due to the dearth of suitable material for the older age and to the feelings of inferiority that arise in children when everyone around them has acquired this skill."[72]

More changes had taken place by 1935. The Walden teachers wrote, "We now set the environment of the six year olds to encourage reading with much flexibility and no pressure. Many children, we feel, show an interest at that age and make a beginning during the course of that year, and reading and writing, greatly vitalized by their activities and interests, are definitely planned as a part of the day-by-day experience of the seven year olds."[73]

Mathematics

In the early years of the school, the younger children learned "number work," mostly through the incidental counting and measuring needed in their activities. By 1935, the staff had also altered the study of mathematics. The teachers engaged in reflective discussions of their own findings, as well as the investigations of experts on the learning processes for arithmetic. The math program that was devised for the lower groups covered less subject matter than that demanded by the conventional schools. It was based on the genuine experiences that children can have with numbers. Walden's math curriculum attempted to eliminate the confusion that results from attempting to teach processes with which the children have no experience.[74]

The Creative Arts and Art Education

Florence Naumburg Cane and Margaret Naumburg both had a deep interest in art and art education. When Florence commented about the art program at the new school, Margaret invited her to become the art director. Florence was directly responsible for the art program; however, during the time that she directed the school, Margaret was involved in the planning and execution of the program. At Walden School both sisters developed and honed techniques that they would use later in other artistic ventures.

In 1916, a reporter wrote in the *Evening Mail,* "Miss Naumburg has her own ideas about teaching youngsters drawing and dancing. The children are encouraged from the first to express their feelings and ideas with paints and crayons. They are given no problems and no models. Their crude scratchings in the beginning are taken seriously. Gradually, as they gain control of their medium, they evolve individual patterns. Children four years old in a few months' time make designs that are clearly stamped with their personality and show a feeling for symmetry, balance, and color harmonies."[75] DeLima, Deming, and Rosenfeld concurred in this opinion of the school. Naumburg, herself, wrote that children develop an interest in color relations and pure design much sooner than she had been led to expect before she tried out her own experiments.

Cane had a unique and radically new way of working with children in art. In 1926, when the coloring book was the only standard art equipment in most schools and a child's proficiency was measured by his ability to copy, she wrote, "The direction of my teaching has been toward the liberation and growth of the child's soul though play and work and self-discipline involved in painting."[76] In their classrooms, the children found open shelves containing piles of drawing paper, boxes of crayons, easels, and paints. They were encouraged to use these materials in an unrestricted manner. When a child showed a drawing to a teacher the teacher refrained from asking the leading question,

"What is it?" It was believed that unless the child volunteers the information, he or she is apt to make up an answer to satisfy the grown-up. As a result, instead of the usual story pictures, a great quantity of abstract designs came from each of the children. Cane never made suggestions of any kind unless she was asked to do so. Even then her suggestions seldom had anything to do with the canvas itself but usually with the child's idea concerning the work."[77] Deming described the utter absorption with which the children worked and the evident satisfaction that they took in the results in this way: "At almost any time that a visitor comes into the school he may see a child here or there intent on a large drawing, studying it critically from time to time and working apparently oblivious of other activities going on in the room. Sometimes a frail, timid little child will use the boldest contrasts of passionate color, putting the energy he does not dare to express in his relationships with other children into this creative form of work."[78]

Under Cane's guidance, the children painted what they felt impelled to paint.[79] Their pictures reflected or expressed their feelings and attitudes toward the environment. Whatever doubts one might have about the soundness of the theory, the results were impressive, and the school's art program was an acknowledged success. The paintings of the Walden School children were on exhibit for successive years at leading art galleries in New York City. They attracted an unusual amount of attention from artists and professional critics, who were astonished at the originality, the design, the feeling for composition, and the richness and color of the work.[80] The teachers of The Walden School learned to read the evidence afforded in the drawings, writings, and other creative efforts of the children and to use it in their ongoing work. Although De Lima felt that some of the teachers carried their psychoanalytic interest and vocabularies to an extreme, she lauded the school for being "unusually perceptive of the subjective growth and needs of its children and unusually responsive to these needs."[81]

The knowledge that Margaret Naumburg gained regarding children's direct use of free art expression and the psychoanalytic interpretation of the art "products" at Walden School formed the foundation for her later moves to art education and further study of and work in the field of psychology. It surely was one of the bases for her pioneering work in art therapy.

MARGARET NAUMBURG AND WALDEN SCHOOL SEPARATE

The birth of her son, the constant need for fundraising, and the increasingly difficult financial picture took their toll on Margaret Naumburg. She turned the executive direction of the school over to two longtime staff members, Margaret Pollitzer (Hoben) and Elizabeth Goldsmith (Hill) during the 1921–1922 school year, and became the advisory director. Naumburg retired as a director in 1924 and, thereafter, is listed as "Founder" in school publications. She was a member of the Board of Trustees for a number of years. Unlike many of the other board members, she did not have the money to contribute from her own funds toward half of the annual salary for teachers who took a sabbatical. However, she did review the philosophical, program, and curriculum development of the school over the years.[82]

The school community began a fundraising drive around 1929, to enable a move to larger quarters. In 1933, the school was moved to the structure at One West 88th Street and Central Park West. The building was renovated and envisioned as "state-of-the-art" to creatively meet the present as well as envisioned future space needs of the school. The school continued to grow until it served children from the ages of 2 through 18 in the nursery and lower school, the elementary school, and the high school.

When Walden School closed in the 1980s, the West 88th Street building (which housed the high school) and another building (which housed the nursery and lower school) were sold to the Trevor Day School, which operates them today. The Trevor School does not follow the Walden philosophy and program.

Although Margaret Naumburg "withdrew" from directing Walden School, in part, because of her unwillingness to continue to fight the economic battles that took up 50 percent of her time, ten years later she returned to speak with the Walden teachers about the economic depression and its effects. She divided her talk into three sections: (1) discussing the direct impact of the depression on Walden teachers and the school itself; (2) dealing with Walden teachers as part of an "advance guard" of progressive educators; and (3) looking at their role among "that international minority who are devoting themselves to bringing the dream of a non-competitive society into the schools of the world." She reminded the teachers that the group of "newer experimental schools to which we belong" came to life through protests of two groups against the old type of education. One was a group of parents "reaching out vaguely for a more liberal and creative type of school." The other consisted of teachers "who were despairing over the lifeless routine of formal education and dreamed of better things." She recounted how Walden made it through its third year of existence, when World War I broke out and many parents withdrew their financial support of the new building. She concluded with exhortations to find the most effective way to develop the progressive school for today and to form an alliance of progressive schools. Her suggestion that educators of all the progressive schools meet to exchange views and to plan for the future led to a lively discussion that spilled over into the following week's teachers' meeting.[83]

Part of the discussion included faculty members' impressions of an address by George Counts to a meeting of the Progressive Education Association held at the Dalton School. Counts "emphasized the imperative necessity for social planning as a result of the present social order." Teacher Ann Elias stated that she "would like to see observation come from the children rather than have finished ideas presented to them." Naumburg responded; "That is the correct approach, but the teacher must have her point of view clearly thought thru [*sic*]." Oscar Nimetz said, "Progressive schools are impotent in bringing about a revolution thru [*sic*] children. They can only contribute to the revolution outside of school." Naumburg remarked, "Preparation for a new social order seems necessary." At the next meeting, Elizabeth Goldsmith began a discussion of freedom, indoctrination, and socialization. Goldsmith reminded the group that one point on which the school had disagreed with Dewey was that of socialization from the early years on. She stated, "We felt that the child had to develop an integrity from within before he could have any kind of social consciousness in the real sense. For that reason we have not had large auditorium assemblies and large meetings of children."[84]

STRONG FEMALE LEADERSHIP

Margaret Naumburg's life and work, including the development of Walden School, provides another example of how strong female leadership was instrumental in the founding and development of progressive schools. From its inception through its first 30 years, the Children's School (Walden School) had strong female leadership. All of the original administrative positions were filled by progressive women with strong academic or creative backgrounds. Between 1914 and 1922, Margaret Naumburg organized and directed the school. She had very specific ideas about the path she wanted the school to take; her leadership style was to "lead by example."[85] Like Murray succeeding Caroline Pratt at the City and Country School, the directors who succeeded Naumburg kept the traditions of the school, adhered to the philosophy of the founder, but created their own modifications that moved the school forward. In 1931, Elizabeth Goldsmith (Hill) was director and psychologist, and Margaret Pollitzer (Hoben) was advisory director. Cornelia Goldsmith was the assistant director for nursery and lower school (children two through six years of age) from 1928 to 1943. She taught class IV in which the children were five years old (of kindergarten age). Because Elizabeth and Cornelia Goldsmith and Margaret Pollitzer had been hired by Naumburg, her philosophy, principles and ideas guided Walden School long after she left the directorship of the school. Pollitzer and Goldsmith organized and supervised the school by remaining faithful to its psychoanalytic principles, its emphasis on the individual child, and its core curriculum of creative expression through the arts. The last of the original (female) staff members to become director was Hannah Falk (later Mrs. Werner Regli). Other directors who made outstanding contributions to the field of early childhood education in general included Mr. Alvie Nitscheke, Mr. Vinal Tibbetts, Dr. Alice V. Keliher, Mr. Hans K. Maeder, and Dr. Milton E. Akers.[86]

THE PROGRESSIVE LEGACY OF WALDEN SCHOOL

In the introduction to *"Schools of Tomorrow," Schools of Today: What Happened to Progressive Education,* Susan Semel presents the 1990 statement of principles of the Network of Progressive Educators.[87] Detailed below are some of the ways in which Walden School exemplified these principles during its 70-year existence.

1. Education is best accomplished in an environment in which relationships are personal.
   a. Descriptions in the literature demonstrate that there were close personal relationships among students, parents, faculty, and administration at Walden School.[88]
   b. Staff members built deep, strong relationships that lasted through their later professional careers. Dr. Alice Keliher, chair of the Advisory Committee, brought Cornelia Goldsmith from Vassar College to found and lead the Day Care Unit of the New York City Department of Health. Goldsmith, who later became the first full-time executive director of the National Association for the Education of Young Children, was influential in the appointment of Dr. Milton E. Akers as a succeeding executive director.
2. Teachers design programs that honor the linguistic and cultural diversity of the local community. Schools embrace the home cultures of children and their families. Classroom practices reflect these values and bring multiple cultural perspectives to bear.
   a. An attempt was made to have a representative group, including children of "workers," "professional people," and "the wealthy." A percentage of children were on scholarship, "so that parents interested in the purposes of the School, but unable to meet the charges, need not be prevented from sending their children."[89]

   b. Many of the staff had studied or visited overseas, particularly in England, France, and Italy.
   c. The students learned about the diversity of the local community from an anthropological and artistic-cultural perspective.[90]
3. Teachers, as respected professionals, are crucial sources of knowledge about teaching and learning.
   a. There are many descriptions in the literature about the methods used by the teaching staff.[91]
   b. The teachers were highly qualified, holding both academic credentials and personal recommendations. Many of the original staff members moved up to positions of greater educational and administrative responsibility within the school and, later, in the broader field of education.[92]
4. Curriculum balance is maintained by a commitment to children's individual interests and developmental needs, as well as a commitment to community within and beyond the school's walls.
   a. The commitment to individual interests was demonstrated throughout the school's history. For example, Alvin Johnson, editor of the *New Republic,* wrote in 1923, "Naturally there is no room in Miss Naumburg's scheme of instruction for the textbook, the set task, the rigid standard. From the youngest child in the school to the oldest, a wide field is kept open for individual interest and personal choice."[93]
   b. Walden School moved, in the 1930s, from being a child-centered to a community-centered school. Students saw the unemployed all around them. They made a specific contribution to the community by working in child care centers, helping out in settlement houses, and participating in interschool organizations. The high school students participated in "Walden-On-Wheels" trips to permanent landmarks of American history. Students attended a welfare conference, met with a southern NAACP group, spent a night at a settlement camp, and corresponded with people across the country.[94]
5. Students are active constructors of knowledge and learn through direct experience and primary sources.
6. All disciplines—the arts, sciences, humanities, and physical development—are valued equally in an interdisciplinary curriculum.

   Principles 5 and 6 above, are amply demonstrated in the literature.
   a. As noted above, Florence Cane developed an outstanding creative art program.
   b. The school physical and biological sciences laboratories were heavily utilized for student projects from the time of the move to 68th Street and on.
   c. Pioneering work in history, anthropology, and other aspects of the social studies was done both by resident faculty and guest lecturers.
   d. Samples of student writing may be found in DeLima's *Our Enemy the Child.*[95]
   e. The school curriculum incorporated all of the movement and physical development methods that Naumburg had studied, as well as those of the music and movement specialists.
7. Decision making within the school is inclusive of children, parents, and staff.
   a. The children's involvement in decision making is exemplified by a photograph of a school committee formed for a specific purpose.[96]
   b. Parents and staff were involved in decision making up to a certain level.

LASTING INFLUENCE ON EDUCATION

Walden School and comparable independent experimental schools had a documented direct influence on public education. According to Walden records, the New York State Education Department, Division of Elementary Schools, incorporated methods pioneered at the independent schools into its plans and assessment criteria beginning in 1935. Among the methods and procedures worked out at Walden and later widely disseminated were the following:

1. Children and teachers participate in selecting subject matter and planning activities.
2. The formal recitation is modified by conference, excursion, research, dramatization, construction and sharing, and interpreting and evaluating activities.
3. "Social studies" is substituted for the separate study of geography, history, and civics.
4. Emphasis is placed on construction and creative expression in the arts and crafts.
5. Special attention is given to the needs and interests of individual children.
6. Discipline is self-control rather than imposed control, and constructive and social behavior rather than "order."
7. The school has a psychologist.
8. More general evaluations of students are used, rather than precise report cards.
9. Parent participation in schools is sought and encouraged.
10. Movable, rather than fixed, desks provide flexibility in room arrangement.[97]

Many of the innovations in policy, curriculum, and method made at the Children's School and Walden School by Margaret Naumburg, her colleagues, and her successors have found their modern expression in the Developmentally Appropriate Practices currently articulated and supported by the National Association for the Education of Young Children.[98]

Margaret Naumburg had a strong and distinct influence on the progressive school movement and other contemporary schools, upon her peers, and on those who followed in her educational path. The modern heirs to Naumburg continue to build on the strong foundation laid down by this "founding mother" of the progressive movement.

ACKNOWLEDGMENTS

The author wishes to acknowledge the research assistance of Ms. Malinda Wolfgang during the Spring 2000 semester. Ms. Wolfgang was a recipient of the undergraduate research scholarship of The College of New Jersey Chapter of Phi Kappa Phi for her work with the author, leading to her paper, "Margaret Naumburg and Caroline Pratt: True Progressive Educators" and ABH/ABC-CLIO 2001 prize-winning bibliography. The author gratefully acknowledges the assistance of the archivists and librarians in the Department of Special Collections, Rare Book, and Manuscript Library, Van Pelt Dietrich Library Center, University of Pennsylvania, Philadelphia, Pennsylvania, particularly Nancy Shawcross, John Pollock, and Amey Hutchins.

NOTES

1. Judith Rubin, "DAYENU: A Tribute to Margaret Naumburg," *Art Therapy* 1, no. 1 (October 1983): 4.

2. The list was described in a paper by Mary Cane Robinson, one of Florence Cane's twin daughters. Mary Cane Robinson, "Statement on Florence Cane for the AATA Meeting on October 28th 1983" (paper presented as part of the symposium "Roots of Art Therapy" at the annual meeting of the American Art Therapy Association, Chicago, Ill., 27 October, 1983), TMs (photocopy), p. 2, personal Collection of Thomas and Kate Frank.
3. Patricia Buoye Allen, "The Legacy of Margaret Naumburg," *Art Therapy* 1, no. 1 (October 1983): 6.
4. Robert Holmes Beck, "American Progressive Education, 1875–1930" (Ph.D. diss., Yale University, June 1942), p. 166. Robert Holmes Beck interviewed Margaret Naumburg on 15 March 1941, and Florence Cane on 2 February, 1941, in New York City. See also Blythe Hinitz, "Margaret Naumburg (1890–1983): A Childhood Educator Who Brought Freudian Theory to the Early Childhood Classroom" (paper presented at the 22nd World Congress of the World Organization for Early Childhood Education [OMEP], Copenhagen, Denmark, 15 August 1998).
5. Lawrence A. Cremin, *The Transformation of the School: Progressivism in American Education 1876–1957* (New York: Vintage Books [Random House], 1961), p. 211.
6. Margaret Naumburg, *The Child and the World: Dialogues in Modern Education* (New York: Harcourt, Brace and Company, 1928), pp. 31–32.
7. Patricia Albjerg Graham, "Community & Class in American Education, 1865–1918," in *Studies in the History of American Education,* eds. H. J. Perkinson and V. Plannie (New York: John Wiley & Sons, 1974), pp. 173–174.
8. "The Walden Story: forty years of living education: Walden School—1914–1954." (New York: privately printed, 16 March, 1954), p. 4, Margaret Naumburg Papers, box 13, folder 783, Special Collections, Van Pelt Dietrich Library Center, University of Pennsylvania, Philadelphia, Pa. (hereafter cited as Naumburg Papers).
9. Margaret Pollitzer Hoben (1895–1983) was a teacher at and later codirector of Walden School during the 1920s.
10. Margaret Pollitzer Hoben [n.d.–1980s] draft notes for autobiography, TMs (photocopy), p. 19, property of Polly Greenberg (originals in the archives of Teachers College, Columbia University library).
11. Maxine Emelia Hirsch, "Caroline Pratt and the City and Country School: 1914–1945" (Ed.D. diss., Rutgers: The State University, October 1978), p. 95, ftn. 33 (Hirsch interview with Jean Murray 27 February 1976).
12. Robert H. Beck, "Progressive Education and American Progressivism: Margaret Naumburg," *Teachers College Record* 50 (1958–1959): 202.
13. Ibid.
14. Beck, "American Progressive Education, 1875–1930," p. 166.
15. Margaret Naumburg, "Maria Montessori: Friend of Children," *The Outlook* vol. 105 (13 December 1913): 798–799.
16. *The Walden Story,* p. 4; Margaret Naumburg, "Montessori Class: The House of Children" (New York: privately printed, [n.d.–1914]), p. 3; idem, Margaret Naumburg, "The Walden School," in *Twenty-Sixth Yearbook of the Society for the Study of Education: Part I: The Foundations and Technique of Curriculum-Construction,* ed. Guy Montrose Whipple (Bloomington, Ill.: Public School Publishing Company, 1926), pp. 333–334.
17. Beck, "Progressive Education," p. 201; Margaret Naumburg, "A Direct Method of Education," 1917, TMs pp. 2, 3, 5, 7, Naumburg Papers, box 15, folders 839–840; idem, "A Direct Method of Education," in *Experimental Schools* bulletin (New York: Bureau of Educational Experiments, 1917), Naumburg Papers, box 15 folder 837. Reprinted in *Experimental Schools Revisited: Bulletins of the Bureau of Educational Experiments,* Charlotte Winsor, ed. (New York: Agathon Press, 1973), pp. 39–45 (page references are to the reprint ed.).
18. Waldo Frank journal entries from 1914, Waldo Frank Papers, Department of Special Collections, Van Pelt Dietrich Library Center, University of Pennsylvania, Philadelphia, Pa.

(hereafter cited as Frank Papers); Alan Trachtenberg, ed., *Memoirs of Waldo Frank,* with an introduction by Lewis Mumford (Boston: University of Massachusetts Press, 1973), p. 64.

19. Florence Naumburg Cane, Eagle Nest, New York (Camp Burnt Rock, Adirondacks, Blue Mountain Lake, New York), to Waldo Frank, 23 and 29 August 1914, in the hand of Florence Naumburg Cane, Frank Papers, box 60.
20. Although she stated in print that she resigned after six months, in actuality the experiment lasted from 19 April 1915 to 1 February 1916. Clippings from: *New York Times,* February 1916, *New York Call,* 5 February 1916, *Tribune,* 5 February 1916, *Sun,* 5 February 1916, and *Evening Sun,* 5 February 1916, Naumburg Papers, box 13, folders 749–752.
21. Margaret Naumburg to George J. Gillespie (correspondence 1915–1916); correspondence from the Board of Education, City of New York to Margaret Naumburg, Naumburg Papers, box 13, folders 749–752.
22. Trachtenberg, *Memoirs of Waldo Frank,* p. 206. The published dates of Margaret and Waldo's wedding differ. The entry in *Who's Who of American Women* (1964), written by Naumburg, states 29 June 1916. The timeline at the beginning of Waldo's memoirs states 20 December 1916. as does the biography in Rosenfeld. Paul Rosenfeld, *Port of New York* (New York: Harcourt, Brace and Company, 1924, reprint, Urbana: University of Illinois Press, 1961) (Page citations are to the reprint edition.) Although Waldo married again several times, he kept every letter and telegram that Margaret Naumburg, his first wife, sent to him.
23. Margaret Naumburg, Gary, Indiana, to Waldo Frank, 18 May, 1916 and 24 May, 1916, in the hand of Margaret Naumburg, Margaret Naumburg Frank Letters, Frank Papers, box 60, folder 1916.
24. Margaret Naumburg, "'City Must Serve Its Children,' Says William Wirt Expounding Workings of the Gary System," *Evening Mail* [1916]. Clipping in Naumburg Papers: "Henry's Parents Visit Gary" TMs; "Interview with Mr. Wirt" TMs; "Mr. Wirt on the Teacher of the Future" TMs; and "Socializing the Teachers in Gary" TMs, Naumburg Papers.
25. Margaret Naumburg, Gary, Indiana, to Waldo Frank, 27 May 1916 and 30 May 1916, in the hand of Margaret Naumburg, Margaret Naumburg Frank Letters, Frank Papers, box 60, folder 1916.
26. Trachtenberg, *Memoirs of Waldo Frank,* p. 206.
27. Interview with Dr. Thomas Frank, 5 July 2000, Cambridge, Mass.
28. Ibid.; Amey A. Hutchins, "Biography" in *Register to the Margaret Naumburg Papers,* Special Collections, Van Pelt-Dietrich Library (Philadelphia: University of Pennsylvania, December 2000), pp. iv-v.
29. Trachtenberg, *Memoirs of Waldo Frank,* p. 250 note 14.
30. Interview with Kate Frank, July 5, 2000, Cambridge, Mass.; and Hutchins, "Biography," pp. iv-vi.
31. Margaret Naumburg, *Writings 1970–1974,* in the hand of Margaret Naumburg, pp. 3–4, Naumburg Papers.
32. Thomas Frank, "Margaret Naumburg: Pioneer Art Therapist (1890–1983): A Son's Perspective" (paper presented as part of the symposium "Roots of Art Therapy" at the annual meeting of the American Art Therapy Association, Chicago, Ill., 27 October 1983) TMs (photocopy), p. 7, personal Collection of Thomas and Kate Frank.
33. Naumburg, "Montessori Class," pp. 1–4, Margaret Naumburg, "Montessori Class: Second Year," 1915. Naumburg Papers, box 13, folder 748.
34. Some documents state that 1914 was the first year because it was the first of two years of Montessori classes conducted at The Leete School. The Walden School catalogs date the beginning of the school from 1914. Cremin, *The Transformation of the School,* and Hal May, ed., *Contemporary Authors* (Detroit, Mich.: Gale Research Company, 1983), p. 109, cite 1915 as the founding date.

35. Hoben, draft notes for autobiography, TMs (photocopy), p. 19 and insert 2 to p. 19 in the hand of Margaret Pollitzer Hoben (photocopy), property of Polly Greenberg, originals in the archives of Teachers College, Columbia University library.
36. Margaret Naumburg, "The Dalcroze Idea: What Eurhythmics Is and What It Means," *The Outlook* 106 (17 January 1914): pp. 127–131.
37. Rosenfeld, *Port of New York,* p. 119.
38. Rosenfeld, *Port of New York,* p. 118; Alvie Nitscheke and Elizabeth Goldsmith, "The Walden Nursery School," in *Twenty-Eighth Yearbook of the National Society for the Study of Education: Preschool and Parental Education,* ed. Guy M. Whipple (Bloomington, Ill.: Public School Publishing Company, 1929), p. 226; Hutchins, "Biography," p. iii; and Margaret Naumburg, "Statement of Philosophy," [n.d.], TMs, p. 7, Naumburg Papers, box 13, folder 788. In *The Child and the World,* Naumburg refers to Franklin B. Sanborn and William Torrey Harris, *A. Bronson Alcott: His Life and Philosophy,* 2 vols. (Boston: Roberts Brothers, 1893). Alcott was a major figure among the transcendendentalists.
39. Nitscheke and Goldsmith, "The Walden Nursery School," p. 227.
40. Ibid., p. 223.
41. Ibid.
42. Ibid., p. 228; Margaret Pollitzer, "Foundations of the Walden School," *Progressive Education* 2 (January-February-March 1925): 17.
43. Nitscheke and Goldsmith, "The Walden Nursery School," p. 225.
44. Ibid., p. 224; Rosenfeld, *Port of New York,* pp. 121, 124, 132.
45. Rosenfeld, *Port of New York,* pp. 121–124.
46. Naumburg, "The Walden School," pp. 334–335, 338–339.
47. Margaret Naumburg, "A Challenge to John Dewey," *The Survey Graphic,* 60 (15 September 1928): 598–600; idem, *The Child and the World,* pp. 102–122; idem, "The Crux of Progressive Education," (25 June 1930), TMs, pp. 1–6. Naumburg Papers, box 15, folder 835 (published in *The New Republic* 63:145); and John Dewey, "How Much Freedom in the New Schools?" *The New Republic* 63 (9 July 1930): 204.
48. Naumburg, "A Challenge to John Dewey," p. 599; idem, *The Child and the World,* p. 110.
49. Laurel Tanner, *Dewey's Laboratory School: Lessons for Today* (New York: Teachers College Press, 1997), p. 26.
50. Naumburg, "A Challenge to John Dewey," p. 599; idem, *The Child and the World,* pp. 111–112.
51. Naumburg, "A Challenge to John Dewey," p. 600; Rosenfeld, *Port of New York,* p. 130. For a discussion of the "project method," see V. Celia Lascarides and Blythe F. Hinitz, *History of Early Childhood Education* (New York: Routledge Falmer [Garland], 2000), pp. 210–211, 230–231, notes. 111–119.
52. Naumburg, *The Child and the World,* pp. 121–122.
53. Pollitzer, "Foundations of the Walden School," pp. 17–18; Naumburg, *The Child and the World,* pp. 121–122.
54. Naumburg, "The Walden School," p. 335; Nitscheke and Goldsmith, "The Walden Nursery School," p, 225; Naumburg, "A Challenge to John Dewey," p. 600; and Pollitzer, "Foundations of the Walden School," p. 17.
55. Naumburg, "The Walden School," pp. 335–336; Rosenfeld, *Port of New York,* p. 132; and Florence Cane, "Art in the Life of the Child," *Progressive Education: A Quarterly Review of the Newer Tendencies in Education* 3, no. 2 (April-May-June 1926): 155–162.
56. Naumburg, "The Walden School," pp. 337–338.
57. Lucile C. Deming, "The Children's School," *Experimental Schools.* bulletin (New York: Bureau of Educational Experiments, 1917), Naumburg Papers, box 15, folder 837. Reprinted in *Experimental Schools Revisited: Bulletins of the Bureau of Educational Experiments,* Charlotte Winsor, ed., (New York: Agathon Press, 1973), p. 48; Naumburg, "The Walden School," p. 335.

58. Nitscheke and Goldsmith, "The Walden Nursery School," p. 224–226.
59. Margaret Naumburg, "How Children Decorate Their Own School," in *Creative Expression: The Development of Children in Art, Music, Literature, and Dramatics,* 2d ed., ed. Gertrude Hartman and Ann Shumaker (Milwaukee: E. M. Hale and Company, 1939), pp. 50–51. (Originally published in *Progressive Education* [1926].)
60. Ibid.
61. Naumburg, "The Walden School," p. 335; Nitscheke and Goldsmith, "The Walden Nursery School," p. 226.
62. Naumburg, "The Walden School," p. 337.
63. Pollitzer, "Foundations of the Walden School," p. 16.
64. Naumburg, "The Walden School," p. 338; Nitscheke and Goldsmith, "The Walden Nursery School," pp. 225–226; Pollitzer, "Foundations of the Walden School," p. 16.
65. Naumburg, "The Walden School," p. 334; Nitscheke and Goldsmith, "The Walden Nursery School," pp. 224–225, 228.
66. Rosenfeld, *Port of New York,* p. 124.
67. Naumburg, "The Walden School," pp. 337–338; Rosenfeld, *Port of New York,* pp. 121, 124, 126–128; Agnes DeLima, *Our Enemy the Child* (New York: New Republic, 1926; reprint, New York: Arno Press and The *New York Times,* 1969), pp. 204–210 (page references are to the reprint ed.); and Nitscheke and Goldsmith, "The Walden Nursery School," p. 223.
68. Naumburg, "The Walden School," p. 338.
69. Rosenfeld, *Port of New York,* pp. 125–126.
70. Pollitzer, "Foundations of the Walden School," p. 16.
71. Deming, "The Children's School," p. 47.
72. Schauffler, Marjorie Page, ed., "*Schools Grow: A Self-Appraisal of Seven Experimental Schools,*" TMs draft (photocopy), pp. 2–3 (New York: Associated Experimental Schools, 1937) RG, 2.5 box 2, folders 6 and 7. City and Country Archives, New York, N.Y.
73. Ibid. This curriculum statement is probably one of the reasons Margaret Naumburg was unhappy with the direction in which the school was going.
74. Schauffler, "Schools Grow."
75. Marion Weinstein, "Life in Schools Means Freedom," *Evening Mail* (New York), 23 November 1916.
76. Walden School catalogs. "*The Walden Story,*" p. 7.
77. DeLima, *Our Enemy the Child,* pp. 203, 211.
78. Deming, "the Children's School," pp. 47–48.
79. Beck, "Progressive Education," pp. 201–202; Cremin, *Transformation of the School,* pp. 213–214.
80. Rosenfeld, *Port of New York,* p. 130.
81. DeLima, *Our Enemy the Child,* p. 214. See also Cane, "Art in the Life of the Child," pp. 155–162; Florence Cane, *The Artist in Each of Us* (New York: Pantheon Books, 1951); and Blythe Hinitz, "Margaret Naumburg and Florence Cane: Progressive Sisters for Creativity in Schools" (paper presented at the annual meeting of the History of Education Society, Chicago, Ill., 30 October 1998), pp. 6–7.
82. Naumburg, *The Child and the World,* p. vii.
83. Margaret Naumburg, "Talk to Walden School Teachers," TMs (photocopy), 17 May 1932, property of Thomas Frank.
84. "Discussion following Margaret Naumburg's Introductory Paper on the Problems and Position of Walden School Teachers in the Present Crisis," TMs (photocopy), 17 May 1932, pp. 1–7; "Teachers Meeting: Continuation of Discussion of Previous Meeting," TMs (photocopy), 24 May 1932, pp. 1–12, both property of Polly Greenberg, (originals in the archives of Teachers College, Columbia University library).
85. Naumburg, *The Child and the World,* p. vii.

86. Walden School catalogs. "*The Walden School: Nursery Through High School,*" (New York, 1931), p. 1, Naumburg Papers, box 13, folder 778; "*The Walden Story,*" pp. 4, 11, 14, 22, 26; and "Walden School on its 50th Anniversary: Its Raison D'Être: Its Educational Pioneering: Past, Present and Future." (New York, April, 1964), pp. 16–17, Naumburg Papers, box 13, folder 786. See also Cornelia Goldsmith, *Better Day Care for the Young Child Through a Merged Governmental and Nongovernmental Effort: The Story Of Day Care in New York City as the Responsibility of a Department Of Health 1943–1963 and Nearly a Decade Later—1972.* (Washington, D.C.: National Association for the Education of Young Children, 1972); Cornelia Goldsmith, "The New York City Day Care Unit: An Interview with Cornelia Goldsmith," interview by James L. Hymes, Jr. (Boston, November 1970), *Early Childhood Education Living History Interviews,* Book 2: *Care of the Children of Working Mothers* (Carmel, Calif.: Hacienda Press, 1978), pp. 57–72; and Blythe Hinitz, "Early Childhood Education Managers," in *Historical Dictionary of Women's Education in the United States,* ed. Linda Eisenmann (Westport, Conn.: Greenwood Press, 1998), pp. 139–141.
87. Susan Semel, "Introduction," in *"Schools of Tomorrow," Schools of Today: What Happened to Progressive Education,* eds. Susan F. Semel and Alan R. Sadovnik (New York: Peter Lang, 1999), p. 18.
88. De Lima, *Our Enemy the Child,* pp. 212–214; Nitscheke and Goldsmith, "The Walden Nursery School," pp. 224–226; and "The Walden Story," pp. 19–20.
89. Nitscheke and Goldsmith, "The Walden Nursery School," p. 227; and "The Walden Story," p. 18.
90. See Walden School catalogs, e.g., "A Child's Own World," p. x.
91. Deming, "the Children's School," pp. 46–48; Naumburg, "A Direct Method of Education," pp. 41–45; De Lima, *Our Enemy the Child,* pp. 205–206; and idem, *The Child and the World* (entire book).
92. "The Walden Story," pp. 20–24.
93. Alvin Johnson. TMs (article). *The New Republic* (28 March 1923). Naumburg Papers, box 13, folder 768.
94. Walden School catalog. "The Walden Story," pp. 17–19.
95. De Lima, *Our Enemy the Child,* pp. 215–237.
96. "The Walden Story," p. 14.
97. Walden School catalog. "The Walden Story," pp. 21–22.
98. Sue Bredekamp and Carol Copple, eds. *Developmentally Appropriate Practice in Early Childhood Programs,* revised ed. (Washington, D.C.: National Association for the Education of Young Children, 1997).

Chapter Four

CAROLINE PRATT AND THE CITY AND COUNTRY SCHOOL

Mary E. Hauser

> I began the adventure innocently enough when at sixteen I became the teacher of a one room school not far from Fayetteville.

With this simple, direct statement, Caroline Pratt introduced her autobiography, *I Learn from Children.*[1] However, the simplicity of the written statement masks the breadth and depth of her experience as a progressive educator and a radical social thinker. It is the purpose of this chapter to examine Caroline Pratt's life work to understand the nature of this woman, who contributed so much to educational thought and practice. Her approach to progressive education has survived, with few changes, for almost a century at the school she founded in Greenwich Village, New York, the City and Country School. What kind of a woman was able to craft a philosophy that has provided such a solid foundation for children to learn? How was her personal and professional experience shaped by the radical-thinking, activist women who were part of her professional and social circles? What was the administrative style of this woman whose school was so much an extension of herself? How did the political and social climate of the times influence her views?

These questions have guided my research into the life of Caroline Pratt, a woman whose story I became acquainted with more than eight years ago. When I read her autobiography as part of my quest to understand the historical underpinnings of early childhood education, I made a strong connection with her thinking about teaching young children. As I have journeyed through her writings and the writings of others about her and her school, I have come to know and respect this complex woman, who lived and worked in a time of dramatic social and educational change.

Once she founded the Play School (the original name of the City and Country School) in New York's Greenwich Village, in 1914, Miss Pratt, as she was known, focused almost all of her energies on the development and operation of her school. It became her all-consuming passion. This chapter draws on sources that provide insight into her early life and development, her thinking about her school and its students, and her administrative practices and style. Although most of the work published about Pratt has been done within the context of the progressive movement in society and in education, I will also consider her life and work in the light of the feminist movement in the early part of the twentieth century. After briefly establishing a context for understanding her work as progressive as well as feminist, both perspectives will be considered as the adventure of her life and work unfolds. I have drawn on Pratt's own writing as much as possible, as one way that her pragmatic thinking is reflected is in her very direct way of expressing herself.

THE PROGRESSIVE CONTEXT

Caroline Pratt's place in the private school progressive movement is well established. She is cited by leading documenters of progressive education[2] as having developed a school that "best reflects the tenets of the progressive ideal."[3] Although those tenets were never easily agreed upon by their proponents, most progressive educators were committed to a child-centered pedagogy that meant children had great freedom in classrooms in order to maximize the development of their creative potential. Activities connected to the world outside the classroom engaged the children in purposeful experimentation and problem solving, rather than in simulations of skills or thinking that would be required later in life.

The Play School embodied a pragmatic orientation to education, which was consistent with the progressive movement of the early part of the twentieth century. This form of progressivism was less focused on social reform than progressive thought had been at the turn of the century. Rather, the emphasis was on conditions that would foster individual expression.[4] Pratt's views on the importance of expression in her educational scheme were clearly influenced by this aspect of progressivism:

> One of the contributions which I hope the Play School may make to education is in art expression. . . . The failure to develop art impulse in education is partly due to the non-recognition of the fact that that the free play of children is art. . . . As the children play with drawing materials, with plasticine, with blocks and toys, with words, with dramatics, the emotions are freed and in a primitive way art is produced. The emotional processes in the children's play are identical with the processes we call art in adult life, and which, with an acquired technique, give us art production. It is what the modern school of artists in their simplified methods of expression try to realize.[5]

In addition to the development of the individual, there was a strong sense of development of community, as the children worked together to represent their experiences through block play. In an article in *Progressive Education,* Pratt presented her views on community: "Children have to learn, then, how to get along with each other, first to attain their own ends and finally because there is a reward in having friendly relationships with one's peers or, on the other hand, in fighting what is disapproved."[6]

The curriculum that she devised was based, for younger children (ages three to seven), on exploration and expression through play, especially block play. "Play with these materials," she wrote, "is an organizing experience. At three or four, children come to block building, for example, after a good deal of experiencing with their bodies. They themselves have been everything . . . cows, animals of all kinds, engines, everything that moves."[7]

Curriculum in the classes for older children (ages 8–13) revolved around practical jobs that each grade level performed for the good of the school

> . . . so that the school eventually functioned as a self-sufficient community. For example, the VIIIs (eight year olds) ran the school store, the IXs the post office. The Xs produced all of the hand-printed materials for the VIIs such as flash cards and reading charts, while the XIs ran the print shop and attended to all of the school's printed needs: attendance lists, library cards, stationery, and so forth. The XIIs first made toys, then weaving, until they finally settled on the publication of a monthly publication called *The Bookworm's Digest* which reviewed new children's books as well as 'Old Favorites,' a particularly popular column in the journal. As students performed jobs they also learned basic academic skills as well as more sophisticated principles of economics, for example. What emerged from this model was a community of independent young children who were actively engaged in their learning, while contributing to the life of their school community.[8]

Both the block play activities and the jobs also provided the context for the numerous field trips that served as laboratories for the social studies, the central curricular theme. Pratt described excursions with the young children: "From our corner of Fourth and Twelfth Streets we could make our journeys, none so long as to tire young children; we could go out and find the answers to the children's questions . . . the six children and I spent a great deal of time at the docks. The river traffic, endlessly fascinating brought good simple questions to their lips. . . ."[9] This description in a school bulletin shows the connection between the jobs and field trips: "The IXs interest in the materials sold in the store leads them to investigate their source of supply—an investigation which is further stimulated and augmented by trips and individual library and laboratory research. They study the spelling of words used in their own writings, as well as those used in store transactions."[10]

At all levels, children's activities were the core of the curriculum, and life experience provided the orientation for learning. Pratt put it this way: "The attempt in the Play School has been to place the children in an environment through which by experiment with that environment they may become self-educated."[11] She elaborated on this basic idea in a symposium published in *The New Republic* in 1930: "The new types of curriculum, as shown in what are termed, 'progressive' schools, have in common more and more opportunities for experiencing inquiring, experimenting, and less stress on subject matter and memorizing. More attention is given to the processes of learning and study of growth habits by teachers and less to teaching and the development of teaching methods. Experimenting means experimentation by children and not experimentation with children. In fact, the zeal for teaching is giving place to the effort to provide an environment in which children can gradually take over their own learning processes."[12]

THE FEMINIST CONTEXT

In contrast to the wealth of information that connects Caroline Pratt and her school to progressive education, her relationship to feminist practices in the first 20 years of the twentieth century has not been clearly established. "Feminist" was a new word then, recently imported from Europe, although women's issues had been debated for years before in terms of the right to vote and own property.

As is the case today, feminism at the turn of the twentieth century lacked unified,[13] widely held, mutually shared understandings. "What distinguished the Feminism of the 1910s was its very multifaceted constitution, the fact that several strands were all loudly voiced and mutually recognized as part of the same phenomenon of female avant-garde self-assertion."[14] The strands that were included in this multifaceted conception were women's claim for full citizenship, equal wages for equal work, psychic freedom and spiritual autonomy, sexual liberation, and the independence of wives. Another women's historian described the groups in the developing movement in this way: "There were radical feminists who saw the woman's problem from the point of view of class struggle and class analysis; others who took a sexist viewpoint and considered all women similarly oppressed; still others were pragmatists who worked for women's rights within women's organizations."[15] Obviously none of these ideas was new, but it was the combination, named as feminism, and the fact that it contained an open-endedness and acceptance of internal contradictions that was the standard around which its adherents congregated.

A significant influence on feminism in New York was the Heterodoxy Club,[16] which met in Greenwich Village between 1912 and the early 1940s. It was an organization of women who consciously regarded themselves as pioneers in forming a "new" feminist theory and practice. Their focus was not primarily on the issues of suffrage that related to the legal and political emancipation of women but dealt instead with ideas about sex solidarity, sexual emancipation, and a search for female role models and for a female based system of values and ethics.[17] The personal and private dimensions of liberation were paramount. Speaking to a public forum in 1914, Rose Young articulated the personal dimension that virtually all of club members valued: "To me feminism means that woman wants to develop her own womanhood. It means that she wants to push on to the finest, fullest, freest expression of herself. She wants to be an individual. . . . The freeing of the individuality of woman does not mean original sin, it means the finding of her own soul."[18]

Caroline Pratt was not a member of the Heterodoxy Club, but several of her friends and professional colleagues were active participants in the organization. One member, Fola LaFollette, was an actress who was associated with the City and Country School between 1926 and 1930. She taught French and English, as well as being in charge of the school library. In a book of testimonials for Pratt, LaFollette wrote, "It would take a book—and I have only a page—to express my thought about your work and my feeling for you. This can only tell you on your birthday that I love you always, that I miss the daily association with you."[19] Another Heterodoxy Club member, Anne Herendeen, a writer and editor, in a 1934 note to LaFollette, reported that she (Herendeen) "was doing some collaborations with Caroline Pratt and loving it."[20] Pratt's 1933–1934 journal had entries that mentioned time spent with Herendeen, as well as professional contacts that she had with Elisabeth Irwin, the founder of another Greenwich Village progressive school, The Little Red Schoolhouse. Irwin and her partner, Katharine Anthony, were both Heterodoxy Club members. Although Caroline Pratt's philosophy and practice were not intentionally feminist, the influence of relationships with these women and with her lifelong companion, Helen Marot—a leader in women's labor organization and investigation—cannot be discounted. These women were clearly advocating for, as well as enacting, the advancement of the new feminism of the twentieth century.

It is important to recognize at the outset of this chapter that some readers of Caroline Pratt's life may argue that making a distinction between her as a feminist practitioner or a progressive practitioner is unnecessary. However, the relevance of using a feminist perspective to understand Caroline Pratt and the other women who are the subjects of this book is found in the contemporary feminist viewpoint that without women's history and a feminist consciousness, the achievements of women in public and professional life can easily be erased. Unless women's part in the development of progressive education is recorded, that influence will be lost to our understanding. Although John Dewey, William Heard Kilpatrick, and Harold Rugg are well established as the primary exponents of the progressive movement in education, the work of Pratt and similar female educators also has a place in the progressive education record. Their work provides important examples of strongly grounded practices of progressive education. Finally, using a gendered lens to understand Pratt's life and work recognizes and makes explicit the role of the sociocultural context in constructing human knowledge. A variety of women activists and men and women artists who had creative approaches to their lives and work were part of the social and professional milieu that shaped Pratt. This chapter, then, will include exploration of the connections between Caroline Pratt and both progressive education and early twentieth century feminism to add depth to our understanding of her as a person and of her important work in education.

EARLY YEARS

Caroline Pratt was born in 1867 into a middle-class family of social and political stature in Fayetteville, a small agricultural town near Syracuse in upstate New York. Her grandfather and father were prominent businessmen, active in the political life of the community in which several generations of Pratts had lived. Her mother was described as the typical Victorian lady, whose sphere of influence was primarily family, church, and charity. She was active in religious, social, and self-improvement clubs that were popular during the end of the nineteenth century.[21]

Caroline Pratt grew up as an independent, capable woman. There is ample evidence that she valued those qualities in herself and that others saw them in her as well. She related that "at ten, my great aunt used to say that I could turn a team of horses and a wagon in less space than a grown man needed to do it."[22] When her great uncle Homer suggested that she become a teacher, she didn't hesitate to "pin up my braids" and do just that. Later, at Teachers College, at Columbia University, in New York City, to which she had earned a scholarship because of her promise as an educator (after five years of country school teaching), her independence was also evident. She openly questioned the Froebelian training methods to which she was being subjected. Originally, she planned to study to be a "kindergartner,"[23] but she soon questioned that decision:

> . . . the more I learned of the newest Kindergarten methods of the day, the more uncertain I became.
>
> Little children, we were taught, should begin the school day by sitting quietly in a circle. They could sing or have a story read, but the sitting in a circle was the important thing. This would give them an awareness of the unity of human life. There was a good deal of this mystical fol-de-rol, and to my practical mind it was more like learning to walk a tightrope than to teach any children I had ever met.
>
> It had a practical value, as I have since come to understand, though not one with which I could have the slightest sympathy. You taught children to dance like butterflies, when you knew they would much rather roar like lions, because lions are hard to discipline and butterflies aren't. All activity in the Kindergarten must be quiet, unexciting. All

> of it was designed to prepare the children for the long years of discipline ahead. Kindergarten got them ready to be bamboozled by first grade.[24]

She talked about this period as being her "rebellious twenties." She wrote, "My first act of rebellion . . . was to go to the Dean and announce that kindergarten was not for me. Guessing rightly that country living had given me a capable pair of hands, he suggested Arts and Crafts. Soon I was happily hammering and sawing in the Manual Training Shop."[25]

Kindergarten and manual training were at this time the two current innovations in education. Kindergarten was becoming popular in public education, and industrial art (tool work) and domestic art (homemaking) were newly developed extensions of high school content into the elementary level. Caroline soon became disillusioned with her shop training, however: "In the shop I had been learning to use one tool after another (someone had decided which were easier and which harder, and the tools were given us in that order), and to perform one kind of operation after another (someone had also decided the order of difficulty for these). I sawed to a line; I planed; I chiseled. I made joints, one after another, until I reached the crowning achievement, the blind dovetail. But I never made a single object!"[26]

These events at Teachers College at Columbia University are significant in understanding Pratt as a progressive educator and as a feminist. In this formal teacher training both as kindergartner and as manual training specialist, she encountered practices that she could not accept. She saw them as being not only disconnected from the experience of the student but also unrelated to the reality of the task to ultimately be accomplished in the world outside of the classroom. Opposition to both of these conditions was embodied in principles of progressive education. An experiential child-centered curriculum, in which children learn from doing real work, is central to progressive philosophy.

The pragmatism of the progressive philosophy is also evident in her thinking. She had taught successfully for five years before coming to New York, using observations of her students and their interests to guide her teaching; and when "experts" did not tell her anything that she could use, it further intensified her pragmatic thinking about teaching. It is probable that the idea that she could use her own abilities to teach, to observe children, to see what they did and what they wanted to do was strengthened during her unsatisfying time as a student.

Feminist thought during this period of Pratt's professional development included advocating for women's individual choices and opposing the self-abnegation historically expected of women.[27] Although she did not invoke feminist ideology to defend her position, Pratt clearly did not feel that she had to rely on the word of college teacher experts to determine how she should think about the teaching and learning process.

ADVENTURES OF THE MIND IN PHILADELPHIA

After earning her two-year certificate, her first job was teaching manual training at the Philadelphia Normal School for Girls. She became a special instructor in woodworking—training teachers to be proficient in skills such as gauging, squaring, sawing, chiseling, planing and boring, and doweling and chamfering. It was the perspective of the school administration that even though schools employed specialists in manual training, classroom teachers needed to understand the value of manual training in the curriculum and how it could be related to other school subjects. She was to implement this practice on a larger scale in her own school.

It was in Philadelphia that Pratt met Helen Marot. Her association with Marot significantly influenced the direction of her life and helped her to focus her independent thinking. Helen Marot (1865–1940) was described as a social investigator, writer, editor,[28] and feminist, who did not seem to care a lick for money, clothes, or fast cars. Her world was that of art, ideas, and reform.[29] Raised in an affluent Quaker household, her education took place in private schools and also at home. She was encouraged by a father who told her at age 14, "I want you to think for yourself—not the way I do."

Helen Marot and a friend had organized a small library in Philadelphia that was a haven for politically liberal and radical individuals. People went there not only for the available materials but also for the opportunities for discussion. Reflecting on her experience there, Pratt recalled, "I took to spending a good deal of time there myself. With my own adventures in learning ever on my mind, I saw there still another aspect of education. Listening to these people, many of them graybeards, as they argued and studied, I began to see education not as an end in itself, but as the first step in a progress which should continue during a lifetime."[30] The library was a place in which she and other women could engage in self-development, in contrast to the practice of self-sacrifice, which was expected of so many women at the turn of the century.

In 1899, assisted by Pratt, Helen Marot undertook an investigation of custom tailoring trades in Philadelphia for the U.S. Industrial Commission. Deeply affected by this experience, Pratt wrote, "It was for me a bitter eye-opener, that experience. The work was done in the home, with no limit to the hours the people work and no check on working conditions—which were also living conditions, and which from both points of view were appalling. The contrast with educational practice as I knew it was painful. Helen and I often discussed the futility of trying to reform the school system, if after leaving school human beings had to earn their living under such conditions as these."[31]

THE ADVENTURE CONTINUES IN NEW YORK CITY

The investigative work galvanized both women into social activism: Marot in labor organization and Pratt in education. In 1901, Pratt moved to New York City with Marot. Marot found employment with the Association of Neighborhood Workers of New York City, where she investigated child labor. Her work led to the 1903 New York Compulsory Education Act.[32] Meanwhile, Pratt combined three jobs to earn her living—one in a small private school and two in settlement houses. The records of Hartley House, one of the settlement houses in which she worked, show that Pratt was paid $750 for a series of 250 carpentry lessons.[33] It is significant that in describing these jobs in her autobiography she makes a point of saying that all of them allowed her a free hand in running the manual training shops.

These three different teaching environments provided Pratt with three different "learning laboratories" in which she could work on the many questions that she had about the methods and materials needed to help children learn. In particular, she was concerned with devising materials that would be effective for children to use for a wide variety of purposes. They had to be both simple and flexible. Lawrence A. Cremin and Robert Beck both refer to her report of observing the child of a friend engaged in working with a miniature railroad system that he was constructing with found materials—"blocks, toys, odd paper boxes, and any material he could find."[34]

In retrospect, this observation was an important turning point in her thinking about teaching and her role in the process. It prompted her to quit her teaching jobs and attempt to produce materials that would be suitable for children to use. She wanted toys that would not limit the children's thinking; what we would refer to today as "open-ended materials." Pratt wrote, "I would make toys which could be used in dramatic play . . . play which would reproduce the children's experience with their own environment. . . . I carefully kept the toys simple in construction so that they could be used as models if the children desired to make others along the same line."[35] However, she was not successful. Her brief ascerbic comment encapsulates this failed venture: "Mr. Castleman and I worked together long enough to wish that we had never met, for the toys were a dismal failure."[36]

Nevertheless, her more well-known invention of the unit blocks, although not a money-making proposition, was, and remains, extremely successful. Generations of children have begun their formal learning in kindergarten in the block corner of the classroom. Blocks had been used since Froebel's "invention" of the kindergarten. But as "kindergartners" began to broaden their concept of kindergarten practice through the scientific study of children as a result of the observations in the growing field of psychology, the complex Froebelian block activities were seen as restrictive to children's development. In contrast, the unit blocks are deceptively simple, but it is their simplicity that provides the versatility needed for children's expression of their experiences. Pratt wanted her blocks to be tools for representing the experiences that children had, both inside and outside of the classroom.

STARTING HER EXPERIMENT IN EDUCATION

In 1913 Caroline Pratt began a two-month experiment with young children from which the Play School developed. With borrowed funds and borrowed classroom space in the Hartley House, she conceptualized a setting in which children would be free to use the materials she collected to construct their knowledge about the world. She went into the Hartley House neighborhood and recruited six five-year-old children to work with her handmade blocks, do-with toys that she had designed and made, crayons, paper, and clay. Her recollections of this first group of children were more about what they taught her than what they, themselves, had learned:

> With my heart filled with gratitude to them for justifying my faith, I was kept busy checking theory against practice. It was so clearly right that play was learning, that this voluntary, spontaneous play-work was far too valuable to be ignored as our schools ignored it, or relegated to spare 'free periods' in the school day, or to the home where a child could work out such play schemes when parents were too busy or too wise to interfere. . . . They were also showing me that children learn to work harmoniously with each other the more quickly and effectively if there is little or no adult interference.[37]

This experiment validated Pratt's vision. In the fall of 1914, with additional funds from her friend Edna Smith, Pratt set up her school in a three-room apartment at Fourth and Twelfth Street in Greenwich Village. She again went out into the neighborhood to recruit her students from among the working-class residents. The first child enrolled was the daughter of the janitress in the building where Pratt had rented quarters for the school. Two girls and four boys, three of whom were from immigrant families, were enrolled in the school.

The events of the 15 years leading up to the opening of the Play School included activities of entrepreneur, inventor, and school founder and head. In the early years of the twentieth century, these were not included in the realm of women's typical endeavors. The work that she did certainly was outside of the boundaries of prescribed gender roles at the time. In her autobiographical writings about this period of her life, she does not question whether these are appropriate endeavors, nor does she wonder how others will receive them. Stories such as these, related years after they happened, tend to be structured to emphasize how the story creates or maintains an identity that is desired by the storyteller.[38] It is interesting, although certainly unintentional, that the identity Caroline Pratt constructed to share with her readers is consistent with feminist thought and action of the early twentieth century.

HER EDUCATIONAL ADVENTURING CONTINUES

So successful did she feel about her work that Pratt began to plan for a larger school the following year, before her first year was even over. The curriculum that she devised during this year was centered on the interest of the children and was designed to enable them to make meaning of their environment. She wrote, "We are not willing to be dominated or have the children dominated by subject matter. We want them to develop strong habits of first-hand research and to use what they find; we want them to discover relationships in concrete matter, so that they will know they exist when they deal with abstract forms, and will have habits of putting them to use; we want them to have a full motor experience, because they themselves are motor; and to get and retain what they get, through their bodily perceptions."[39]

The scope and sequence of the school curriculum, briefly described previously in The Progressive Context section of this chapter, has been well documented in Pratt's autobiography, as well as in a number of publications that she wrote or coauthored with City and Country School teachers.[40] It is unique within the progressive tradition and has remained viable since its inception with few changes. Incorporating computers into the curriculum is an example of the kinds of changes that have been deemed necessary to keep the school viable.

Unfortunately, her original idea of establishing a school for the children of poor immigrant families who lived in Greenwich Village was short-lived. The free tuition was a draw, but the fact that the school continued to emphasize play beyond the kindergarten year was a deterrent to these families, who believed that their children's success in their adopted country depended on education that they understood in a traditional sense. Pratt was unable to convince them of the value of exploration and play in learning. But this perspective did appeal to another segment of Greenwich Village residents—the artistic, creative families who sought the low rents available in this neighborhood for studios and living spaces. As a result, the emphasis that the school placed on creative expression and exploration flourished. The future of the school was guaranteed with the support of these families. Had Pratt begun her school in a different part of New York, one can only speculate about what its fate may have been. As Cremin observed, the history of the school was colored by the fortuitous association of Caroline Pratt and the Greenwich Village intelligentsia.[41]

As described earlier, an important aspect of the school day of children was artistic expression. Paints, crayons, and clay were standard equipment in all of the classrooms for the children. So much did Pratt value artistic expression that a special art teacher—in many cases a practicing artist—was hired to teach art as a specialty. The first such teacher was the sculptor William Zorach, who was a strong advocate of the City and Country philosophy.

Both Cremin and Beck saw the school as an exponent of the expressionism that was flourishing in the early twentieth century in major cities in bohemian type enclaves such as Greenwich Village in New York. The arts were a counter to both the superficial aestheticism of the Victorian period and to the bleakness and spiritual aridity of the Industrial Revolution.[42]

Pratt was surprised at the attention that was soon visited on the school. Evelyn Dewey visited in 1915 and included a section about the school in *Schools of Tomorrow*, which described schools that demonstrated progressive philosophy. However, whether or not the school was considered progressive was not a concern for Pratt. She was adamant about not being placed in any category of educational philosophy or practice: "My resistance to anything in the nature of a blueprint was instinctive and desperate." She added, "All my life I have fought against formula. Once you have set down a formula, you are imprisoned by it."[43]

When Beck interviewed Pratt in 1941 for his dissertation, she told him that she did not evolve her ideas about education through play from the teachings of either John Dewey or William Heard Kilpatrick.[44] Beck thus described her as having fundamental educational beliefs but seeming "to ride no hobby. It was not a school of Dewey or of Rousseau or of anyone else. Practice was altered as experience and imagination suggested the modification."[45]

The following comments by Charlotte Winsor, a colleague of both Pratt and Lucy Sprague Mitchell, describe the kind of controversy that Pratt avoided:

> When I first knew her, even at that time, the progressive 'experimental schools in New York City' which were vanguard movements, had already established themselves with a wide range of philosophical background and ideas. It's hard to believe that in 1924 they were already fighting with each other about conceptual things and what we mean by progressive education. And the City and Country School, which I think was a profound reflection of these new ideas, was at that time being attacked by some other progressive school crowd as being over-intellectualized, as being too concerned with contacts. This was not true freedom for the child. City and Country was not really progressive because they had a curriculum. That we had a curriculum and that [of] all the dreadful things, teachers were expected to write out their curriculum goals at the beginning of the year—this distressed the critics.[46]

Did the school founded by a radical thinker such as Pratt demonstrate what we now term gender equity? Was the curriculum what we now consider gender neutral? Her discussion of this topic in *Experimental Practice in the City and Country School* is evidence that the school was trying to counter the conventional gender stereotypes prevalent at the time:

> We make every effort to get all the children to play with these adaptable materials. [Note: She referred earlier to 'materials which they can work with, which they can dominate, and feel their power over.'] If the children come to us young enough, they work on floor schemes without much urging. The best work is done by the boys although there are good girl block builders as well. We should like to find out why girls are not so ready to organize on a broad basis as boys. Sometimes we may be informed that it is because of the biological difference in the sexes, but in the meantime, we are working upon the supposition

> that it is social. Whatever her future is to be, we can see no reason why it is not as valuable for a girl as for a boy to use materials which lead out into ever wider experiences instead of remaining centered in less active interests. In our own experience we find that we treat girls differently from boys. We cannot help ourselves. If we do it, we who are trying above all else not to make a distinction while the children are so young, how much more is the distinction made in casual contacts? Society cannot get used to 'boyish' reactions in our girls, even very little ones. Excusable roughness in boys still remains questionable in girls. They are pigeonholed by the most tolerant of us, and by others they are victimized by deep-rooted habits of thought. Nevertheless, while we encourage both boys and girls with suitable materials to carry out domestic experiences, we also encourage girls as well as boys to follow any lead out into the bigger world. There are many references to this difference between boys and girls in our school records.[47]

There is also ample anecdotal evidence of the conscious awareness of gender in the work of the students. She reported, "In all these activities (shop work and cooking and sewing), boys and girls were given equal opportunities, and generally their interest in one or the other was not markedly based on sex."[48] She followed this statement with a story about how a holdout—a boy who was "stubbornly masculine"—was treated when he refused to sew an apron, on the grounds that it was girls' work and he wanted none of it. He could not even be convinced by his peers. But when he saw the products of their work, which ironically were from the cooking class, he quickly produced an apron so that he could join them in cooking.

LEADERSHIP PRACTICES

Caroline Pratt was in charge of her school from the day it opened, and, according to some, she remained in charge even after she retired in 1945. How her colleagues, former students, and their parents remember the nature of her influence leaves us with a somewhat paradoxical picture of her leadership. Although some remembered it as a small school that "felt like a family with a strong leader—you had to have that," others likened the way the school was run to National Socialism in Russia.[49] She was always very clear about what should happen at the school. A former teacher recalled that Pratt had a strong voice and penetrating eyes that showed a determination to get things done the way she thought they should be done.

The City and Country School was administered as a teachers' cooperative school, receiving the last such charter in the state of New York. All the teachers who had completed one year of service were members.[50] A staff member recalled that one of the influences of this system was that the staff meetings were about teaching. For example, every new toy that appeared on the market was evaluated by Pratt and her staff before it was allowed to be included in the school. What could it do for the self-image, confidence, and creativity of the children? Was it manageable? How much learning was inherent in the material?[51] Another former teacher's comment indicated that the influence of the school's cooperative organization extended even beyond Pratt's tenure at City and Country. She reflected that when she began teaching in the mid-1950s there was more openness about the program than there is now. She commented, "Of course, then it was a teachers cooperative and it had an effect on practice."[52]

Under the conditions of the cooperative, the school operated under the direction of a five-member executive committee, over which Pratt presided. She remained the ultimate authority over all facets of the educational program.

Her manner was not always authoritarian, however. Another retired teacher commented, "One of her strong points was that she didn't hinder the ideas of others as long as it didn't interfere with the children's education. What teachers did with their political lives outside of the school was OK." An example of a situation in which Pratt saw a teacher's politics interfering is found in her personal journal (1933–1934 school year). She wrote about being very direct in opposing a teacher who wanted a change of assignment for the following year, " . . . whereupon I fell upon her and said I would never trust her with the twelves because of her radical interpretation of history . . . and toward present day economics." Nevertheless, several journal entries later, Pratt acknowledged that she acquiesced to the change that the teacher wanted.

Her adversarial relationships with parents became legendary; even her obituary in the *New York Times* (June 7, 1954) mentioned that she trusted children but not their parents. There are at least two perspectives that can be used to understand her often turbulent relationships with parents. Because she was so definite about the importance of children being in charge of their own learning and adamant that this occur with relatively indirect support by adults, she often saw parental behavior as thwarting this process. She began a letter to a parent in this way, "Regarding our conversation yesterday, I am so positive that I am right in feeling that you should not have anything to do with A's work that I am going to put it down on paper."[53] She often felt that parents were inhibiting the development of children's independence.

Pratt also saw the unique nature of her school as a deterrent to parental involvement. She wrote, "Every progressive school offers a challenge to adults because in order to understand what is going on in it, the grown-up must forget all his own early schooling. He must forget that he depended upon the teacher to keep him straight." She went on to remind the grown-up of other things that he or she must forget: " . . . the furnishings, that it was a sin to look out of the window, or to whisper a remark in some one's ear.[54]

However, portions of one journal entry provide an example of how she did take parents' concerns into consideration: "Had luncheon with Miss F. A good deal of criticism came out. . . . She has a child who hates blocks, doesn't even like the feeling of them . . . thinks we ought to have other materials and tried to suggest something but was as nonplussed as we always are when we face the necessity for other adaptable materials. She thought some children might never want to use blocks. . . . She finally admitted the possible value of blocks but that some children must be eased into them through something else. We had a very friendly time and I thought she gave us something to think about."

A former teacher recalled in a 1999 interview, "Parents were not welcome in the school. The teachers and Pratt ran the school. I think she was afraid of parents . . . she knew children, but not parents." Pratt herself wrote about being afraid of parents, but for another reason. She explained, "For my own part I was afraid that they would get in our way, that they would attempt to curtail our freedom of action, try to steer us closer to the more familiar, more comfortable kind of school."[55]

Caroline Pratt's leadership did not include being a good fiscal manager. The same former teacher related that Pratt had little money, herself, and no sense of it for running the school. She believed so strongly in her school that she just knew it would be supported financially. Lucy Sprague Mitchell's comment echoes that view, "Pratt had such intense belief in her experiment that she felt the world owed it support."[56] For the first three years of her experiment, Pratt received financial support from Edna Smith and thereafter from Lucy Sprague Mitchell. Both women had extensive personal resources and were able to provide the needed financial base for the school. The entire Mitchell family became interwoven with Pratt and the school. A part of the Mitchell property was used as the third location of the school; Lucy taught in it, and her children were enrolled there. However, this very strong connection began to disintegrate by about 1928. By 1930, Mitchell was engaging in her education work elsewhere, and she made arrangements to end her financial support of the school.

Records of a fundraiser in 1936 indicated that George Gershwin contributed the music and was in the cast, George S. Kaufman wrote a skit, Dorothy Parker contributed a scene, and Gypsy Rose Lee made a guest appearance. Pratt may not have been able to balance a budget, but she was able to capitalize on the artistic and creative community of the City and Country School after her benefactors left.

There is ample evidence that by most standards Pratt could not be considered an example of feminist leadership according to its current manifestations in educational practice. For example, Carol Gilligan's work describes female networks of caring relationships as characteristics of feminist leadership. Pratt cared passionately about the education of her students, but her leadership did not did not openly demonstrate a caring network. She did value affective qualities in her teachers and talked about trying to choose teachers "for their emotional and intellectual maturity, their qualities as human beings, rather than for academic eligibility."[57] Other feminist perspectives characterize female leadership as more humane and democratic, and less authoritarian than that of males. Such descriptions do not fit the leadership practice of Pratt according to most contemporary accounts. However, one of the ideas of feminism that was cherished by the women of the Heterodoxy Club was the freedom to be "willful women, the most unruly and individualistic females you ever fell among."[58] Certainly, Caroline Pratt was that.

CONCLUDING THOUGHTS ABOUT CAROLINE PRATT'S ADVENTURE

How do we understand the life and work—the adventure—of Caroline Pratt? How do the descriptions of "a tart spinster,"[59] "a blue-eyed smiling person,"[60] "a dominant personality with great personal charm and a twinkling sense of humor,"[61] "a kind of genius"[62] contrast with "she was not a warm person, not demonstrative with her emotions," "a figure of authority and often feared,"[63] and a formidable physical character?[64] What do we learn from this strong, engaging, radical woman?

City and Country is one of the few current progressive schools in which the curriculum and pedagogy continues to reflect the philosophy of its founder. That the school has retained this philosophy and the practice through which it is enacted while still adapting to changing trends in education is a tribute to the strength of both Caroline Pratt's ideas about how children should be educated and to her personality.[65] Her unflagging faith in the ability of children to control their own learning has stood the test of time because it is based on observing what children do—not on abstractions of children or on aspects of their development, but on children's lived experience.

Grounding new knowledge in one's experience rather than the authority of an expert reflects a feminist way of knowing.[66] This framework was used to shape the school organization as well as the way each classroom was run. The curriculum and the way it was enacted with each successive age group was developed from observations of children. Modifications were made based on children's actions within it. In the individual classrooms, teachers supported rather than directed the efforts of their students to represent the meanings of their experiences through the use of classroom materials. Learning through collaboration, which was valued as a feminist way of knowing, was encouraged at City and Country. Another characteristic of a feminist classroom is validation and integration of the personal. Evidence for the presence of this quality in the City and Country School is found in the recollection of a former student: "City and Country gave us confidence and stimulated a thirst for learning, evaluating, and arriving at our own conclusions. There were no real discipline problems although it must be remembered that the feeling we had of being in charge, of having a voice in what we did individually and collectively, would tend to keep us from antisocial behavior."[67] Pratt's ideas about what boys and girls should do were also feminist, in that she believed that gender was not biologically constructed and worked to make that apparent at the City and Country School. Countering the biological construction of gender was a central topic of the early twentieth century feminists.

Caroline Pratt lived and worked in a social context that included both the expressionist progressives and the radical feminists of Greenwich Village. The woman she was and the school she shaped with such passion and energy are products of both of these environments. Although her independent thinking about the learning process is compatible with the basic ideas espoused by leading progressive educators, this chapter has added a feminist perspective to the understanding of her life and work. Just as we honor progressive ideology as one of our historical antecedents, so must we validate the feminist thought and practice that also underlie our contemporary thinking about education. Our understanding of Pratt and her educational practice is richer as a result of considering both of these perspectives.

NOTES

1. Caroline Pratt, *I Learn from Children* (New York: Harper and Row, 1948/1970), p. xiv.
2. See Robert Beck, "American Progressive Education 1875–1930" (Ph.D. diss., Yale University, 1942/1965), idem "Progressive Education and American Progressivism: Caroline Pratt." *Teachers College Record* 60, no. 3, (1958): 129–137; Lawrence Cremin, *The Transformation of the School: Progressivism in American Education 1876–1957* (New York: Vintage Books, 1961); Patricia A. Graham, *From Arcady to Academe* (New York: Teachers College Press, 1967); and John and Evelyn Dewey, *Schools of To-Morrow* (New York: E.P. Dutton, 1915) for their analysis of Caroline Pratt's progressive school practice.
3. Dewey and Dewey, *Schools of To-Morrow.*
4. Beck, "Progressive Education and American Progressivism."
5. Charlotte Winsor, ed., "Experimental Schools, The Play School by Caroline Pratt and Lucille Deming," in *Experimental Schools Revisited: Bulletins of the Bureau of Educational Experiments,* Bulletin no. 3 (New York: Agathon Press, 1973), p. 29.
6. Caroline Pratt, "Making Environment Meaningful," *Progressive Education* 4, no. 1 (1927).
7. Ibid., p. 105.
8. Susan F. Semel and Alan. P. Sadovnik, "Lessons from the Past: Individualism and Community in Three Progressive Schools," *Peabody Journal of Education,* 70, no. 4 (1995): 69. I appreciate the editors' willingness to share text from this and other sources about Pratt and the City and Country School to aid in my preparation of this chapter.
9. Pratt, *I Learn from Children,* p. 42.
10. "Bulletin of the City and Country School," n.d. (Note: In *I Learn from Children,* Pratt reports that the VIIIs operated the school store and the IXs provided the school mail service. Other publications indicate the opposite. It is not clear when the sequence changed.)
11. Pratt and Winsor, p. 23.

12. Caroline Pratt, "The New Education, 10 years after," *The New Republic* 63 no. 2 (2 July, 1930): 172–176.
13. Nancy Cott, The *Grounding of Modern Feminism* (New Haven: Yale University Press, 1987).
14. Ibid., p. 49.
15. June Sochen, *Movers and Shakers: American Women Thinkers and Activists, 1900–1970* (New York: Quadrangle), p. 8.
16. Caroline Pratt, "As to Indoctrination," *Progressive Education* (Jan.–Feb. 1934): 106–109.
17. Kate Wittenstein, "The Heterodoxy Club and American Feminism, 1912–1930." (Ph.D. diss., Boston University, 1989).
18. Judith Schwarz, *The Radical Feminists of Heterodoxy* (Norwich, Vt.: New Victoria Publishers, 1986), p. 25.
19. Pat Carlton, "Caroline Pratt: A Biography," (Ph.d. diss., Teachers College, 1986), p. 14.
20. Schwarz, *The Radical Feminists,* p. 102.
21. See Carlton, "Caroline Pratt: A Biography," Ph.d. diss., for extensive information on Pratt's family genealogy and activity in Fayetteville and the surrounding communities.
22. Pratt, *I Learn from Children,* p. xv.
23. This term was used at the time to describe kindergarten teachers. Nowadays it refers to a student in kindergarten.
24. Pratt, *I Learn from Children,* p. 11.
25. Ibid.
26. Ibid., p. 12.
27. Cott, *Grounding Modern Feminism,* p. 6.
28. Edward T. James, ed. *Notable American Women, 1607–1950: A Biographical Dictionary* (Radcliffe College, 1971).
29. Lewis Mumford, *Sketches from Life* (New York: Dial, 1982), pp. 108, 109.
30. Pratt, *I Learn from Children,* p. 14.
31. Ibid., p. 15.
32. James, *Notable American Women.*
33. Jennifer Wolfe, *Learning from the Past: Historical Voices in Early Childhood Education* (Mayerthorpe, Alberta: Piney Branch Press, 2000), p. 311 (Excerpt from Hartley House Board minutes, 1 March 1905).
34. Pratt, *I Learn from Children,* p. 19.
35. Ibid., p. 20.
36. Ibid.
37. Ibid., pp. 36, 37.
38. Jill Kerr Conway, *When Memory Speaks* (New York: Vintage, 1998).
39. Caroline Pratt, *Experimental Practice in the City and Country School* (New York: E.P. Dutton, 1924) p. 32.
40. In addition to Pratt's autobiography, *I Learn from Children,* see *Experimental Practice in the City and Country School,* (1924); Pratt, "Learning by Experience," *Child Study Magazine* 11 no. 3 (1933): 69–71; and idem, *Before Books,* with Jesse Stanton (New York: Adelphi, 1926).
41. Cremin, *The Transformation of the School,* p. 203.
42. Ibid., p. 204.
43. Dewey and Dewey, *Schools of To-Morrow;* Pratt, *I Learn from Children,* pp. 56, 64.
44. Beck, "American Progressive Education," 1942/1965, p. 151.
45. Beck, 1958, p. 137.
46. Columbia University Rare Book and Manuscript Dept., Lucy Sprague Mitchell Documents, box 4.
47. Pratt, *Experimental Practice,* pp. 7, 8.
48. Pratt, *I Learn from Children.* p. 71.

49. Susan Semel, "The City and Country School: A Progressive Paradigm," in *"Schools of Tomorrow" Schools of Today,* eds. Susan Semel and Alan Sadovnik (New York: Peter Lang, 1999), p. 131.
50. Ibid.
51. Frank and Teresa Caplan, *The Power of Play* (1973).
52. By 1981, it had become clear that no matter how dedicated it was to the school, the executive committee could no longer maintain the school's financial solvency while attending to pedagogical concerns. A restructuring process resulted in the creation of a board of trustees and the dissolution of the teacher cooperative. Semel, "The City and Country School," pp. 137–138.
53. Letter to Mrs. M.B., 1 June 1938. City and Country School Archives.
54. Untitled typed draft, City and Country School Archives, RG 3.2.4 box 6, folder 12.
55. Pratt, *I Learn from Children,* pp. 182, 183.
56. Lucy Sprague Mitchell, *Two Lives* (New York: Simon & Schuster, 1953), p. 412.
57. Pratt, *I Learn from Children,* p. 64.
58. Schwarz, *The Radical Feminists,* p. 4
59. Mumford, *Sketches from Life,* p. 218.
60. *New York Times,* (7 June 1954).
61. Ibid.
62. Mitchell, *Two Lives,* p. 413.
63. Carleton, "Caroline Pratt," p. 11.
64. Interview, Virginia Parker, 1999.
65. Semel and Sadovnik, "Lessons from the Past," p. 70.
66. Mary Field Belenky, et al., *Women's Ways of Knowing* (New York: Basic Books, 1986).
67. Recollections by Elizabeth Jones Greenhall, class of 1928, the second graduating class. City and Country School Archives.

CHAPTER FIVE

HELEN PARKHURST AND THE DALTON SCHOOL[1]

SUSAN F. SEMEL

Helen Parkhurst, progressive educator, was the founding mother of the Dalton School, an independent, child-centered school located in New York City. In its early years, the school survived because of Helen Parkhurst. Her vision and force of personality engendered great loyalty from her faculty, school parents, board of trustees, and students. Her particular strand of progressive education, which came to be known as the Dalton Plan, was adopted in places as distant as Japan. However, Helen Parkhurst, the woman, was an anomaly. Her competence as a progressive educator was unquestionable, but on a personal level she exhibited a single-minded persuasiveness, a driving ambition, and an unparalleled ability to use people to achieve her own ends. I believe that her entrepreneurial approach to education, her forceful personality, and her single-minded determination were responsible for The Dalton Plan taking root in the Children's University School, renamed the Dalton School in 1920.

PRE-DALTON YEARS

Parkhurst's life and the history of the Dalton School are intertwined from the years between 1916 and 1942. It is important to examine Parkhurst's early years if we are to discover how her educational philosophy was formed. She was born in Durand, Wisconsin, in 1887 and excelled in high school, graduating as her class valedictorian. In college, at Wisconsin State, she completed four years in two, graduating in 1907, with the highest professional honors ever awarded.

Although no direct documentation exists, it is probable that Helen Parkhurst came under the influence of Frederic Burk and Carleton Washburne, the latter having developed the "individual system,"[2] which allowed students to progress through their studies at their own pace. Parkhurst, however, dubbed the philosophical underpinnings of her approach, the Dalton Laboratory Plan, and emphasized the laboratory, or "lab," as her students and faculty came to call it, as her own unique creation and one of the focal concepts of her plan. She claimed to have discovered the lab as a teacher in a normal school at age 16. Small, frail, and knowing little about mathematics, Helen Parkhurst had to cope with teaching this subject to a group of farm boys, some of whom were far larger and knew more than she. Parkhurst grouped the students around a table in the rear of the classroom and, in an early instance of peer instruction, had the older, more knowledgeable students help and instruct the younger ones.[3] Thus, the laboratory was born out of pressing need and ingenuity.

In 1914, while on leave as director of the Primary Training Department of Central State College in Stevens Point, Wisconsin, Helen Parkhurst was appointed by the Wisconsin State Department of Education to report on the Montessori method. Parkhurst studied in Rome with Maria Montessori, and, according to her, became the only person ever authorized by Montessori herself to train teachers.[4]

Helen Parkhurst claims that she never returned to Stevens Point, Wisconsin but directly founded the Dalton School. To the contrary, historical evidence points to the fact that Parkhurst did return to Stevens Point, where she taught until 1915.[5] Around 1915–1916, her rather fortuitous relationship with Mrs. W. Murray Crane began, the woman who was to provide Helen Parkhurst with the financial support that she needed to found her school. Mrs. Crane, a midwesterner by birth, transplanted by marriage to a pastoral factory town in Dalton, Massachusetts, was the second wife of one of the wealthiest men in America. The Crane family business, founded in 1801, was a paper-printing enterprise: paper for stationery and paper for the U.S. mint. The Cranes had three children—two boys and one girl—and Louise, the youngest, was born in 1913. It was the education of Louise Crane that Helen Parkhurst was invited to Dalton, Massachusetts, to supervise. There is little documentation regarding just how and precisely when Helen Parkhurst began her school for Louise Crane and three or four of her friends in the Crane family house in Dalton. Of particular significance, however, is the relationship begun by the two women: Mrs. Crane, a rich, powerful and important patron of the arts; and Parkhurst, a practitioner of modest means and a messiah of progressive education.

According to Winthrop Crane, Parkhurst's school in Mrs. Crane's house lasted for one year;[6] she then went on to implement the Dalton Plan in the public high school in Dalton, Massachusetts, which also lasted for one year. Although difficult to document, it has been suggested that the relative brevity of the experiment was due to parental resistance, perhaps portending the fate of similar experiments in the Dalton School in the public high schools in New York City.

THE DALTON SCHOOL: IMPLEMENTING THE DALTON PLAN

After this abortive experiment in the public sector, with the encouragement and financial support of Mrs. Crane, Helen Parkhurst opened the Children's University School in New York City in 1919. It was renamed the Dalton School in 1920; a compromise perhaps, as Helen Parkhurst, according to informants, wanted to name the school after her benefactor who declined the honor (and perhaps the publicity) in favor of the name of the town from which the Crane family originated. Her particular progressive pedagogical plan, originally called the Laboratory Plan, was also renamed the Dalton Plan in keeping with the new name of the school.

The original site of Helen Parkhurst's school was a brownstone on West 74th Street. As the student body of the school began to increase, larger quarters became necessary. Thus, the Lower School was moved to West 72nd Street, and the High School was opened on West 73rd Street. In the fall of 1929, both divisions were moved to one of the school's present locations at 108 East 89th Street.[7]

Helen Parkhurst's early educational efforts attracted a great deal of attention. It was said that John Dewey was a frequent visitor to Dalton;[8] Evelyn Dewey published *The Dalton Laboratory Plan* in 1922, devoting an entire chapter to the Children's University School in operation. But Parkhurst, herself, was her own best publicist. Her book, *Education on the Dalton Plan,* was published in 1922, and within six months of publication it was translated into 14 languages. During the 1920s and 1930s, she traveled by invitation to such places as England, Japan, Russia, China, Chile, Denmark, and Germany, lecturing on her educational philosophy.

Helen Parkhurst was, above all, concerned with providing students with a better way to learn, a way that would permit them "to pursue and organize their studies their own way."[9] She wanted to create "a community environment to supply experiences to free the native impulses and interests of each individual of the group. Any impediments in the way of native impulses prevent the release of pupil energy. It is not the creation of pupil energy, but its release and use that is the problem of education."[10]

The guiding principles of her plan were freedom and cooperation. By freedom, Parkhurst intended the student to work free from "interruption . . . upon any subject in which he is absorbed, because when interested he is mentally keener, more alert, and more capable of mastering any difficulty that may arise in the course of his study."[11] To this end she abolished bells, for she was also cognizant of the fact that students acquire knowledge at their own rate and that they must have time to learn thoroughly. As Parkhurst dramatically suggests, "Freedom is taking one's own time. To take someone else's time is slavery."[12]

By cooperation, Parkhurst meant "the interaction of group life."[13] Concerned with preparing students to live in a democracy, she attempted to create an environment in which there would be maximum cooperation and interaction between student and student and between student and teacher. Believing that "education is a co-operative task,"[14] she set out to implement her principle through the work problem: "Under the Dalton Laboratory Plan we place the work problem squarely before him [the student], indicating the standard which has to be attained. After that, he is allowed to tackle it as he thinks fit in his own way and at his own speed. Responsibility for the result will develop not only his latent intellectual powers, but also his judgment and character."[15]

In September, students were confronted with the year's work in each subject. They were required to discuss their plans of action with their teachers, for it was essential to Parkhurst that both students and teachers perceived their tasks. Later, students might discuss their plans with their fellow students, they might modify their plans on peer recommendations, or they might even abandon their plans and start over. Students would have participated, nevertheless, in planning their studies with both faculty and peers, interacting with the community in a spirit of cooperation.

In addition to planning, cooperation could be achieved through student activities, such as clubs and committees, or the House system, which was a particular Dalton phenomenon. Helen Parkhurst conceived of the House system, particularly in the High School, as the arrangement of the student population into advisory groups, representing all four grades and meeting four times a week for a total period of 90 minutes with a teacher-advisor.

House meetings might consist of students planning their work with advisors, discussing problems of scheduling appointments with teachers, or perhaps assembling as a group to discharge their responsibility to the rest of the community. House discussions might approach a more personal level, as well. Student attitudes, habits, and experiences might be considered by the group, as they had a definite bearing on community life within the school. Thus the House system fostered the spirit of cooperation among students; however, it also served to develop the qualities of independence and social awareness.[16]

Other important components of the Dalton Plan were the contract system and the assignment. In the Children's University School and, in the early stages of the Dalton School, the curriculum was divided into "jobs" encompassing 20-day time periods. Students "contracted" for their tasks and actually signed contracts to that effect.[17] The tasks appeared as the assignment—"an outline of the contract job with all its parts."[18]

To free the children—to assist their growth in independence and responsibility—large blocks of time were set aside each morning from nine to twelve o'clock called lab time. Each teacher had a lab, and students were expected to utilize the resources of their teachers to help them fulfill their contracts. Lab could either be a group or individual experience; lab rooms were stocked with textbooks and "adult books," as well, to facilitate learning.

Student progress was recorded by the graph method. At the beginning, there were four graphs: an instructor's laboratory graph, a pupil's contract graph, a house graph, and an attendance graph. The instructor's graph was posted in each lab, and students marked off how many units of work they accomplished, with 20 units being equal to one assignment. Teachers and students alike could use the instructor's graph to measure progress. For the teacher, it served as a means of determining when a "conference" might be called in which a group of students reaching the same level might come together to discuss common problems; for a student, the graph allowed the measurement of his or her achievement in relation to the group as a whole. The pupil's contract graph permitted a student to record his or her progress in all subjects; later, it became known as the unit card for, like the instructor's graph, it contained space for 20 units divided into four sections for each week of the 20-day period. The house graph emphasized the entire number of units of work completed and contained a space for individual house members to record their progress. This served mainly as a tool for the house advisor. The attendance graph was posted on a bulletin board, which was accessible to the students; it was their responsibility to record the time that they arrived each morning.

Teachers under Parkhurst's plan were referred to as "specialists," for she believed that they knew their subject matter intimately. Training and credentials, however, were not important. In fact, one of her most eminent and talented teachers, Elizabeth Seeger, author of *The Pageant of Chinese History*, only completed high school. To her credit, Helen Parkhurst saw the teacher, "not as a peddler of facts, but one who helps to recreate personality. Every child has a very definite personality, but a great deal of work needs to be done on all personalities. It takes time to make adjustments. It is necessary for us not only to extend our own personalities, but to consider the school a garden where many personalities are to be encouraged."[19]

Flexibility, then, was the keystone of the plan. Conferences were called as needed. Classes met, too, when necessary, and consisted of grade meetings to discuss problems common to a particular age group. During Helen Parkhurst's time, the school exuded a quality of informality, spur-of-the-moment decision making, enormous energy, high-level engagement on the part of both faculty and students, and the element of surprise. Harold Thorne, a former teacher, now deceased, was fond of saying, "You could come back to your classroom on Monday and someone else would be teaching there."[20]

Andre Malraux once said that "artists theorize about what they would like to do but they do what they can do." Helen Parkhurst's Dalton Laboratory Plan, as she conceived it, was far from perfect. Former students complained of lack of structure, conferences called without warning, and the disruption of precious time needed to complete projects. Teachers had to be reeducated in Dalton ways; often, because of the emphasis on process, they were insecure with regard to curriculum. For the student to realize his/her potential as an individual and to be a contributing member of a community remained a problem that was largely unsettled.

The Dalton Plan flourished in the Lower School, particularly from grade four through eight. Because of college requirements, it was not introduced in the High School until 1933, when Dalton became part of a group of independent schools participating in the Eight Year Study.

Helen Parkhurst founded a school with a particular philosophy—a special environment in which teachers had lab rooms and classrooms and used educational jargon indigenous to the institution. Perhaps Parkhurst's greatest contribution to education, however, and the *raison d'être* of her school was her emphasis on process rather than product. She saw the Dalton Plan as "a vehicle for the curriculum, i.e., a new way of school living, permitting children to acquire flexible habits to put behind ideas."[21] She sought to instill in her students "good habits" because she realized the necessity of being able to adjust to new situations with facility as the primary principle of living. She believed that she was creating an environment in her school that would allow her students to make "adjustments and do things in terms of principle . . . to get the right mental habits for life."[22] As Helen Parkhurst stated in her Caxton Hall lecture in England, "the trouble with the education given to us in the past is that it really was not a preparation for the particular kind of living that we are enjoying today. Statesmen, teachers, all professions are baffled by problems that we face but do not solve. We must have flexible individuals in the future, who can do their tasks which we, in our ignorance, are unable even to discern today."[23]

Granted this description of the Dalton Plan, one might usefully locate it within the larger context of the Progressive Education movement. Parkhurst was not an isolated visionary but, rather, one of many people who established child-centered day schools in New York City. In their book, *The Child-Centered School,* Harold Rugg and Ann Shumaker, placing this movement in a national context, mentioned that during the years before World War I, "Carleton Washburne began his experimentation in the practical individualization of instruction in Winnetka, Illinois,"[24] within a large, public setting.

Parkhurst's Dalton Laboratory Plan is original in its language only; it is a synthesis of the ideas of leading progressive educators of her day, and, as is demonstrated below, perhaps had its origin in the "individual system" of Carleton Washburne. In giving credence to her ideas, Parkhurst, herself, credits John Dewey with having directly inspired her. As a medieval scholar might refer to Aristotle as authority, so Helen Parkhurst invoked John Dewey. She made the connection easy for her reader to grasp, by demonstrating that the second principle of the Dalton Plan—cooperation or the interaction of group life—was taken from Dewey's *Democracy and Education* and quotes the passage for emphasis.[25]

In *Education on the Dalton Plan,* her chapter entitled, "Inception of the Dalton Plan," is particularly relevant to this discussion, both for what is included and what is omitted. Helen Parkhurst credits Edgar Swift's book, *Mind in the Making,* as influencing her profoundly, maintains that she owed her "first conception of educational laboratories" to it;

she quotes extensively from this text on such matters as working "with the students, inspiring them to delve into things for themselves."[26]

She continues to explain her "philosophical evolutionary process" with a brief allusion to her experience in Italy with Maria Montessori, and then moves on to mention Frederic Burk, who made it possible for her to make "a practical test of her laboratory plan upon a selected group of one hundred children, between the ages of nine and twelve."[27]

I believe that to understand fully the origins of the Dalton Plan, we should look to a name that Parkhurst omitted—Carleton Washburne—a man who became associated with the "individual system," or the Winnetka Plan and whose schema bears resemblance to Helen Parkhurst's Dalton Plan.

It is interesting to note that both Parkhurst and Washburne have in common as a mentor, Frederic Burk. In *The Transformation of the School: Progressivism in American Education 1876–1957,* Lawrence A. Cremin describes Burk's work at the San Francisco Normal School: "President Frederic Burk had begun as early as 1912 to redesign the curriculum of the model elementary school to allow greater freedom for each of the 700 students to progress through his studies at his own particular pace. Each child was given a copy of the course of study for each subject on his program. Class recitations were abandoned, as were daily assignments. Provision was made for testing and promoting pupils as soon as the work outlined for any grade in any subject was completed."[28] Cremin traces the outcome of Burk's "individual system" as culminating in the work of three people: Willard W. Beatty, Carleton Washburne, and Helen Parkhurst.

For the purpose of this discussion, it is necessary to look at Carleton Washburne vis-à-vis Helen Parkhurst. As he stated in his book, *Winnetka: The History and Significance of an Educational Experiment,* Washburne worked, "under the aegis of Frederic Burk for five years." Then in May 1919, he assumed the position of superintendent of the Winnetka Public Schools. There, "in 1919 . . . what came to be known as the Winnetka educational system had its origin. The source was three streams—the educational desires of Winnetka, Burk's work in San Francisco, and my own early experiences followed by the rigorous training under Burk. Therefore all three influences merged into one continuous stream."[29]

Washburne's "individual system" and Parkhurst's Dalton Laboratory Plan have much in common. Both are based on the idea that the child should progress at its own pace, and both used the method of individualized instruction to this end. Both emphasized the idea of group life through projects, student government, clubs, and committees. Both had similar content in their projects; for example, Greek Olympic Games, which became institutionalized at Dalton, were part of the curriculum for fifth graders in Winnetka. Both had some standardized form of measurement for student progress: Parkhurst used unit cards or progress charts; Washburne used goal cards. And both had systematic approaches to individualized work: Parkhurst used assignments that followed a particular set structure, whereas Washburne used "specially prepared individual progress materials."

Obviously, contemporaries of both had difficulty discussing the differences between the plans set forth by Washburne and Parkhurst—so much so that Washburne addressed this problem in a footnote in his book:

> The 'Winnetka Plan' has often been confused with the 'Dalton Plan.' The 'Dalton Plan' first worked out in her Children's University School in New York by Helen Parkhurst, was tried in the public high school in Dalton, Massachusetts, beginning in 1920. It had, at first

> one thing in common with our work in Winnetka. It did, during the early years, provide the individual progress of the children. I shall not attempt to describe the Dalton Plan—Miss Parkhurst has done so fully in books and periodicals, with translations in several languages. It had many merits and it had very wide vogue, especially in the British Commonwealth (its first fame was in England), in the Netherlands, in the early days of the USSR, and in the Orient. It differed markedly from the work in Winnetka in that it was never characterized by research, by the preparation of self-instructive teaching materials, by the scientific construction of curriculum ('The Dalton Plan,' Miss Parkhurst said, 'is a vehicle for any kind of curriculum.'), or by techniques for group and creative activities.[30]

I certainly would agree with Washburne that his Winnetka Plan was far more "scientific" than Parkhurst's pedagogy, although perhaps I would substitute systematic or rational for the word scientific.

It still remains a mystery as to why Parkhurst made no mention of Carleton Washburne's work. In fact, in a letter to Lawrence A. Cremin, shortly before she died, she protested his linking her with the Burk-Washburne strain and emphatically stated that Maria Montessori was her teacher, in every respect.

In the final analysis, however, the degree to which her thinking was original matters not a whit to the institution that Parkhurst founded. Her genius lay in her ability to merchandise existing ideas, to publicize, and to popularize. Like the *philosophes* of the eighteenth century, who publicized and popularized the scientific revolution of the seventeenth century, Helen Parkhurst publicized and popularized the strain of individualized instruction, an important current in progressive education. She was not original, but she was effective as a woman in education with a vision that became a reality with the creation of the Dalton School.

PARKHURST AND THE DALTON SCHOOL: 1919–1942

Helen Parkhurst's personal style set the tone of the Dalton School from the very beginning. A forceful, creative individual, she inspired great devotion—in some, even a fierce loyalty—on the part of her staff, students, and parents. Her authoritarian, paternalistic, and nonrational mode of administration was tolerated by the school community precisely because she was viewed as a great educator and a formidable individual. As will become evident below, Helen Parkhurst went about the business of deciding upon educational philosophy and its implementation, expansion of the physical plant, and financial matters undaunted by guidelines and decisions made by her board of trustees. As one former trustee and admirer stated, "You might decide something at a board meeting, but she wouldn't do it." Parkhurst truly believed that the Dalton School was her creation and, in the end, was inextricably connected with the institution that she had created.

Although the Dalton School had been in existence before 1929, records are sparse and intermittent for this early period. Thus, for our purposes, the history of the school will begin in 1929, when the physical plant was moved to its present location at 108 East 89th Street, and a board of trustees was created that kept recorded minutes. By this time coeducation had been phased out of the High School, and enrollment was estimated at 383 students.[31] The building at 89th Street was underutilized from the very beginning, for enrollment during the Parkhurst years never reached 500 students. Students from these early years constantly cite the unusual amount of space in which to play and the great freedom of movement within the building. The truth of the matter was that Dalton opened its new doors during the onslaught of the Depression and, for many years thereafter, suffered, as did many other independent schools, from underenrollment.

Parkhurst, however, managed to attract a small, if steady, stream of applicants to Dalton through her speaking engagements, publications, and personal contacts. Many parents chose Dalton because of Parkhurst's educational philosophy. Once a child from a multisibling family entered the school, the others were not far behind. A former alumna, the late Elizabeth Steinway Chapin, related that, while visiting Chapin's mother to discuss her older brother, Parkhurst saw her, age two-and-a-half, toddling around the house and announced, "Send that child to school"—whereupon a class of preprimary students was created with the two-and-a-half-year-old as its first member. Dalton became a school for children of the professional class, with a well-balanced religious distribution through the Middle School. Many of the Protestant families, however, withdrew their children from Dalton's more permissive atmosphere after the eighth grade, to exchange it for more traditional modes of education.[32] Ultimately, to counter the loss of male students going off to boarding schools, Parkhurst declared the High School to be the exclusive province of female Dalton students.

The school attracted children of people actively engaged in the arts and education (many of whom were awarded scholarships) as well as those who had physical disabilities. Thus early on, Dalton became known as a "motherly institution," precisely because it attended to individual differences well beyond learning styles.[33]

The Dalton School achieved a great amount of national prominence through a unique opportunity presented to Helen Parkhurst: participation in the Eight Year Study. In 1930, the Progressive Education Association, meeting at its annual convention in Washington, D.C., posed the question, "How can the high school improve its service to American youth?"[34] The Commission on the Relation of School and College, consisting of 26 members was formed under the leadership of Wilford M. Aikin; it was "concerned with the revision of the work of the secondary school and eager to find some way to remove the obstacle of rigid prescription."[35] By 1932, an experimental solution was found through a plan that allowed cooperation between a select group of 28 schools;[36] the colleges and universities permitted the schools to be "released from the usual subjects and unit requirements for college admission for a period of five years, beginning with the class entering college in 1936. Practically all accredited colleges and universities agreed to this plan. Relatively few colleges require[d] candidates to take College Entrance Board Examinations."[37] Dalton School psychologist, Dr. Genevieve L. Coy, served on three subcommittees of records and reports, established by the commission; Dr. Hilda Taba, Dalton School curriculum coordinator, served as an associate on the evaluation staff of the commission, headed by Dr. Ralph W. Tyler.

The Eight Year Study created a very exciting time for the Dalton School community. Former staff members were eloquent about this period in the school's history. Former students, too, note an excitement generated by participating in an educational experiment in which staff, students, and parents frequently came together to discuss the purpose of education and to speculate on what should be taught. Giles et al. characterize this feeling as "the adventure of living,"[38] and, undoubtedly, the Dalton community generated an atmosphere similar to the feeling of adventure in embarking upon previously uncharted waters.

The Eight Year Study for Dalton, however, when placed in proper perspective, might be viewed as an extension of Parkhurst's Dalton Plan into the High School. Curriculum had to be reorganized in an attempt to reflect a particular philosophy of education that was already present in the Lower and Middle schools: (1) the development of many sides of the child's nature—intellectual, emotional, aesthetic, and spiritual; (2) provision for individual differences; (3) the development of the self-discipline in the pupil that makes it possible for him to use freedom; and (4) the growth of an active appreciation of, and concern for, the needs and achievements of other individuals and peoples.[39] The philosophy of the High School incorporated these precepts, placing a strong emphasis on "appreciation of individual differences" and "social awareness" so that the purpose of a Dalton education was to prepare the student to live in a democratic society. Curriculum, Parkhurst stated, should not be presented as subjects in "water-tight compartments" but, rather, "integrated, cutting across subject lines . . . which might help the student orient herself towards large problems of the present world."[40] School objectives were defined as the development:

1. On the part of the student, of a personally formulated and cherished outlook on life. This is the overarching objective of the program, and it is, finally, the *integrating force for the individual.*
2. Of intellectual powers: generalization, consistency and persistency in thinking, planning through attacks on problems; transfer of ideas from one field of action to another, etc.
3. Of intellectual tools; basic concepts and information in significant areas of knowledge, ability to read with purpose, ability to discriminate in reading and observation, facility in oral and written expression, etc.
4. Of a self-awareness that leads to satisfying and joyful living. This is the overarching objective *for the student,* though perhaps never formulated. It will be realized through location and pursuit of interests, self-understanding through social relationships, plan of imagination in all areas of school activity, etc.[41]

The curriculum was organized so that students would "form a unified point of view on some of the problems of modern life,"[42] and the technique of selecting large problems for students to work on was adopted for them to better comprehend their environment. The four-year course of study in the Dalton High School was as follows: (1) grade 9—life in New York City, considered as a metropolitan community; (2) grade 10—the political, economic, and cultural trends that have given character and differentiation to life in the United States today; (3) grade 11—the impact of European culture on our life today; and (4) Grade 12—outstanding, international problems and America's relation to them.[43]

In addition to their prescribed course of studies, ninth grade students were also obliged to care for infants for two weeks in the fall and two weeks in the spring of the freshman year. This program, known as "the nursery," grew out of an inspiration Parkhurst had when a close friend and trustee of the school became pregnant. Helen Parkhurst decided to develop a course in which students cared for babies from disadvantaged families "to make human biology more meaningful."[44] The program began in the spring of 1932[45] on a voluntary basis and was open to all high school students. By 1936, it became an integral part of the High School biology program for ninth graders. In addition to caring for the babies, students were responsible for picking them up and delivering them home in the school car. One informant stated that she learned more about life in her surrounding community from those home visits than from any other school experience. She was particularly impressed with the treatment of the infant as dictated by its socioeconomic circumstances and cultural background. After carefully nurturing the child at school, she was shocked to see that the child was fed mashed spaghetti, meat sauce, and diluted wine at home.

Students in the nursery also studied such topics as digestion, food, diet, disease, bacteria, reproduction, and heredity. Trips to places such as Borden's Milk Plant and New York Hospital were also arranged. A registered nurse supervised the Nursery, and a physician visited daily to examine the babies. This program remained an integral part of the freshman year through 1956.

Helen Parkhurst and her staff worked tirelessly during this experimental period to perfect their notion of Dalton's philosophy of education and to implement it in the curriculum. She delegated the running of the High School to Charlotte Keefe Durham; however, Mrs. Durham notes that when Parkhurst was headmistress "she was central and final in its management—very few ideas came from the faculty."[46] She characterizes Parkhurst as "a benevolent, creative, autocrat,"[47] who "cared passionately about freedom for children and went to great lengths to originate plans for that and student responsibility, e.g., the nursery, Dalton Plan."[48]

That Helen Parkhurst was enmeshed in every aspect of school life was an understatement. She visited the homes of students, befriended parents, and opened previously untapped channels of communication between the school and the home. A believer in conferences as an effective vehicle of communication, she held many during her administration, in which students, parents, and faculty participated. A three-day Parents Education Conference, held in May 1938, dealt with studying the needs of the children and their education "to acquaint parents with the kind of work and activities the students are engaged in throughout the school."[49] During one meeting, students presented their views of what sort of teacher they preferred; parents expressed opinions on the sort of teacher they would like to have as their child's instructor, and teachers reported on what kind of parents they would "appreciate."

Helen Parkhurst was fond of appearing at student assemblies to discuss the Dalton Plan. She would also call students at random to her office as individuals, or in groups, to discuss personal or communal concerns. In addition, she instituted a reporting system, not just for parents of Dalton students, but for nurses and governesses who were concerned with the welfare of their charges. Parkhurst's commitment to the school community was contagious. Parents, too, became so involved with the education of their children that one was moved to write a letter to the student newspaper testifying to that effect:

> Being a parent of a daughter who has been practically brought up in Miss Parkhurst's school, I feel it almost a sacred duty to follow the activities of the school and the principles it teaches. I have watched with pleasure the splendid progress of my child in her studies and in her general development and I often ask myself what are some of the outstanding characteristics that my child has been imbued with through her contact with her teachers and her superiors. One of the most striking of these, I would say, is the clear development of her own logical thinking and her persistent demand for logical reasoning on the part of her parents, also.
>
> A Dalton admirer[50]

The most telling portrait of Parkhurst, however, can be found through reading the minutes of the board of trustees and by interviewing former board members. What emerges is the picture of an administrator who did as she pleased, regardless of what the board mandated. She was, of course, supported by certain board members who were close personal friends, such as Helen Parkhurst's patroness, Mrs. W. Murray Crane, and Mrs. Evangeline Stokowska. Mrs. Stokowska often accompanied Parkhurst on both school-related and personal trips; she also provided financial support when additional funds were necessary. The antique, renaissance-style cabinets that once graced the second-floor corridors were said to have come from Stokowska's apartment; likewise, the elaborate gold-threaded drapes that were transformed into angels' costumes for the Christmas pageant. Helen Parkhurst also received support on the board, first, from Benjamin Buttenwieser, a wealthy and prominent member of the New York German-Jewish community, who had an immense respect for education and believed in Parkhurst's ideas; and later, when Buttenwieser joined the Navy, from his wife, Helen, a lawyer. Parkhurst also commanded the loyalty of trustee Lloyd Goodrich, art historian and curator of the Whitney Museum, whose son was the recipient of a scholarship. It is alleged that those members who forcefully opposed her were pressured until they resigned for either "reasons of health" or "business responsibilities." Parkhurst was able to manipulate the board successfully until the advent of Mr. Richardson Wood as president. Then her style of leadership led to her undoing.

As stated in the minutes of the board for September 26, 1929, the responsibilities of the head of the school were as follows: "The Head of School shall have full charge of the educational policy of the school and its administration, for both of which she shall be responsible to the Board of Trustees."[51] The head of the school was also mandated to "regulate the course of study, school sessions and vacations, admission, suspension, and expulsion of pupils."[52] Only later, in 1936, did the board act to curb the power of the headmistress by creating a small educational advisory board consisting of three teachers, Miss Keefe, Miss Seeger, and Mrs. Mukerji, to act in concert with Helen Parkhurst on matters concerning educational policy "and to accept responsibility of maintaining budget limits established by the Board."[53] Any payments made by the school had to have the signatures of Parkhurst and two of the three teachers on the advisory board. These three teachers, however, were close to Parkhurst and sources interviewed stated that they were likely "to do her bidding."

There is much evidence that Helen Parkhurst acted with calculated insouciance regarding money matters. In 1937, she spent $250 on publicity photographs of the school, although unauthorized to do so by the board. She cavalierly stated that the PTA would pay the bill. The Board, however, objected, stating that the PTA was not responsible for the debt and that Parkhurst should pay for the photographs herself. Yet, the PTA willingly paid the bill, for Parkhurst was quite successful in her ability to coopt the parents.

Although Helen Parkhurst may have carefully thought out her educational philosophy, she certainly did not devote the same amount of meticulous planning to school policy. A particular example of shoddy planning was the infamous merger with the Todhunter School in 1939, "a conservative, fashionable school for girls—totally unrelated to Dalton ideals."[54] According to several sources, the Todhunter School presented a number of desirable elements to Parkhurst. First and foremost, it had $17,000 in its building fund that might be added to Dalton's construction fund. It also had Mrs. Franklin Roosevelt on its board, Bernard Baruch's daughter as a student, and a population drawn from wealthy East Side Protestant families. Dalton teachers were not consulted about the merger nor was the board consulted about Parkhurst's plan until she had informally mustered support from some board members.

"The merger," states Charlotte Keefe Durham, "was a disaster. Confusion and questionable financial plans followed from both sides."[55] Few Todhunter students came to Dalton; those who did left the following year. Parkhurst, nevertheless, managed to keep the $17,000 from the Todhunter building fund and applied it the following year to her fund to expand the 89th Street building.

Yet another incident of thoughtlessness and dictatorial management on the part of Helen Parkhurst was her "New Milford Experiment" or "City and Country," as one trustee referred to this scheme. Apparently no one involved seems to know exactly how it came about, what its purpose was, when it was first implemented, or how it was financed, although there is much speculation, especially regarding the latter. There is, however, much agreement on the fact that the site was chosen because Helen Parkhurst had a country house in New Milford, Connecticut.

The first mention of New Milford is found in an editorial in *The Daltonian* on December l, 1937, entitled "Integration of Work and Play." It comments favorably on Helen Parkhurst's notion of obtaining a small house in the country where some members of the High School might go for a week and combine academic studies with nature studies and winter sports. Students interviewed who had participated in the New Milford experiment were unclear as to why they were there. One viewed it as "a lark"; another stated it to be "the most miserable experience of my life." *The Daltonian* of February 28, 1941, reported that on March l, students from the junior and senior classes and the faculty would go to New Milford for one month in which they would devote their weekdays to academic studies and spend their weekends taking field trips. Apparently the students were boarded in dormitories and fed in a communal mess hall. It was thought that these facilities were constructed to house city children who might have to be evacuated because of the war. As to the ownership of the property, one informant speculated that Parkhurst had purchased the site through remortgaging the 89th Street building. There is no evidence to support this piece of speculation. It is a fact, however, that Parkhurst could not account for the sum of more than $40,000 over the budget, according to board minutes. According to another informant, members of the faculty who participated in this experiment were "good sports"; however, it was felt that there was an undercurrent of resentment for the authoritarian, insensitive manner in which it was implemented. Indeed, it must have been quite difficult for faculty members with families to absent themselves from home for a full month.

Parkhurst's building expansion program, which she undertook in 1940, illustrates her lack of concern for board approval and perhaps serves to underscore her feeling that the school was hers to direct as she saw fit. Parkhurst decided to implement a plan to build additions to the east and west wings of the school. Without giving the board notice, she sent letters to parents asking for contributions, the percentage to be determined by the amount of tuition paid to support her program. Ultimately, she did obtain board approval but only after she procured loans from two parents totaling the full amount of the projected expenditure.

Helen Parkhurst's Achilles' heel was money. She raised it successfully but spent it imprudently. In her 1938 annual message, she reported to the board that the cash deficit of 1929 had been completely eliminated. Yet, in 1940, because of lack of funds, the faculty had to pitch in and donate a percentage of their salaries to the school to prevent the staff from being cut. Two years later, in 1942, after declaring a financial deficit of $91,764.65, the school went into bankruptcy. Where did the money go? Many informants who were interviewed stated that Parkhurst used school funds for personal gain and that she granted scholarships whimsically. Whatever the case, the climax leading to the inevitable denouement occurred in 1942, when the board under the presidency of Mr. Richardson Wood, editor of *Fortune Magazine,* investigated the school's finances. Concurrently, the faculty, particularly in the High School, split into two factions. Many of its newer members voiced strong opposition to Parkhurst's dictatorial approach in determining academic policy. The schism unleashed such passion that for years thereafter, one loyal Parkhurst supporter, Dora Downes, refused to speak to members of the opposition.

According to board minutes, during a private meeting on April 30, 1942, in which Helen Parkhurst, Richardson Wood, Charles D. Hilles, and Stanley Isaacs were present, Parkhurst agreed to resign. Subsequently, a formal announcement of her resignation was made in a board meeting on May 5. A letter written by Helen Parkhurst tendering her resignation was accepted by the board and her contract was to be terminated as of June 1942.

There is much evidence to support the proposition that Helen Parkhurst did not accept her fate lightly. One faculty member reports that, in an attempt to quell the faculty revolt, Parkhurst called an assembly of High School students, locked the faculty out, told the students not to go to classes, assured them that no one would fail, and promised to sign the diplomas of graduating seniors.[56] She also contacted former board members and powerful parents to muster support. During the board meeting of May 13, 35 telegrams and notes from parents were delivered to the chairman, expressing opposition to or disappointment in Parkhurst's resignation. The most pathetic series of events, however, were those that involved students being called into Parkhurst's office to be confronted by the once-powerful headmistress, begging them to go home and tell their parents, "I'm good, really I am."

One former trustee suggested that Parkhurst was guilty of financial "hanky panky." It is more than likely, however, that she viewed the school as her creation—her property—so that by the end of her administration she was incapable of distinguishing between what belonged to her and what belonged to the institution. Benjamin Buttenwieser, a trustee during Parkhurst's administration, said of her, "when she was on the ship she was captain." Her tragedy was that she failed to comprehend that the ship that she commanded was only temporarily entrusted to her.

PARKHURST AFTER DALTON

After her resignation in 1942, Parkhurst went to Yale as the first Yale fellow in education and received her M.A. in 1943. From 1947 to 1950, she became an award-winning radio and television broadcaster, with "Child's World," where children discussed their problems. She had a number of other shows and made over 300 recordings with children on psychological issues.

From 1952 to 1954 she taught at the City College of New York, and wrote three books from 1950 to 1963. Shortly before her death in 1973, she was invited back to the Dalton School by Headmaster Donald Barr. It was the first and only visit that she made to the school after her resignation. While working on a book about Montessori and another about her childhood, she died in New Milford, Connecticut, of a pulmonary embolism at the age of 86.

CONCLUSION

Helen Parkhurst was one of the early twentieth century pioneers of progressive education. Her Dalton Plan became internationally known, and although it had little lasting effect on large, public systems, it has found its way into smaller, alternative and/or charter schools, particularly in urban areas. It has also influenced independent, progressive schools in the United States and abroad. Although the Dalton School in New York City, which she founded, is no longer progressive, it remains one of the most important schools in the New York independent school world.[57] Today, many of the progressive reforms initiated in public education have their philosophical origins in the Dalton Plan, sometimes consciously, other times, not. The University of Wisconsin at Stevens Point has dedicated a lecture hall in Helen Parkhurst's memory.

To recapitulate, Parkhurst's contribution to progressive education is indisputable. However, Parkhurst's autocratic leadership style and her recruitment of elites for the student body remain problematic for progressive educators today. In fact, her style and her school mirror the progressive paradoxes that I have written about elsewhere, particularly democratic education autocratically delivered and democratic education for the elite.[58] Parkhurst's leadership style was not unlike that of many of her contemporaries and, in many ways, not unlike that of heads of start-up schools in the public sector today. Her recruitment of elites for her school, in part due to funding issues, however, speaks to the complexity of social class issues embedded within progressive education and explored in the work of the late British sociologist, Basil Bernstein.[59] Paradoxes notwithstanding, Helen Parkhurst represents an important founding mother who made a lasting contribution to progressive education.

NOTES

1. This chapter is adapted from Susan F. Semel, *The Dalton School: The Transformation of a Progressive School* (New York: Peter Lang, 1992), chaps. 2, 3.
2. Lawrence A. Cremin, *The Transformation of the School: Progressivism in American Education 1876–1957* (New York: Alfred A. Knopf, 1961), pp. 295–96.
3. Helen Parkhurst relayed the incident to former Dalton teacher, Nora Hodges, in 1936 (interview with Mrs. Nora Hodges, January 24, 1979). There is some discrepancy between Parkhurst's account of her physical vulnerability and of those who knew her as a tall, imposing woman. Perhaps at age 16, she had not fully matured physically.
4. Marilyn Feldman, "Helen Parkhurst" (New York: Dalton School Archives, 1977), p. 3 (mimeographed).
5. Former UW-SP teacher honored at dedication," *Stevens Point* (Wisconsin) *Daily Journal*, March 4, 1974.
6. Interview with Mr. Winthrop Crane, August 8, 1991. There is some discrepancy about when Parkhurts's plan was adopted in the high school in Dalton, Mass. (I have chosen Winthrop Crane's account, although Carleton Washburne dates it at 1920.)
7. Marilyn Moss Feldman, ed. *Dalton School: A Book of Memories* (New York: Dalton School, 1979).
8. Interview with Mrs. Nora Hodges, January 24, 1979.
9. Helen Parkhurst, *Education on the Dalton Plan* (London: G. Bell and Sons, 1927), p. 15.
10. Helen Parkhurst, as quoted in Evelyn Dewey, *The Dalton Laboratory Plan* (New York: E.P. Dutton, 1922), p. 136.
11. Parkhurst, *Education on the Dalton Plan,* p. 16.
12. Ibid.
13. Ibid.
14. Ibid., p. 19.
15. Ibid., p. 18.

16. Helen Parkhurst, "Report of the Dalton School to the Commission on the Relation of School and College" (New York: Dalton School Archives, 1937), p. 5 (mimeographed).
17. The younger children were expected to sign contracts; this was not so in the High School.
18. Parkhurst, *Education on the Dalton Plan,* p. 47.
19. Elizabeth Seeger, *The Pageant of Chinese History* (New York: David McKay, 1934); Helen Parkhurst, "Lecture at Caxton Hall" (New York: Dalton School Archives, 1926), p. 9 (mimeographed).
20. Interview with Georgia C. Rice, January 26, 1979.
21. Parkhurst, "Lecture at Caxton Hall," p. 5.
22. Ibid., p. 9.
23. Ibid., p. 10.
24. Harold Rugg and Ann Schumacher, *The Child-Centered School* (New York: Arno Press, 1969), p. 51.
25. See Parkhurst, *Education on the Dalton Plan,* p. 20; John Dewey, *Democracy and Education* (New York: Macmillan, 1916).
26. See Parkhurst, *Education on the Dalton Plan,* p. 10.
27. Ibid., p. 14.
28. Cremin, *The Transformation of the School,* pp. 295–296.
29. Carleton W. Washburne and Sidney P. Marland, Jr., *Winnetka: The History and Significance of an Educational Experiment* (Englewood Cliffs, N.J.: Prentice-Hall, 1963), p. 14.
30. Ibid., pp. 152–53.
31. It is important to keep in mind that the Dalton School of today has an enrollment of over 1,200 students. Thus, it has more than tripled in size since the early years. This increase in size has had significant effects, which are discussed in Susan F. Semel, *The Dalton School: The Transformation of a Progressive School* (New York: Peter Lang, 1992).
32. Interview with Elizabeth Steinway Chapin, February 9, 1979.
33. Interview with Frank Carnabuci, September 29, 1990.
34. Wilford M. Aikin, *The Story of the Eight Year Study* (New York and London: Harper and Brothers, 1942), p. 1.
35. Ibid., p. 2.
36. Later, two more schools in California were added.
37. Aikin, *Eight Year Study,* p. 12.
38. H. H. Giles, S. P. McCutchen, and A. N. Zechiel, *Exploring the Curriculum* (New York and London: Harper and Brothers, 1942), p. 289.
39. Helen Parkhurst, "Report of the Dalton School to the Commission on the Relation of School and College" (New York: Dalton School Archives, 1937), p. 1 (mimeographed).
40. Ibid., pp. 1, 2.
41. Ibid., p. 2.
42. Ibid., p. 5.
43. Ibid.
44. Helen Parkhurst, "Motherhood Training, High School Nursery" (New York: Dalton School Archives, n.d.), p. 1.
45. Evidence conflicts as to the exact date; however, correspondence points to 1932.
46. Ibid.
47. Ibid.
48. Ibid.
49. *The Daltonian* (New York: Dalton School Archives, May 20, 1938).
50. *The Daltonian,* October 30, 1930.
51. Board of Trustees, minutes, September 26, 1929, p. 5.
52. Ibid., p. 7.
53. Ibid., p. 354.
54. Charlotte Keefe Durham correspondence.

55. Ibid.
56. Incident related by Nora Hodges in an interview, January 24, 1979.
57. See Semel, *The Dalton School,* for a discussion of the school's transition from its progressive roots under Parkhurst to its current status as an elite college preparatory school.
58. For a discussion of the paradox of leadership, see Susan F. Semel, "Female Founders and the Progressive Paradox" in *Social Reconstruction Through Education,* ed. Michael E. James, (Norwood, N.J.: Ablex, 1995), pp. 89–108. For a discussion of the paradox of democratic education for the elite, see Susan Semel and Alan Sadovnik, *"Schools of Tomorrow," Schools of Today* (New York: Peter Lang, 1999).
59. For a discussion of Bernstein, see Alan R. Sadovnik, ed. *Knowledge and Pedagogy: The Sociology of Basil Bernstein* (Norwood, N.J.: Ablex, 1995). For a discussion of the social class dimension of progressive education, see Semel and Sadovnik, "*Schools of Tomorrow.*"

CHAPTER SIX

ELSIE RIPLEY CLAPP AND THE ARTHURDALE SCHOOLS

SAM STACK

> A community school foregoes its separateness. It is influential because it belongs to its people. They share its ideals and its work. It takes from them and gives to them. There are no bounds as far as I can see to what it could accomplish in social reconstruction if it had enough wisdom and insight and devotion and energy. It demands all these for changes in living and learning of people are not produced by imparting information about different conditions or by gathering statistical data about what exists, but by creating with people, for people.[1]

Written in Kentucky during the darkest days of the depression, Elsie Ripley Clapp described her pedagogy and her democratic conception of the community school. Deeply influenced by her association with and reading of John Dewey, Clapp believed that the depression only worsened the loss of community in American society and that the school could serve as a means to restore community life. In our own era, where discussion of community abounds, Clapp's work in progressive education provides insight into linking the school and community in preparing children for active participation in a democratic society. Her work presents a challenge to those who see education as merely the imparting of information and learning defined as a point in time, easily assessed by pencil and paper tests. Learning for Clapp was both an individual and a social process, grounded in human experience—the foundation of community and democracy as ethical association. Yet, her work also brings attention to the progressive paradox—the contradiction between democratic theory and actual school practice. As a biographer, my goal is to tell her story in the context of her historical time frame and life experiences. Through studying Elsie Clapp, a teacher and administrator, contemporary educators can gain a clearer understanding of the difficulties women leaders faced. Through this study of Elsie Clapp and other leaders in progressive schools, a renewed dialog about the purpose of American education in a democratic society will ensue with the goal of improving practice.[2]

By 1933, when Elsie Clapp defined her conception of the community school, she was well known in progressive education circles. From 1933 to 1934 she chaired the National Committee on Rural Education for the Progressive Education Association (PEA), was vice president of the PEA from 1933 through 1934, and served as a member of the PEA Advisory and Executive Board from 1924 to 1936.[3]

Although Clapp's experience in rural education was limited to Kentucky, she became extensively involved in rural education in West Virginia, truly learning by doing. Her experiences in rural education are documented in two books, *Community Schools in Action* written in 1939, and *The Use of Resources in Education,* published in 1952, as well as in several articles.[4]

EARLY LIFE AND EDUCATION

Clapp's intellectual maturity and her attempt to integrate theory and practice developed over a number of years from personal, extensive experience in public and private education. Born in 1879 in Brooklyn, New York, Clapp spent her early years in affluent Brooklyn Heights. She described the first 14 years of her life "as incredibly comfortable, protected, in our own world. . . . We lived unostentatiously perhaps, but luxuriously."[5] Later in life Clapp recalled this affluent life-style as highly restricted and never impromptu. Her true education took place in the home. "The real education I received in childhood," she wrote, "came through familiarity with the libraries of my father and grandfather and association with the older members of my family and exposure to their interests."[6] This education also included concerts, theater, and dancing. Educated in local private schools in Brooklyn, Clapp attended high school from 1894 through 1899 at the Packer Collegiate Institute in Brooklyn.[7] Packer challenged Clapp intellectually and socially, giving her some interaction with girls from other social classes and ethnic groups. Much of her energy in high school was spent improving her Latin skills: reading and translating. Following high school graduation, she enrolled at Vassar College; however, she expressed disappointment during her first year because the courses at Vassar were less rigorous than those at the Packer Institute. "Classes may be dull," she recalled, "but at least I have this library."[8]

Unfortunately, during her sophomore year, Clapp was diagnosed with chronic appendicitis and, later, with phlebitis. These conditions resulted in her leaving Vassar, eventually transferring her credits to Barnard College, where she matriculated in 1908 with a degree in English. Prior to completing her degree at Barnard, Clapp accepted a teaching position at the Brooklyn Heights Seminary, teaching seventh and eighth grade English from 1903 to 1907.[9] For five months during the 1908 to 1909 school year she taught at the Horace Mann School at Teachers College in New York. Here she tutored fifth, sixth, and seventh grade children who needed remedial work.[10] While finishing the Bachelor of Arts, Clapp took graduate classes in the English and the Philosophy Departments at Columbia University. She studied the history of philosophy with William Montague, fundamental problems of philosophy and Aristotle with F. J. E. Woodbridge, Plato with Wendell Bush, Kant with Arthur Lovejoy, and took courses in ethics and curriculum with John Dewey. Her association with Teachers College brought her into contact with William H. Kilpatrick, whom Clapp described as a lifelong friend. Following graduation from Barnard, she enrolled in graduate study at Columbia University, where she received a masters degree in philosophy in 1909.

During 1909–1910, Clapp spent most of her efforts in the English Department, although her heart always seemed to be drawn to philosophy. She studied with Dewey, describing a course on Kant as, "a course notable for its clarity and conciseness," and also took Dewey's course in the Philosophy of Education at Teachers College. Known by members of the Philosophy Department, Clapp gradually became one of Dewey's most admiring students. She explained, "One day Dr. Dewey appeared in the Philosophy room with his hands full of papers. Could I, he asked, find time to help him with these? I could and did. Many students were I found failing because they were confused."[11] Clapp suggested that Dewey conference with the students and possibly cut down on lecturing. Apparently, Dewey complied, to the success of his students. During the summer of 1910, Clapp assisted Dewey in his course *Aims and Principles of Education* at Teachers College—an assistantship, Clapp believed, Dewey paid for out of his own pocket.

Clapp continued her studies in English and in philosophy, noting attendance at Dewey's revolutionary lectures on Types of Logical Theory. "Dewey received no support in this endeavor, Clapp recalled, "Montague was a realist, Woodbridge called himself a metaphysical realist, and Dr. Bush was a Platonist. . . . To them, the distinctions against Dewey inveighed were necessary and inevitable. Although his ideas apparently fascinated them, for members of the Department attended most of his lectures [, t]hey found it difficult to grasp his conception of the individual-in-the-world, acting upon it and reacting to it, living and learning."[12] Clapp was attempting to address Dewey's concept of experience, which was central to his pedagogy and to hers.

Although Clapp took graduate courses in English, it was in philosophy and in the philosophy of education that she experienced intellectual stimulation and satisfaction. Continuing study with Dewey, she took two courses, The Logic of Experience and Philosophy and Education in their Historic Relations. In taking these courses, Clapp "discovered what it is really to know a writer, and realized that the insight that discerns significant relation between education's development and the history of thought is the result of both reflection and wide knowledge and experience."[13] In the summer of 1911, Clapp assisted Dewey in preparing for two courses, An Analysis of Experience and Theory of Experience. She appreciated and admired Dewey's patience and "willingness to receive ideas offered by his young assistant and by his generosity in finding in them matter relevant to his own thinking."[14]

Clapp completed all of the course work for a doctorate in English but never completed the dissertation. Describing her study in the English Department, Clapp wrote, "No one guided a graduate student's choice of courses; I now cannot recall my reasons for my selection but, generally speaking I chose courses which I had not already had in high school and college. . . . I worked hard in these years and was happy feeling that at long last I was getting some solid work done and was succeeding in it. I have often said that while I was at the University no one ever praised me; that was true." Clapp's thesis was to examine the theory of English grammar in the sixteenth, seventeenth, and eighteenth centuries, a foreboding task. However, during the preliminaries—what today might be called a qualifying exam—several of the English faculty members began to argue among themselves and after an hour Clapp left the examination in disgust. "I felt outraged," she recalled, "for I knew I had been baited—a fact which Trent [her advisor] did not deny, but also that I had allowed myself to be routed, and of this I was ashamed. My only comfort was a note from Dr. Dewey to whom I had communicated the bare facts."[15] In a note to Clapp, Dewey seemed stunned and claimed, "I thought I knew University life, but find I have it still to learn. The whole situation eludes me completely so that I cannot react in any intelligent way, so far as advice is concerned. . . . All I can make of it is that they thought you needed a little discipline and that requiring a second examination would give it to you."[16] Clapp described her relations with the English Department as formal, casual, and impersonal and she refused to undergo a second examination; thus, she never received the Ph.D.

In the spring of 1912, after leaving Columbia University, Clapp began work as a member of the Committee on Children in the Patterson Silk Workers' Strike Organization. As a member of the committee, Clapp was charged with visiting the homes of workers taking care of the strikers' children. This job opened Clapp's eyes to a different world from Brooklyn Heights. She became well acquainted with the Lower East Side and the uptown tenement districts. "I was amazed," she wrote, "to discover how poor and crowded were the homes of those who had offered refuge to the children of their fellow workers."[17] In her first eye-opening experience to poverty and class conflict, Clapp would meet Bill and Margaret Sanger, Bill Hayward of the Industrial Workers of the World, Elizabeth Gurley Flynn, William Zorach, Carla Tresca, Arturo Giovanniti, and John Reed. Clapp acted mostly as an observer and was particularly surprised and impressed with the workers' ability to work cooperatively in a common cause. Clapp learned a great deal from her observations and participation; if nothing else, a growing sympathy with the plight of the working class, a sympathy that would bode her well in rural Kentucky and West Virginia.

In 1913 Clapp traveled south to Charleston, South Carolina, to teach in an exclusive girls' school, Ashley Hall, where she headed the English Department from 1913 to 1914.[18] While teaching at Ashley Hall and showing a growing political maturation, Clapp and 22,000 women participated in a suffrage parade in Washington, D.C., during President Woodrow Wilson's inauguration. Although police were everywhere, they ignored the "hoodlums who surged out from the sidewalk and forced us to reduce the marching lines to eight abreast. But not a single woman faltered and finally we reached the Auditorium at the top of Capital Avenue."[19]

Clapp left Ashley Hall in 1914 and taught English for one year at Jersey City High School. In 1915 she returned to teach at the Brooklyn Heights Seminary, where she remained from 1915 to 1921, serving as head of the English Department and as executive secretary to the principal.[20]

During the summer of 1921, Clapp once again entered Dewey's life, asking him if she could help him in his summer courses at Teachers College. Clapp's experiences now gave her insight into philosophy of education, claiming "the practical work I had been doing for the past six years seemed to have deepened my understanding of philosophy of education."[21] Leaving Brooklyn Heights in 1921, Clapp moved to Milton, Massachusetts, to teach English and history at the Milton Academy for Girls, where she also headed the English Department from 1922 to 1923. Milton was a pleasant experience for Clapp. She wrote, "Even if I was the only progressive at the school, other staff members cooperated readily with me."[22]

From 1923 to 1924, Clapp taught seventh grade at the City and Country School in New York. Begun by Caroline Pratt in 1914, the City and Country School served as an experimental institution in the middle of Greenwich Village. The school attracted artists and writers, many who were willing to place their children in an innovative educational program.[23] Close colleagues of Clapp, Elizabeth Stanton and Lucy Sprague Mitchell, were also associated with the City and Country School and remained close to Clapp throughout her educational career. While teaching at the City and Country School, Clapp helped Dewey during the 1923–1924 winter sessions at Teachers College. Clapp felt like an outsider at the City and Country School and found the teachers' attitude of discipleship toward Caroline Pratt distasteful. She also believed that Pratt was jealous of her extensive education and her association with Dewey.[24]

After leaving the City and Country School, Clapp served from 1924 to 1929 as principal of the Rosemary Junior School in Greenwich, Connecticut. She described the early years at Rosemary much like rolling a ball uphill. "The children, accustomed to maids and chauffeurs and to a weak and inefficient school, were both bad mannered and indolent and lacking any work habits or interests . . . some parents were hopeful, more doubtful and a few hostile."[25] Gradually, the children became more alive and interested in their work. Clapp commented that the teachers in the secondary school (called the upper school at Rosemary) were "astonished to find that in progressive schools, such as ours, teachers occupied a far more responsible and independent position. . . ."[26] Clapp implied from her experiences at the Rosemary Junior School that a highly selected and trained staff was necessary for a progressive school to work, a belief she carried and applied at the Ballard School in Kentucky and at the Arthurdale Schools in West Virginia. Student teachers from Vassar were baptized at Rosemary in progressive pedagogy under Clapp's direction, and one student, Elisabeth Sheffield followed her to Ballard and Arthurdale. Although successful at the Rosemary Junior School, Clapp wished to apply her progressive pedagogy in a larger community setting, gaining a better understanding of the role of the school in the community. At the time, Clapp wrote, "The Progressive Education Association continually sought ways and means of introducing progressive methods into the public schools in which the majority of the children in our country are educated—a task rendered difficult by the size of their class groups, and especially the fact that most public school teachers then were untrained in progressive ways of thinking."[27]

Such an opportunity came when Clapp was offered the position as principal of the George Rogers Clark Ballard Memorial School in Jefferson, Kentucky. During this time she also lectured at the University of Louisville. Clapp remained at Ballard until 1934, when she accepted a position at Arthurdale, the first federal subsistence homestead project of the New Deal, located near Reedsville, West Virginia. The work at Ballard and Arthurdale are clear attempts to implement progressive pedagogy in more rural settings and in a more public sphere.

THE BALLARD SCHOOL, 1929–1934

The Ballard School was a rural public school in Jefferson, Kentucky, and under the control of the Jefferson County Board of Education. Land for the school was donated by Mrs. Thurston Ballard in memory of her son George Rogers Clark Ballard. The Ballard School comprised several consolidated one- and two-room schools in the county and enrolled 212 students in 1932. During Clapp's tenure, the school consisted of ten grades, eight elementary and two secondary. Seventy-five percent of the children came from poor rural areas with approximately 25 percent from more well-to-do farm families.[28]

Clapp described the children as "backward in reading, they did not seem to comprehend what they were studying, and exhibited no particular interest or curiosity. The plan however, got its start when in order to find what their interests were, we began to share in their outside-school activities."[29] At Ballard and Arthurdale, Clapp began the community schools by first addressing health and recreational needs. This linkage was the first step in bringing together—building a sense of trust between the school and the community. Once this was established, the educators focused on pedagogy, largely through enhancing cultural understanding and self-realization. The traditional focus on teacher and text had kept the children from understanding real life experiences—subject matter being separate from real life experience. Education needed to focus on the interest and ability of the children, giving them a feeling of ownership in the process, the school as the center of the community.[30]

Clapp saw herself implementing Dewey's "My Pedagogic Creed" through her work at Ballard. In "My Pedagogic Creed," Dewey discussed education as being a social process. "The school is simply that form of community life in which those agencies are concentrated that will be the most effective in bringing the children to share in the inherited resources of the race, and to use his [*sic*] own powers for social ends."[31] One could not enter into a community with a preconceived notion of a community school. Being unique by nature, the essence of a community school had to be discovered by inquiry and questioning, coupled with a study of culture, history, belief, values, and the nature of labor. It was through the latter that Clapp began to build the community school.

Clapp received quite an awakening at Ballard due to what she called "her protected childhood on [*sic*] Brooklyn Heights."[32] Considered by many in progressive education circles to be an expert in rural education, the situation in Kentucky proved an enormous challenge. She expressed her concern in *Community Schools in Action* saying, "Unfortunately, or fortunately, we knew nothing of rural education. All this we had to learn."[33]

Clapp viewed Kentucky as the "mingling of the old and new, that today makes her rich educationally, for her children still can see the things around them, the whole history of her growth and can, through it, understand the history and development of this country."[34] An understanding of the past was necessary for children to comprehend their role in the present. In essence, cultural understanding provided the foundation for self-realization, the first step in understanding the role of the individual in a democratic community. The school as a social institution could and should enhance and nurture this understanding. Clapp explained the role of the school and community in a 1933 article published during the Ballard years. "The fact that schools are schools of communities implies reciprocal and cooperative responsibility on the part of the community schools—cooperative endeavor for community affairs. It involves shared responsibility, the community for the school and the school for the community and this involves action of the school in community life because of its interests and investment of interest and activity as well as of money by the community in the school."[35] In *Community Schools in Action* (1939), Clapp wrote that it was in Kentucky "that we came to an understanding of the nature and functioning of a community school . . . answering the needs of the children and families of the school district, that brought us the realization that a public school in a rural area is necessarily a socially functioning school."[36]

The location of Ballard, near the Ohio River and the city of Louisville, afforded a unique opportunity for learning. The school used as its subject matter what it found in the surrounding environment. For example, due to the limited resources in science, teachers embarked on geological study, along with the study of trees, birds, flora, natural resources, and local sources of power and energy.[37] Clapp's belief in the social role of the school in the community was clearly evident at Ballard. She sought to demonstrate to educators a "socially functioning school using the agencies at hand and where necessary creating these, also demonstrating the organization of subject matters for use in social education."[38] This involved close association with agricultural agents and medical and dental specialists at the University of Louisville and in the community. Ballard also included a sliding-scale lunch program, cooking and sewing for girls, home visits, proper planting, and public health education issues. As an example of the social role of the school, during the summer of 1930, of 218 children in the Ballard school, 140 received health examinations. "The results of these examinations," Clapp recalled, "which confirmed our worst fears, came as a surprise to everyone. Of the 140 children examined, 109 had posture defects, 73 suffered from acute malnutrition, and in this undernourished group, 30 were threatened with infantile tuberculosis . . . unless ways and means could be found to meet these conditions half the children of Ballard would neither learn nor develop."[39]

Although taking care of the health needs of the children was crucial, so was continuing to build upon the cultural heritage of the students. Clapp, an accomplished artist, viewed art as instrumental in helping children integrate present experience with cultural past; therefore, art was emphasized as a significant aspect of a child's education. "Art has been used with all ages in the school," Clapp reported, "both [*sic*] as a means of realizing what they have been experiencing. It is freely and continuously used and is greatly enjoyed and appreciated. It has been a means of discovery, to the child's beauty in their surrounding and has constituted a personally satisfying way of uniting the child and his learning."[40]

Clapp wrote of a Christmas play, directed by teacher George Beecher, which combined school and community participation: a social effort that stimulated growth of knowledge, interests, and appreciation. Clapp described it as a rural gathering and a sharing of talent and interest to benefit the school and community.[41] Much like Dewey in *Art as Experience,* Clapp seems to comprehend that art by its very nature embodies democratic traits such as freedom of expression, open communication, creativity, and imagination. Art served as a means to involve the whole community and the school.[42]

In linking experience with interest and cultural understanding, Clapp supported studies that linked the past and present; however, rarely from a critical perspective. Clapp discussed an eighth-grade play that emphasized the history of Kentucky during the Andrew Jackson era. The children decided that a play might be the best means to investigate this part of Kentucky history. Due to the economic importance and geographical location of the Ohio River, this led to a study of transportation, including riverboats, railroads, and wagons. The children made costumes and furniture with community assistance. Clapp recalled, "the success of the plays, the intense interest, the engagement, the growth, were due to the fact that they filled a need for all the children, satisfied desires, gave meaning to familiar things around them, as in their own past."[43] The play provided a means for the children to better understand their place, as well as a means to involve the community in the life of the school.

The focus on cultural history served as the foundation for other grades at the Ballard School. The first grade studied farm life while the second grade studied village communities. The third grade concentrated on Native American peoples, who had inhabited the Kentucky area prior to the arrival of the white man. The fourth grade studied pioneer life, and the fifth grade transportation. The sixth grade focused on the coming of the French, the English, and the Spanish.[44] Unfortunately, from Clapp's description of these activities it is obvious that the perspective taken is Eurocentric. In her writing, Clapp referred to Native Americans as Indians, which was typical of the time. However, unfortunately there is no discussion in Clapp's writing that teachers discussed the conflict between the white man and the Native American over land and culture. The traditional depiction of native peoples as inferior or savage is present even among these progressive educators. At Ballard this was evident in Clapp's description of a fourth-grade play written and acted out by the students. The students decided to write a story about the American folk hero Daniel Boone and his family. In the play, Boone's daughter strays from camp and is captured by the "Indians." Only through the bravery, cunning, and honor of the white man is she saved. The eighth-grade study of the Jackson era also failed to discuss the "Trail of Tears," which forced native peoples to move from the Southeast to Oklahoma in a treacherous, deadly, and costly journey. For Clapp and colleagues, this was a serious failure to understand the diverse dimensions of democratic community and that community cannot exist in an environment of classicism, racism, and inequality. These activities failed to stimulate reflection, inquiry, and questioning—essential traits of the democratic citizen. The inclusion of the Native American experience could have strengthened their understanding of the democratic community and its diverse nature as well as the exploitation and oppression within their own Appalachian experience. Although the democratic rhetoric is strong, actual practice reveals some troubling observations; an issue for contemporary issue educators.

The Ballard School did influence the lives of the people in the community. In a study of wells by the first-grade class, it was discovered that some wells were contaminated by sewage. The discovery resulted in moving several wells to higher ground, which improved the drinking supply. This was a good example of a community school-people working together to improve the conditions of the community and gain a new response for the value of education as meaningful in the lives of the people. "It can," Clapp wrote, "because of its position and the value of its work penetrate and affect the lives of the people. It may improve conditions in the village and in the homes, means of health, ways of living. It teaches whatever is taught about leisure . . . and improved ways of living."[45]

THE ARTHURDALE YEARS, 1934–1936

Concern about the loss of community intensified among progressives following the crash of the stock market in 1929. In north central West Virginia, signs of coming economic collapse struck the coalfields by the mid-1920s. Particularly hard hit were coal camps called Scott's Run, Jere, and Osage. Many coal mines had shut down, putting people out of work and leaving them without means to provide for their subsistence. Activity of the United Mine Workers and the National Miners' Union increased as workers sought support in the region, with some activity bordering on worker revolt. The Hoover administration provided some aid to feed children through the efforts of the American Friends Service Committee; however, this aid was far from enough. Following the election of Franklin Roosevelt, Eleanor Roosevelt traveled alone to Scott's Run, an area wracked by poverty, violence, and disease.[46] Enticed by Lorena Hickok, an Associated Press reporter, Mrs. Roosevelt to came to observe the devastation and try to help the people. Upon returning to Washington, she began to push for aid to the people, which would come through the National Industrial Recovery Act, passed in 1933. Under Section 208 of the Act, the President and Secretary of the Interior, Harold Ickes, were given $25,000,000 to begin a federal subsistence homestead program. Due to the devastation of the area and Mrs. Roosevelt's insistence on helping the people of the devastated area, the first federal subsistence project would begin in north central West Virginia, officially called the Reedsville Experimental Project, and later called Arthurdale.

Subsistence homesteads were designed to be model communities, where people, given four to seven acres of land, could provide for their food subsistence. Ideally, the communities were to be located near industry or other possible wage-earning opportunities.[47] The subsistence homestead was to be a community in which people could gain a sense of ownership and make ends meet; to ideally be in control of their destiny. The homesteaders selected for Arthurdale were a highly selective group. Potential homesteaders were tested on their knowledge of farming, carpentry, dairy farming, and so forth, and chosen on their ability to make the project a success. However, because of the politically charged atmosphere, prejudice, and Jim Crow laws, African Americans were not allowed to settle in the community, despite Eleanor Roosevelt's insistence. At that time in West Virginia, separate schools were required due to Jim Crow laws.[48]

Foremost in the mind of Mrs. Roosevelt and federal planners was a special type of education—experimental in nature—just like the subsistence homestead. Eleanor Roosevelt described her thoughts about the project and attempted to respond to criticism that the project was too expensive. "The Reedsville project from my point of view was an entirely different thing. It was from the start a laboratory in every way . . . the place where new types of rural schools might be tried out as an object lesson to communities of a similar kind through the country."[49] With the advice of Clarence Pickett of the American Friends Service Committee, Eleanor Roosevelt offered Elsie Clapp the job of Principal of the Arthurdale Schools and Director of Community Affairs.

Clapp officially became an employee of the project on July 7, 1934.[50] She accepted the position because, "John Dewey wished to have worked out for education use a plan of community education—a cooperative enterprise of a community in and through a school."[51] Clapp viewed the project at Arthurdale as an ideal opportunity to utilize the school as a tool to restore community life. The school could grow just as the community was being built around it and with it. She did show concern that "no one including myself really knew the function of a school in a homestead community project." So, as in Kentucky, they began to address health and recreational services to build community trust. There is no doubt that the Ballard experience helped Clapp and her staff as they began their work at Arthurdale, another opportunity where the school could serve as the center of community life. She learned from Ballard that education needed to be a cooperative endeavor, a skill so necessary for problem solving during the depression. Work needed to be shared by all: parents, students, teachers, and other members of the community.[52]

To get a sense of the people and their culture, Clapp visited the Scott's Run area prior to the 1934–1935 school year. She described her visit in *The Use of Resources in Education.* Coal operators had cut off the electricity, but allowed the miners to remain in the houses for fear the empty houses would become firewood.[53] If anything, the plight of the people had worsened by the summer of 1934. She saw in the people frustration, alienation, and hopelessness but believed, like the federal planners, that progressive education in the new homestead and community school could free them from their misery and alienation. Discouraged and displaced miners complained to Clapp on her visit, "You ain't never going to make nothing of us. We're like them old apple trees out there, all gnarled and twisted."[54] Clapp believed that the coal camp existence had "bred habits of complaint, suspicion, obedience to the boss and, resource to the relief of excess and also casual and irresponsible living."[55] Although there is a degree of progressive paternalism here, Clapp did understand that the community school needed to be built on trust—a concept not fostered in the coal camp environment of competition and self-survival. She sensed the difficulty in overcoming this alienation and believed that the subsistence community and progressive schools could give the people a greater sense of place, self-realization, and ownership.

With the advice of the Arthurdale Schools Advisory Committee, Clapp earnestly began work. Members of this committee included Eleanor Roosevelt, John Dewey, E. E. Agger of the Resettlement Administration, Fred Kelly of the Office of Education, Lucy Sprague Mitchell of Bank Street College, Clarence Pickett of the American Friends Service Committee, and W. Carson Ryan, a well-known figure in progressive education.[56]

Clapp worked closely with the Arthurdale Advisory Committee and project architects, Eric Gugler and Steward Wagner, in designing the school aesthetically and geographically as a central facet in the lives of the people.[57] However, when school opened in the new homestead in the fall of 1934 there were no school buildings, books, or furniture. Although disappointed, Clapp and her staff saw this as an opportunity to involve the community in the life of the school. Young children made use of the Arthur mansion while the high school students held classes in the community center and two sheds. Clapp believed that these problems brought the people together and benefited the school children and the community at large. For example, fathers of the children built classroom tables, benches, and blackboards for the school. For Clapp, this type of activity helped the people see the school as a community school, that it was theirs, and that they should have a voice in it. This clearly involved putting aside self-interest for the benefit of the community.[58] Clapp believed that "A community school is not provided—it grows by concurrence and consent. It is a function, never a system. It is a joint production, the result of living and learning, shaped and guided by many events, as well as by ideas and purposes and by the feelings and responses of a large number of people, above all by the desires and the needs of the people whose school it is."[59]

Clapp considered the nursery school as the best example of integrating the school and the community. The nursery school catered to preschool children and served as a social function through a well baby clinic, health work, and parental education. Mothers were taught about sanitation and how to prepare nutritious meals. Designed by architect Steward Wagner, with input from Clapp and Elizabeth Stanton, the nursery school portrayed high ceilings and large windows for ventilation. Each group had its own playground and sleeping porch, designed to allow the children to play outside in inclement weather. Teachers in the nursery school made home visits—then quite prevalent, to learn about the children and their environment.

As at Ballard, Clapp believed that the involvement of teachers in the community at Arthurdale brought a sense of responsibility and solidarity to the teaching corps. Teachers were to share in the activities of the community, to experience life like the homesteaders. Neighborliness was viewed as an educational asset and a means to build trust.[60]

Clapp and colleagues sought to develop a curriculum adjusted to the special needs of the community—a curriculum not hampered by traditional schooling through the grading and grouping of students. "This means that the real learning experience of the school will come chiefly through the vocational life of the community."[61] Examples of this approach to curriculum abound in Clapp's *Community Schools in Action* and *The Use of Resources in Education* and include carpentry, metal work in the forge, constructing musical instruments, and subsistence gardening. Yet these innovations in curriculum did create concern among the homesteaders at Arthurdale.

The high school at Arthurdale did not receive state accreditation until 1938, two years after Clapp's departure. Some homesteaders believed that their children were not receiving as strong an education as other children in Preston County, particularly those in the Reedsville-Masontown areas. They viewed education as they had experienced it—painful and thorough domination of the teacher and the text. Unfortunately, Clapp and her staff did not communicate with the parents as well as they thought about what they did and why they did it. This lack of understanding and the failure to communicate by Clapp and her staff proved a detriment to building a true community school. Yet, in all fairness to Clapp and her progressive colleagues, the failure to practice a critical/reflective democratic pedagogy must be understood in light of the powerful political, economic, and social forces surrounding Arthurdale and the federal homestead program.

The philosophy of the Arthurdale Schools stated a faith in democracy and the belief that people are capable of governing themselves. Democratic procedures and processes were to govern the formation and the administration of the school. They defined learning beyond the accumulation of information, but "implying rights and aptitudes and appreciations for all kinds of useful labor and a sincere regard for moral and social values."[62] Unfortunately, those ideas guiding pedagogy at Arthurdale proved more difficult to put into practice. Because of the politically charged nature of Arthurdale, Clapp and government officials were constantly concerned about the ability of the homesteaders to make judgments and at times overruled the wishes of the homesteaders.[63] Any failure at Arthurdale was pounced upon by Republicans, who immediately claimed failure of the New Deal policies and branded Arthurdale as socialist. Elsie Clapp found herself in the middle of a political nightmare over which she had no control.

Clapp and her staff did seek to link culture and subject matter to help the children and adults better understand who they were and their place in the community, ideally overcoming alienation. For Clapp, the curriculum could eliminate much of this confusion and frustration through a study of culture. "In rural communities it is the school that introduces what is called culture. It brings the children into contact with facts and ideas which it would never know without its aid—literature, history, science, and world events."[64] Through study of the past, Clapp believed the homesteaders and children could gain a sense of security, "a sense of belonging here, a feeling that the land was theirs."[65] Although the school was instrumental in bringing the people together in common cause and advanced cultural understanding, it did not attack social prejudice with democratic community viewed through a myopic lens.

Clapp and her progressive colleagues were right to link culture with identity—identity being necessary for democratic community to exist. They began their work in the folkways, arts, and historical traditions of the past. One of the outstanding examples of this took place under the direction of Fletcher Collins, the director of drama and music at Arthurdale.[66] Collins described the traditional culture of the homesteaders: "While still basic to them, and very much in their blood, it [culture] came to Arthurdale layered over by coal dust; their experience in the mine camps brief though it had been, had obscured to them their cultural heritage; and being in the shadow of Morgantown, they had also been disturbed by the radio, movies, and by bourgeois cultural standards."[67] Collins voiced a concern familiar to contemporary educators as to how the capitalist media influences beliefs, values, and perhaps even a consumerist and individualist mentality. For an educator whose focus was democratic community, these influences are an obvious concern. As the director of drama and music, Collins attempted to use Appalachian culture, largely Scotch-Irish as he and others saw it, to enhance self-realization—the first step to community. These activities included square dancing, fiddle playing, guitar playing, ballad singing, quilt making, writing plays, and nursery songs, dancing jigs, and playing the mouth harp.

According to Collins, the square dance provided a forum to meet together, to communicate—an opportunity for the homesteaders to express themselves through song and dance. He viewed it as a type of activity that was communal, embodying shared interest and expression; in essence, a democratic type of activity. Collins recalled, "At Arthurdale, square dancing had also the added social values of bringing the homesteaders into easy, natural contact with the people of the region through the sharing of a cultural expression which was inherent in both groups."[68]

However, this wonderful social and educational experiment in community planning lasted only two years. From the onset, the Arthurdale Schools were funded by federal, state, and private sources, and by the spring of 1936 the failure to attract industry to employ the homesteaders led to the demise of private funding. Although the homesteaders were able to produce enough food for their families and often much more, they still needed wage labor to purchase goods and services. Consistent employment did not occur until the economy was stimulated by the World War II; many of the men went into service or back into the mines. Much of the private funds for the school and staff came from Eleanor Roosevelt and Bernard Baruch. Clapp and her staff were largely paid through private funds and so the progressive education experiment came to an abrupt close. When the schools opened in the fall of 1936 they were under the jurisdiction of the local Preston County Schools, with curriculum becoming more traditionally teacher and text centered.

Clapp left Arthurdale with deep sorrow and regret, yet she may have experienced some personal relief because the project continually attracted media attention and political rhetoric from the Republicans, who dubbed it Eleanor's baby. Although the school was now in traditional hands, the social project of Arthurdale would continue through the war and end in 1947 when the federal government divested and sold the homes.

CONCLUSION AND CHALLENGE

During the last year of her work as editor for the journal *Progressive Education,* Clapp's book, *Community Schools in Action,* was published, documenting her extensive experience from 1929 to 1936 in the Ballard and Arthurdale Schools. Clapp made special note of one review of her book by Samuel Everett of Northwestern University. Although Everett saw practical benefits for the book, he did not believe that the programs from Ballard and Arthurdale were "sufficiently intellectualized. Issues such as those of race, farm tenancy, unionization, conservation of human and natural resources, unemployment, paternalism, dictatorship, nationalism, the maldistribution of wealth and income—in a word, the sickness of an acquisitive society—are noticeable for their almost entire absence."[69] Everett challenged Clapp for not bringing to bear an analysis and more critical attitude, based upon her extensive experience. Interestingly and somewhat ironically, Dewey came to Clapp's rescue and responded to Everett. Dewey stated, "that the educational policy of the school as she [Clapp] conducted it is a beautiful concrete exemplification of what I stated in very general terms—the necessity of beginning with the local face to face community and developing its potentialities and resources, human and otherwise, if any serious attack is to be made upon the problems of the larger society."[70] Apparently Dewey believed that Clapp had done a good job in linking the school and the community and, to a point, she had. Even Everett saw the practical pedagogical benefits in terms of curriculum innovation and development, but Clapp failed to push community into society at large. Arthurdale failed to adequately deal with race, and although it did attempt to remedy unemployment and poverty, political power and profit motive kept industry from moving into the area to provide wages for the people. The homesteaders succeeded at what they were asked to do yet had little understanding and control over the social, political, and economic forces dominating their lives. Clapp's pedagogical successes are tempered by this failure. As contemporary educators, we must take to heart the necessity of linking school and community with self-realization and democracy yet also attempt to understand how this might be corrupted by the "maldistribution of income" and what Everett calls the "the sickness of an acquisitive society."[71] Everett's problems plaguing society are alive and well today, implying that we have yet to seriously address them.

Through the work at Ballard and Arthurdale, Elsie Clapp, as a school administrator, sought to nurture an environment fostering a sense of belonging, and interdependence; in essence, a sense of community. Current feminist pedagogy suggests that women administrators are more democratic, less authoritarian, more caring and nurturing, and often at odds with male-defined leadership paradigms. Elsie Clapp did care about her professional colleagues and associates and the people she worked with in Appalachia. Yet, perhaps it is Elsie Clapp's progressive ethic that has much in common with the contemporary ethic of caring, so well-articulated by Carol Gilligan and Nel Noddings. Elsie Clapp's ethic of caring was grounded in the democratic trait of respect and tolerance for the other and what she understood of democracy as ethical association and the basis for community. She worked closely with the teachers at Ballard and Arthurdale and spent much time with the homesteader committees at Arthurdale. She did not attempt to control her teachers, but, in all fairness, did not have to because most were experienced progressive educators and were compatible ideologically. It is unfortunate that the educational experiment at Arthurdale lasted only two years. Elsie Clapp's desire for the community school foundered in the political and economic morass of depression politics. As teachers sought to empower students, federal planners feared that their parents might become too empowered and damage the credibility of the Democratic Party. Clapp was aware of this and often found herself in the middle—caught between the homesteaders and the male-dominated federal planners who could only see the bottom line and the next day's headlines. Yet, to this day the children of the original homesteaders speak fondly of their education at Arthurdale and consider their education a "highlight of their lives.[72]

Through historical study and analysis of this type of social and pedagogical experiment, we can gain insight into the hurdles that need to be overcome and adequately addressed in democratic society. There is a need for a serious dialog on the purpose of education in American society. Clapp and her colleagues had a sense of purpose, and although they made mistakes—largely misunderstanding the power of the school in social reform—they were guided by a belief in democratic community, a community grounded in shared interests for a common good. Today, the common good is rarely addressed seriously with efficiency, competition, and an individualistic mentality guiding many pedagogical reform efforts. These characteristics are not conducive to a caring ethic and challenge a democratic one. An enormous challenge is before us as historians and as educators preparing students who enter the teaching profession. Perhaps Dewey implied and Elsie Clapp understood: A creative ethical democracy was and is still the task before us.[73]

NOTES

1. Elsie Ripley Clapp, "A Rural School in Kentucky," *Progressive Education* 10 (March 1933): 128.

2. See Ronald Cohen and Raymond Mohl, *The Paradox of Progressive Education: The Gary Plan and Urban Schooling* (Port Washington, N.Y.: Kennidat, 1979), and David B. Tyack, *The One Best System: A History of American Urban Education* (Cambridge: Harvard University Press, 1974), pp. 130–132. These works attempt to articulate the paradox between democratic theory and practice.
3. U.S. Department of Interior application to Arthurdale, employment and biographical data. "Elsie Ripley Clapp Papers" (hereafter designated ERCP), 1910–1943 Series 1, Collection 21, Special Collections, Morris Library, Southern Illinois University. Collection 38 contains photographs from a scrapbook that Clapp maintained.
4. Elsie Ripley Clapp, *Community Schools in Action* (New York: Viking, 1939); and *The Use of Resources in Education* (New York: Harper and Row, 1952). Although Clapp lived until 1964 these projects were considered by her to be her best work and the culmination of a productive career as an educator.
5. "Elsie Ripley Clapp Memoirs: 1879–1964." [hereafter designated ERCM], p. 54. These are in the possession of the author. The memoirs are autobiographical and end when Clapp began the Ballard work in 1929. The last 20 years of her life are difficult to chronicle. This may have been due to declining health as well as the declining fortunes of progressive education.
6. ERCM, p. 46. Clapp seemed to identify more with her father, seeking his comfort and solace, and did not feel comfortable in the traditional female role, exemplified by her mother. This created a tense relationship with her mother that Clapp did not resolve until adulthood.
7. There is a discrepancy in the Clapp papers regarding the actual date of her birth. The most consistent date is 1879.
8. ERCM, p. 70.
9. For a brief biographical sketch of Clapp, see Barry Westfall, "Elsie Ripley Clapp and the Arthurdale School: 1934–1936" *Vitae Scholasticae* (Spring 1993): 53–64. Due to illness, Elsie lacked 26 credits to finish Vassar.
10. Clapp remained in this position for five months, but due to its interference with classes, left to work as secretary for the *Journal of Philosophy, Psychology, and Scientific Methods,* sponsored by the Philosophy Department at Columbia University. F. J. E. Woodbridge served as editor of the journal at this time. She received a stipend of $50 per month.
11. ERCM, p. 105.
12. ERCM, p. 109. This gives some insight into how pragmatism was not intellectually accepted in American philosophy departments.
13. Ibid., p. 111. At this time she received aid through the Curtis Scholarship in English, $50 from the journal, and a summer assistantship of $75 from Dewey. For a broader discussion of the Department of Philosophy see George Dykhuizen, *The Life and Mind of John Dewey* (Carbondale: Southern Illinois University Press, 1973), pp. 116–152.
14. Ibid., p. 114.
15. ERCM, pp. 131–132. This seems to be further evidence that Clapp was struggling with her identity and self-esteem in an academic world dominated by males.
16. Ibid., p. 132.
17. ERCM, p. 137. This was quite a change from the affluence of Brooklyn Heights. Not noted in her memoirs, Clapp had made an ideological transition to the liberal left.
18. While in Charleston, Clapp spent time exploring the city and interestingly sought out the socialist local. ERCM, p. 150.
19. ERCM, p. 162.
20. In 1916, once again Clapp began to experience health problems, this time diagnosed with a cataract. In 1917, feeling a need to make a contribution to the war effort, Clapp took a job with the Red Cross Canteen, which had charge of all soldiers and sailors passing through the city on their way to and from Europe.

21. ECRM, p. 189.
22. Ibid., p. 193. Clapp spent the summer of 1922 assisting Dewey in a course entitled, Special Problems in the Philosophy of Education, and painting with William Zorach.
2.3 See Lawrence Cremin, *The Transformation of the School: Progressivism in American Education 1876–1957* (New York: Vintage, 1964), pp. 204–205. The City and Country School is also discussed in John Dewey and Evelyn Dewey's "Schools of Tomorrow," in *The Middle Works 1899–1924,* vol. 8, 1915. (Carbondale: Southern Illinois University, 1979), pp. 285–290. For a recent discussion of the City and Country School, see Susan Semel's "The City and Country School: A Progressive Paradigm" in *Schools of Tomorrow, Schools of Today,* eds. Susan Semel and Alan Sadovnik (New York: Peter Lang, 1999), pp. 121–140.
24. During her short stay at the City and Country School, Clapp became friends with Jessie Stanton and Harriet Johnson, who later assisted her at Arthurdale, largely offering expertise in early childhood education. Apparently there was a close network of women educators in progressive schools, and they kept in contact.
25. ERCM, pp. 228–229. Elsie knew these children because she had grown up much like them in Brooklyn Heights.
26. Ibid., p. 242.
27. ERCM, p. 271. Clapp is conveying a concern among progressive educators that their success was limited largely to the private sector and lab schools. This is clearly indicated in the Eight-Year Study. Dewey was also aware of this limitation and pushed Clapp to explore this option in Kentucky and West Virginia.
28. Elsie Ripley Clapp, "Social Education in a Public School," *Childhood Education* 9 (October, 1932): 24.
29. ERCP, Notebook-Manuscript on the Ballard School (1929–1930), Series 4, p. 2.
30. Ibid., p. 7.
31. Ibid., p. 12. See also John Dewey, "My Pedagogic Creed," in *The Early Works 1882–1898: Early Essays,* vol. 5 (Carbondale: Southern Illinois University, 1975), pp. 86–96.
32. ERCP, Manuscript on the Ballard School, p. 6.
33. Ibid., p. 4. Clapp's limited expertise is again evidence of her lack of understanding and experience in rural/public education.
34. Ibid., p. 18.
35. Clapp, "The Teacher in Social Education," in *Progressive Education* 10 (May 1933): 283.
36. Clapp, *Community Schools in Action,* p. 3.
37. Clapp, "Social Education in a Public Rural School," p. 24.
38. Ibid., p. 26.
39. ERCP, Manuscript on the Ballard School, 2. See also Clapp, *Community Schools in Action,* p. 12, for further documentation of this activity.
40. Ibid., p. 25. Clapp studied with William Zorach (1887–1966) and seems years ahead of Dewey in understanding "art as experience" and its communicative capacity. See William Zorach, *Art Is My Life: The Autobiography of William Zorach* (Cleveland: World Publishing, 1967).
41. ERCP, Typed copy of Christmas experience of school and community by George Beecher, Series 3.
42. See John Dewey, *Art as Experience* (New York: Perigee, 1934).
43. Elsie Ripley Clapp, "Plays in a Country School," *Progressive Education* 8 (January 1931): 38.
44. Clapp, *Community Schools in Action,* p. 22.
45. Elsie Ripley Clapp, "A Rural School in Kentucky," *Progressive Education* 10 (March 1933): 127.
46. Eleanor Roosevelt made the trip on August 18, 1933. By the time of her visit, Scott's Run had been nicknamed "bloody run" due to violence in the area. The Roosevelt

administration was clearly aware of the conflict in the area and wanted to make a political and social statement.

47. Lois Scarf, "First Lady/First Homestead," in *A New Deal for America,* ed. Bryan Ward, (Arthurdale. Arthurdale Heritage, 1994), p. 105. This is the most notable publication on Arthurdale and its place in the larger federal subsistence homestead program of the New Deal.
48. See Dan Perlstein, "Community and Democracy in American Schools: Arthurdale and the Fate of Progressive Education," *Teachers College Record* (Summer 1996): 625–650. Critiques Arthurdale's failure to deal with race and other issues of diversity.
49. Eleanor Roosevelt, An address called "Subsistence Farm Steads-Reedsville Project," National Archives, Record Group 48, Department of the Interior, 1–277, box 53, pp. 3–4.
50. U.S. Department of the Interior application for employment. Clapp described herself as 54 years old, 5'6" tall, and weighing 147 pounds. She listed John Dewey, William Heard Kilpatrick, and George Arps as references.
51. Ibid., Dewey's papers are virtually silent on Arthurdale. The Center for Dewey Studies located a letter from Dewey following his visit to Arthurdale in April 1936. He describes the Arthurdale Schools as one of the best public schools in the nation. John Dewey to J. A. Rice, April 16, 1936, Black Mountain College Papers, Center for Dewey Studies, Southern Illinois University, Carbondale.
52. ERCP, Notes on the Arthurdale School, 1934–1935, Series 3. Most of Clapp's staff at Ballard joined her at Arthurdale. For a list of these teachers, see Clapp, *Community Schools in Action,* pp. 395–398.
53. Clapp, *The Use of Resources in Education,* p. 4.
54. Clapp, *Community Schools in Action,* p. 116.
55. Ibid., p. 122.
56. Ibid., p. 397.
57. Alice Davis, along with faculty from the Department of Education at West Virginia University, government officials, Eleanor Roosevelt, and homesteaders, had input on the type of school desired. See Clapp, *The Use of Resources in Education,* p. 8.
58. Clapp, *Community Schools in Action,* p. 67.
59. Ibid., p. 124.
60. Dewey reiterates the role of the teachers in this type of community in his foreword to Clapp's *Community Schools in Action,* p. x. Due to housing shortages in Arthurdale, the teachers tended to live outside the community.
61. Clapp, *Community Schools in Action,* p. 74.
62. Ibid., pp. 72–75.
63. See Holly Cowan, "Arthurdale" (Masters thesis, Columbia University, 1968), p. 40. Bruce Beezer, "Arthurdale: An Experiment in Community Education," *West Virginia History* 36 (October 1974): 33.
64. Ibid., p. 217.
65. Ibid., p. 126.
66. Collins interview; also see his discussion of activities in Clapp, *Community Schools in Action,* pp. 218–272.
67. Fletcher Collins, in Clapp's *Community Schools in Action,* p. 219.
68. Ibid., p. 221.
69. Samuel Everett, "Review of Elsie Clapp's Community Schools in Action," *Curriculum Journal* 2 (March 1940). A copy is also in ERCP.
70. John Dewey to Samuel Everett, April 29, 1940. ERCP, Series 2.
71. Ibid.
72. Interview with Glenna Williams, Jettie Eble, Joe Roscoe, and Annabelle Mayer, August 18, 1997. Arthurdale teacher, Fletcher Collins, was also present at this interview. For a discussion of feminist leadership theory, see Narotan Bhindi and Patrick Duigan, "Lead-

ership for a New Century," *Educational Management and Administration* 25, no. 2 (April 1977): 117–132. Also see Barbara Pazey, "The Ethic of Care: An Essential Link for the Administration of Special Education," *Journal for a Just and Caring Education* 1, no. 3 (April 1997): 296–310. For discussion on the ethic of caring, see Carol Gilligan, *In a Different Voice: Psychological Theory and Women's Development* (Cambridge: Harvard University Press, 1982) and Nel Noddings, "Stories in Dialogue: Caring and Interpersonal Reasoning," in *Stories Live Tell: Narrative and Dialogue in Education,* eds. Carol Witherell and Nel Noddings (New York: Teachers College, 1991), pp. 157–170.

73. John Dewey, "Creative Democracy—The Task Before Us," in *The Essential Dewey: Volume I. Pragmatism, Education, Democracy,* eds. by Larry Hickman and Thomas Alexander (Bloomington: University of Indiana, 1998), pp. 340–344.

CHAPTER SEVEN

CARMELITA CHASE HINTON AND THE PUTNEY SCHOOL

SUSAN MCINTOSH LLOYD

Carmelita Chase Hinton was a lifelong adventurer, possessing the daring, the visionary heedlessness, and the organizational acumen that any adventure requires. She was born in Omaha, Nebraska, the daughter of Clement Chase, a financial editor, newspaper owner, and book retailer, and Lula Belle Edwards, a full-time mother to her four children, a devoted Episcopalian, and philanthropist, who hailed originally from an old Kentucky family. She is chiefly known as the founder, in 1935, of the Putney School. Innovative for its time, the Putney School continued during and long after her 20-year tenure as head to inspire other schools, both public and private. Like most progressive schools, Putney reflected the originality of its founder, in particular, her swings between a practical, self-preserving bent and a dashing, mercurial idealism. And while the school appeared often to be standing in opposition to prevailing cultural and political assumptions, its students came in surprising variety.

Carmelita Chase was told by her father that she was *born* husky and active; her boundless energy only grew as she moved through childhood. It was a time of profound optimism among American middle-class women; Carmelita absorbed this optimism early and entered the twentieth century with the confidence that she could shape her life in any way she chose. Her mother, no feminist, tried lovingly to steer her toward the conventional young womanhood befitting the daughter of one of Omaha's "first families." But Carmelita would not be steered. Instead, she followed her older brother on his pony rides and camping trips. She never forgot a visit with a group of Pawnee Indians out on the prairie on one of these trips, where she spent the night near the Indian camp, cooking pancakes for breakfast in the wind. Her father loved her questing spirit, having long favored women's rights.[1] He had had adventures of his own, including a four-year stint in Washington, D.C., as secretary to a Nebraska senator.

Carmelita cared little for school after sixth grade, but her after-school activities at least partly matched her energies: the troop of Campfire Girls, which she joined, and then helped to lead; and her intense work/play as volunteer helper to a neighbor with six small children. Her alliance with her father strengthened as he made her his assistant in his bookstore. There, she alternately served customers at the counter and read the entire works of Charles Dickens, sitting on the floor. She remembers her father striding into the family home after work saying, "I'm going to Chicago next week. What little girl here would like to go to Chicago?" At that time Carmelita was the only little Chase girl available, and several times she accompanied him on his publisher's railroad pass.

A bustling prairie town when Carmelita was born, Omaha had become a fast-growing city as she entered her teens. Opportunities widened for every youngster. Carmelita became a competitive tennis player, as though to compensate for the dullness and lack of purpose she found in her classical education at Omaha's Episcopal School for Girls. Her tennis technique was unrefined, but she won repeatedly by wearing her opponents out. Again and again during her later career, she would win other kinds of contests with the same endurance.

Carmelita's four years at Bryn Mawr College were much more stimulating than high school had been. She found the faculty's commitment to women's equality—if not to their *superiority*—congenial and warmed to the idealism of many of her classmates. Though athletics, music, and theater appealed to her far more than the scholarship prized by the college, she realized by the end of her junior year that she would have to prepare herself for some kind of career, as most of her classmates were doing.

In many ways, her choice was conventional: a path to usefulness that thousands of women had taken before her. Carmelita joined Bryn Mawr's teacher education program for seniors, and for the first time found college texts that seemed worth reading. Her most memorable course was entitled, The Psychology of Instinct, Emotion, and the Will, three essential ingredients of Carmelita Chase's personality. Speeches she wrote much later enthusiastically quote both William James and Friedrich Froebel. More than once she cited her reading of one of James' precepts: "If you find yourself deeply stirred by what you read and see, it is harmful for you to let this emotion dissipate itself without resultant action. *Do* something."[2] She began to own the progressive movement's confidence that a "scientifically based" education could transform the nation's future, and to wonder whether she might not turn her love of children into a vital profession.

It was at Bryn Mawr that Carmelita found her most congenial medium of communication: speech making. As president of the college Athletic Association, she remembered that at a "Sports and Health" assembly, "I got so excited about how to have the best possible health that I said, 'If you keep eating just any old thing and go to bed at any old time, some morning—*some morning*—you will wake up dead!' This brought down the house."

The rich student friendships that flourished among whole groups of Bryn Mawr students offered Carmelita powerful emotional support and may have become her model for the "gangs" of Putney School girls and boys whom she encouraged over the prevailing tendency to pair up or "go steady." It was the rare Bryn Mawr student who had anything close to a "date." Carmelita had already learned to count on the friendship of her older brother and father; her Bryn Mawr experience taught her ease and happiness with other women as well, including her mother, whose good works she began to admire and share. The rewards of friendship may have seemed the greater to her because she so often got her way with both men and women.

Fervent spirits at Bryn Mawr attached themselves to causes of every stripe on the progressive spectrum. Although Carmelita was not a political person, her father greatly admired Teddy Roosevelt, the president throughout her adolescence. These were the hopeful years before World War I and the Russian Revolution, years when all problems seemed soluble by the enlightened application of human creativity and organizational skill. Teachers could be heroes in a movement that saw education as "an engine for social betterment," in the words of sociologist Frank Tracy Carlton.[3] It took just a year of make shift postgraduation work on her father's weekly newspaper for her to realize that she would have to move away from Omaha if she was to act on her dream of becoming a heroic teacher, and claim an adulthood of her own.

As she contemplated traveling to Chicago and finding some safe and interesting place to live, she recalled Jane Addams, whose broad vision of an educational community had inspired many Bryn Mawr students. Eager to know her, Carmelita Chase persuaded Addams to allow her to live at Hull House in 1913 as her part-time secretary. Addams made her immediately useful, convincing her that "You can't just bathe babies. You've got to go down deep and change things, whether people like it or not."[4] Carmelita entered eagerly into the life and work of Hull House, including the passionate dinner table conversations that the residents enjoyed.

Among the many education reformers who came to dine and talk with Hull House residents was Neva Boyd, a renowned leader of the American nursery-kindergarten movement. As one of Hull House's playground volunteers, Carmelita discovered how much more she had to learn about teaching; late in 1914, she enrolled in Neva Boyd's two-year playground course. Boyd's conviction that "the play group is society epitomized,"[5] brought the larger contexts of teaching and learning into focus for Carmelita, who already knew the power of children's natural curiosity. She was inspired by Boyd's way of teaching, recalling later that the best way to prepare for teaching children carpentry or music was to learn carpentry or music yourself. Discussing her daily lessons at the dinner table with school reformers such as Edward Yeomans, she realized that both carpentry and music are stifled unless the grinding inequalities imposed by urban-industrial society are also addressed. She loved talking with Yeomans' friend Sebastian Hinton, a patent lawyer, backyard engineer, and passionate amateur naturalist; they spoke fervently of how human affairs might be ordered by enlightened social science, or of how absurdities become realities, courtesy of Thorstein Veblen's "leisure class." She could hardly wait to make her own small, firm step toward a new society by starting and running a nursery school.

Just before the playground course ended, Carmelita Chase married Sebastian Hinton. Almost immediately, she opened her own nursery playschool, using the couple's Chicago apartment and the outdoor park across the street. So compelling was Mrs. Hinton's teaching that the school's clientele doubled yearly throughout its six-year history, interrupted only briefly by the birth of the three Hinton children. Thus, she won her first "public" in the progressive families of Chicago. Dorothea Smith Ingersoll, daughter of the founder of the new North Shore Country Day School, attended the playschool as a child. She remembered "Mrs. H." as "a dynamic person," bringing "joyful noise." Ingersoll recalled, "She would come into a room and it was an explosion. But it was a happy occasion. She could sweep people up and carry them off to Mars. . . . I can see her talking to (parents), talking to children, and the children all looking at her, spellbound."

Like all of Mrs. Hinton's teaching experience, this school was a family affair. The school went with them when the family moved out of Chicago to Winnetka, fulfilling a dream of country living that had been nurtured by both parents' distress with city life for children. Sebastian Hinton found some of his happiest moments inventing and constructing playground equipment (taking out a patent on the first "jungle gym"), instructing occasional groups of children on how to care for a wild raccoon or teaching them shop. Each of the three Hinton children began in this first school their crucial role as their mother's pupils, her inspiration, and her goad; for in her mind no one else's school ever seemed as appropriate to their needs as her own schools were. One of her most steadfast contributions to progressive education was a parent's understanding that children need and welcome discipline, as well as boundless opportunity to experiment.

Sebastian Hinton often struggled with depression. Determining at last to find the best possible treatment, in 1923 he went for an extended stay at the Riggs Clinic in Stockbridge, Massachusetts. A few weeks after he arrived, he hanged himself. His sudden death left his wife and children stunned. It is impossible to know fully what it meant to Mrs. Hinton, who seems to have spoken about the tragedy with only one person outside of her own brothers and sister: Jane Arms, for years her only intimate friend at Putney School. Even the Hinton children did not know of their father's suicide until I (the author) naively asked them how they would feel about my revealing the fact in her biography, having learned of it by way of the only person whom Jane Arms ever seems to have told.

Did it matter to her subsequent career? Surely it did. There is no doubt that Carmelita Hinton's strength and sense of responsibility—perhaps her very self—quickly expanded to fill the empty space in the family. She carried into her later elementary and high school teaching a dread of inactivity, a key factor, she believed, in adolescent and adult depression. The tenacity with which she embraced her secret led to deceptions. Putney School "senior discussion" groups, for example, heard much of the mysterious Mr. Hinton's "death from smoking." More pervasive was her tendency to keep others at a slight but steady distance, no matter how jolly and friendly—even affectionate—she was with students, parents, and colleagues; a tendency her husband's suicide may have accentuated. The few people who crossed this gap, including Jane Arms, found themselves enriched and perplexed by an often stormy relationship with a needy, passionate, and endlessly intriguing friend.

Perry Dunlap Smith offered Mrs. Hinton (called "Mrs. H." by almost everyone from this time forward) a kindergarten job at the North Shore Country Day School, and she gratefully took it; not because she needed money—since her husband's patented jungle gym design was bringing in a small, steady income—but because she needed activity. It was a healing two years; nevertheless, she began to think of leaving the Midwest. Winnetka was becoming a "society suburb," she later said. Shady Hill School in Cambridge, Massachusetts, had an opening for a second grade teacher and was ready to accept Carmelita Hinton, sight unseen.

In the summer of 1925, the Hinton family "hit Cambridge like a cyclone," according to one family friend.[6] They settled into a house with a large barn, filled it with animals, and were soon riding (or being pulled by their ponies on their sleds) to Shady Hill. There was more to Shady Hill than exciting artistic and intellectual challenges. Mrs. H. found the cheerful puritanism of Shady Hill congenial and her faculty colleagues remarkably supportive of her desire to teach math and reading through active, absorbing projects. One year found her much-envied pupils designing and building small houses to make a town, each with carefully measured door and windows, and ceramic dishes crafted by its child carpenter and fired in the school's kiln. A lot of arithmetic went into these houses, and into the post office and railroad station built by teams. One night the children even "slept" in their houses on the rugs they had woven out of wool from their class sheep.

Even before Mrs. H. moved her family out to a Weston farm, she was boarding children whose faraway families wanted their children to attend Shady Hill. On the little farm, she included students who were ready to ride their horses to the progressive Cambridge School in Weston with her oldest daughter Jean (Cambridge School had only a tiny boarding department). On weekends and in summers, there were camping expeditions to the beach and the mountains, some of them on horseback. Edward Yeomans told the family that "You should write a book about yourselves, and call it 'The Indestructible Hintons' or 'The All-Weather, Heavy-Duty, Uncollapsible Hintons.'"

Other forecasts of the future Putney School lay in Mrs. H.'s lectures to her children and the nine boarders on the dangers of coffee, tea, and alcohol: her insistence that really noble boys do not feel sexual urges ("Now when my Billy feels *that way* he goes and chops down a tree");—and the manner in which the amused boarders nodded their heads in agreement because it was just so much fun to live there.

The two most exciting Hinton summer trips were a result of Mrs. H.'s learning of the Experiment in International Living. In 1933 she took 15 youngsters (8 of them related to her) on a bike and hosteling trip through Austria and Germany. Naive about politics, she was deeply impressed by the rugged, uniformed youth-camp life with its songs, its drills, and the German campers' sense of purpose. Her children made a few Nazi friends among the families and counselors they met; they also pulled off some comical rebellions against a German coleader who insisted that everyone cycle in an orderly line. The German youngsters were as delighted as the Americans when the coleader quit in disgust.

It was the success of the coeducational boarding and overseas camping experiments, as well as her son Billy nearing high school age, that impelled Mrs. H. to begin thinking about starting a coeducational boarding school of her own, a school free of the "shadows of the old pedagogy" that doomed even progressive high schools like Cambridge School to put college preparation and competitive sports foremost—shadows that fell between adolescents and the real world, obscuring both their responsibilities toward that world and their joy in learning.[7] She had spent eight wonderful years at Shady Hill; as her ninth year progressed, she set out to plan a school of her own.

It was natural for Mrs. H. to begin fashioning her "castle in the air" after the progressive elementary schools in which she had taught. It was also indicative of her self-preserving practicality that she was glad to accept the advice of educators who had far more experience than she with older students. In early 1934, her old Hull House friend, Ed Yeomans, found a place for her school. Yeomans had recently begun his own elementary school in Ojai, California, and had been consultant to the founders of Antioch, Bennington, and Black Mountain colleges. Two philanthropic sisters who owned a farm on the edge of the town of Putney, Vermont had read his latest design for "A Small Rural Private School" and had offered to help him bring it to life on their family estate with its 25 purebred cows; its 300 acres of pasture, timber stand, and meadow; and its long view of Mt. Monadnock across the Connecticut River. Two other ladies had built a dormitory wing and assembly hall to adapt the farm for a summer drama school but could not find a winter clientele. Following a trial year, these four generous women were willing to turn over the entire place and all of its equipment to a nonprofit Putney School in return for $20,000 and a seat on the board of the corporation.

Here, at last, was the ideal site. Mrs. Hinton followed John Dewey in admiring rural life as a natural induction into adult usefulness. Dewey, however, had regretfully concluded that farm life had become an anachronism, whereas to Mrs. H., a working farm made the perfect setting in which adolescents could build the competence and independence needed by useful, critical citizens of a complex urban-industrial society. A college preparatory curriculum would only be enriched by the social discipline of cooperative work that a farm demands, by the responsibility of caring for animals, by the self-discipline required to produce the music, dance, visual arts, and theater through which an isolated community would entertain itself, and by the healthy exercise natural to outdoor life.

Friendly critics tried to dissuade her. The West Hill site was swept by winds, it was entirely inaccessible, except by sledge, in heavy snow and mud, and the buildings were inadequate; the whole idea of a fully coeducational New England boarding school was unprecedented, impossible. But Carmelita Hinton, with her children at her side, accepted the ladies' offer with delight. She asked for and received a year off from teaching at Shady Hill to make detailed plans.

With a great deal of help from Shady Hill colleagues and parents, a curriculum was put together, along with a daily schedule that was equally hospitable to the arts, a rigorous academic program, and outdoor work and play. A winter house party was arranged to seduce wavering prospective students to commit themselves. The six initial applicants and the Hinton children became the chief characters in a film designed to persuade reluctant parents and likely teachers to sign on. A cook, a housekeeper, a teamster, and an all-round herdsman/craftsman, all born to New England farm life, were already working for Elm Lea Farm and agreed to stay on. A woman who managed a working farm at Framingham State Prison answered an ad for a Putney School farm manager, and after one more Elm Lea Farm house party scheduled to show off the loveliness of May in Vermont, six more extraordinary teachers joined the outdoor teaching staff already in place.

As for students, girls were the first to sign up—almost invariably the daughters of teachers or professionals who had come to know Mrs. H. at Cambridge, in Chicago or at Bryn Mawr, as teacher, experiment leader, or fellow alumna; their brothers' educations, it seemed, could not be risked to any boarding school more innovative than Deerfield or Exeter. Yet Mrs. H. insisted that the number of boys should at least equal the number of girls in an age when nearly everyone assumed that coeducation in a 24-hour setting would unman little men. "We had a few weird boys at Putney in those days," says an early Putney pioneer " . . . including one who always wore an Indian headband and walked around on tiptoe." For many parents of daughters, a thoroughly coeducational boarding school seemed madness: they feared an untimely *pregnancy,* above all else. It would be better to send their daughters to one of the Quaker boarding schools, whose long practice of partial coeducation at least symbolized women's equality.

Yet volunteer students-to-be and teachers joined the Hinton family in sowing the fields and planting the gardens in the spring of 1935, for a working farm cannot wait. The first Putney Labor Camp found the same group—and many intrigued newcomers—cutting hay, weeding endless rows of vegetables and berries, transforming a barn into science classrooms and a ceramics studio, and preparing for the opening of school. Open it did in early September, with 54 students who immediately found themselves needed for the harvest—a part of an all-ages, working enterprise.

Clearly, the 1930s offered a constituency for progressive education, as the early success of Bennington and Black Mountain colleges suggested. Almost all of the first Putney parents were college-educated, internationalist liberals who had caught the spirit of the New Deal or even helped to design it: "professors, lawyers, doctors, government workers, musicians," wrote Mrs. H. for *The Vermonter* magazine, "and businessmen," she added, almost as an afterthought.[8] A growing group of parents in the largely impoverished town of Putney, including George Aiken, who was about to become Vermont's governor and then one of its senators, began to see Putney School as a viable alternative to Brattleboro High School for their children. These youngsters, and the $80 that the town contributed in lieu of paying each year's tuition costs at Brattleboro High, became one source of stability for the school and helped to root it within its surrounding community.

The seven-year-long careers of Ursula and Hugh MacDougall at Putney School give a sense of what Putney asked of and received from its early teachers. Ursula was a poet and not at all enamored of the rugged outdoor life; and Hugh, a mechanical engineer whose budding career as a Massachusetts Institute of Technology (MIT) teacher and research scientist had been interrupted by the Great Depression. Early salaries of less than $1000 a year could not discourage a couple with one small child in an era when many boarding school teachers counted themselves fortunate to be working for their food and shelter. Never mind that at Putney, you might have to help build your shelter before you could live in it! Every morning the family left their tiny apartment in the main house, ate in the common dining room (first with the summer "work campers," later with the 55 Putney School students) and then separated. Hugh, builder of suspension bridges, oversaw the construction of the boys' dormitory that he had designed within the roof and frame of the old carriage shed; while Ursula worked on her own writing, advised the student editors of the *Putney Magazine,* or taught English to as many as six different grades in her exacting and sensitive way. Hugh, along with two other gifted science teachers, gave life to Carmelita Hinton's assertion in her 1934 "*Prospectus of the Putney School*" that "To think scientifically is to master the machine age." Mrs. H. opened the way for the MacDougalls' and other early teachers' increasingly passionate investment in the school's success by according them a great deal of responsibility in leading the overall academic program, and for the most part, taking their advice.

Thus, the depression helped Putney School as well as hindering it. In a more prosperous economy, it might have been impossible to attract faculty who could give such energy and skill to the building of the school, to say nothing of being able to buy and hold the land and the buildings. These holdings included the 900-acre neighboring farm, which Mrs. H. bought herself, by mortgaging, and then selling the Weston house. Finally, she had the seven to ten acres per energetic adolescent that she had long dreamed of as the ideal educational space.

The Great Depression had energized both teachers and students to come to Putney on low pay or high tuition ($1400 yearly, until the optional summer language learning term was dropped), live in spartan quarters, and labor in barn or gardens to put food on the table. Wealthier parents, aware that they were paying for the children of poorer parents as well as for their own, were often inspired by 1930s social reformers' determination to make the world anew. In a decade when capitalism itself was on the barricades, educational experiments with a socialist tinge could thrive. Carmelita Hinton often told students, "I didn't begin my school to help you to ADJUST to the world; I began it so that you would CHANGE the world." Generations of young people took her seriously; or, much as her boarders in Cambridge and Weston had done, they laughed at her behind her back—especially at her condemnations of radios, caffeine and heavy petting—but then went ahead and tried to live according to her wishes. Art, music, and theater were allowed so much time in the weekly schedule of classes and evening activities, and led by such resourceful and independent individuals, that disciplined self-expression could find full voice beside communal projects and enthusiasms. Music was so important—and the music director so compelling, irascible, and ingenious—that the first-year tradition of herding every student and most teachers into the Assembly Hall for "Friday

Sing" would continue for 65 years more, launching countless alumni's musical vocations or avocations in spite of the coerciveness of it all.

In addition to the time they spent on solitary craftsmanship, arts, or music practice, the young gained respite from the pervasive pressure to cooperate with others through the simple fun they shared on long Saturday walks or horseback rides in the surrounding woods and hills—with perhaps a fugitive cigarette to go with the sandwich lunch one had packed after the last Saturday class—or even from the lone 6 A.M. trek from dormitory to barn in winter's dark to greet and feed one's "own" calf, sow, or workhorse. A few carefully vetted older boys who had proved their skill on the logging or carpentry crews were allowed to build and live in their own cabins in the woods. "Utopia was at hand," wrote Carmelita Hinton when she later recalled the school's opening years and the intense philosophical planning for its future, which had involved students and teachers alike.[9]

By 1940, the student enrollment had reached 120, on its way to 180, once a new boys' dorm was added after the war. Putney's fitness as a college preparatory school was no longer in question, with admissions officers at some prestigious colleges (especially Harvard and Radcliffe) having discovered their faith in the preliminary findings of the Eight Year Study vindicated in the remarkably strong academic records of Putney graduates whom they had accepted. It seemed that Putney students' immersion in outdoor work and the arts, and their school's refusal to push any team except the ski team or to sanction academic competition in the form of grades and class rankings (which were kept on file for colleges but never shown to students) had not harmed them. On the contrary, Putney's version of progressive practices may actually have helped them grow into more original and independent (though often quirky) learners.

Mrs. H. continued throughout her tenure to prize forceful character over scholarship and to support the odd but promising "low-tester" by allowing him (it was usually "him") to stand for the "general diploma" instead of the much more usual academic diploma. Classroom teachers sometimes paid for her indulgence in the dizzying range of academic competence that they confronted in their classes; but students loved the zany variety of their schoolmates.

Teachers came and went, especially during the war years, arriving full of energy and admiration for Putney School's ideals, and then moving on exhausted after a year or two, as even idealists need sleep, some time to themselves, and some minimum of financial security. Again and again, however, alumni celebrate these short-term teachers as "among the best I ever had, including all my teachers in college and graduate school," and praise the academic freedom that allowed their teachers full scope to explore their deepest interests with their students. "Mrs. H. had the extraordinary ability to pick wonderful people with whom she couldn't agree," said one of Putney's most gifted and longest tenured chemistry teachers.[10] Only in the school catalog did Putney's curriculum cohere.

Certain adults (a few of them more privileged than the rest in their dormitory duties and housing arrangements) helped Mrs. H. to anchor the school throughout its first dozen years. They included Ed and Mabel Gray, original farm manager and housekeeper for Elm Lea Farm and among the few ready to stand up to Mrs. H. when she pushed her luck too far in asking sacrifices of other outdoor and classroom teachers. She had a way of wanting to start a new project (a youth hostel, a student travel camp, a nursery school for faculty children) the minute the last set of projects had reached some tenuous equilibrium. Mrs. H. seemed to have an endless capacity for leading recovery from crises like the loss of the first large barn in a spectacular fire and the hurricane of 1938.

Such disasters remained valuable stories with which to build the Putney epic, telling of the valor of both young and old in the face of dramatic challenges.

By 1948, however, the daily emergencies that brought the community together during its founding struggles had receded; the difficult war years had passed. An ever-growing group of teachers liked being at Putney School, thought of it as their place, and wished for a slightly more formal school governance that would ensure them a steady say in how things were run rather than leaving the structure of authority to Mrs. H.'s whim or to the faculty friendships that she happened to be enjoying at a given moment. Ed and Mabel Gray, whose salaries (pared down by Mrs. H. to match the local rates for "hired help") had been stuck for years at $50 a month, felt no desire to disparage their classroom colleagues' wish for a few more comforts. The classroom faculty hoped for higher salaries and roofs that did not leak, a limit on dormitory and evening duties, some system for airing grievances without fear, even a pension plan, all of which private schools of similar size were beginning to guarantee their teachers. They appreciated the freedom that their principal granted them to govern the academic program through faculty meetings and to teach as they thought best—and they wanted a commensurate share in the governance of the school as a whole. Some of the most capable and experienced of the newer faculty had taken over heavy leadership responsibilities from others who had left to fight in the war, only to be demoted when the warriors returned.

About 1945, a group of parents pleaded with Mrs. H. to substitute a board of trustees for the closed corporation that she wholly dominated. At last she did appoint a faculty Executive Committee to share some administrative decisions and to consider whether the corporation's stock (nearly all shares controlled by Mrs. H. herself) should be turned over to an independent board of trustees. But Mrs. H. had appointed her oldest, closest faculty friends to the Executive Committee, and the harder it worked, the more excluded others felt.

Mrs. H. often hired neophyte teachers on inspiration, certain that they could turn the experiences of an unusual life into a capacity to teach. Most such inspirations proved remarkably successful, but in 1947–1948, one of them, Edwin Smith, was daily losing ground with his economics and history students, in spite of his decade of work as one of FDR's first appointees to the National Labor Relations Board, or his subsequent career as a Congress of Industrial Organizations (CIO) labor organizer and director of the National Council of American-Soviet Friendship. When Mrs. H.'s oldest daughter Jean (also a labor organizer) first met Smith and persuaded her mother to give him an interview, he was running just one stride ahead of the packs of "internal security" hounds who were gathering strength in Congress and the FBI. Putney School seemed a haven for Ed Smith and his personable wife Marion, a skillful secretary. Yet he was unable to translate his experience even to his bright and eager seniors, who recall him as "an unbelievably poor teacher" who "did nothing but lecture and read aloud, and that badly." The students took his fulsome reading list (Hazen's *Europe Since 1815,* Marx's *Communist Manifesto,* and Moon's *Imperialism and World Politics,* among other works) and taught each other after class. Mrs. H. called Smith in after eight months to offer him the help of a proven teacher who shared many of his interests and work out a plan of improvement, but told him that if he could make no progress within a further trial year, he would have to leave in June of 1949.

There were so many discontents among the faculty as a whole that Ed Smith cannot be blamed for what happened next. Twenty-six of the 35 indoor faculty, including a few members of the secretarial staff in close contact with the students, formed a Faculty Association the next fall and began to plan excitedly for the truly democratic school that they thought Carmelita Hinton had originally envisioned. Students had a great deal of say in school governance; why could adults not have more administrative responsibility? Full of hope, a delegation of teachers, all highly respected by both students and colleagues, presented the Faculty Association's purposes to Mrs. H. They handed her a letter, which began,

> We know that you as an educator realize that there exist among any group of teachers certain problems which can best be solved by an orderly presentation of them to the administration . . . (including) such matters as the exercise of their professional responsibility in serving the best interests of the school, salary scale, tenure, etc.

The letter went on to say that the new Association would "from time to time," present proposals addressing these problems to her, all in "the best interests of the Putney School." It ended by expressing confidence that the Association would thereby be able to "contribute actively to the morale and the spirit of this community."

Carmelita's immediate response was to burst into tears. She was not comforted by learning how the Association proposed to bring about its hopes: a system of elected department heads, for example, who would work together to coordinate academic offerings for a fixed term of years and have a voice in the hiring of teachers. Association members wanted to help with the orientation of new faculty. They wanted to have some part in all of the decisions that Mrs. H. had made entirely on her own (consulting only those whom she chose to consult), and they wanted some form of deliberate, inclusive due process in the release of shaky teachers like Ed Smith.

Smith, with his CIO tactics in mind, was almost certainly behind the fateful rule that the Faculty Association drew up as soon as Mrs. H.'s panicked response to the letter became known: No association member was to speak privately to any nonmember on any subject related to the association's issues. Thus, the constant knockabout exchanges that had hitherto edged Putney School forward were absolutely forbidden, as was association membership for all teaching members of the present Executive Committee. As the lines hardened and Mrs. H. demanded the resignations of all adults who felt they could not "work harmoniously within . . . the framework" in place when they were hired, the prohibition of school-related conversation extended de facto to all conversation. A music teacher had a baby; her father died. Only one of her old friends, an association member, crossed the line to express her support and sympathy.

Mrs. H. soon rescinded her "resignations" ultimatum, tearfully apologizing for her hasty memo, but the damage was done. One beloved young teacher of English, Spanish, and flute resigned over Christmas vacation. Mrs. H. fired a dormitory counselor whom she accused of homosexual advances on the word of a single student. Acting on her usual yearly timetable, she warned two teachers that their contracts would not be renewed. One was a substitute for a theater teacher on leave of absence; the other was Edwin Smith.

By the middle of January, "It was War—all War, and no holds barred," one observer recalled. Mrs. H. made the most of her access to parents by trying to enlist their support. She rejected the Association's demand that all teacher dismissals be decided by a majority vote of a new committee, consisting of the principal and four elected Association members. "How can teachers hire and fire each other?" Mrs. H.'s daughter Jean wanted to know.

Many of the older students, deeply saddened by the conflict, struggled without success to reconcile their faculty friends with each other. A wry joke went round the school dining room, passed from one association teacher to another: "At Putney School, the dogs are treated like children, the children are treated like adults, and the adults are treated like dogs." The wonder is that school went on, with each adult remaining loyal to the students in his or her charge, ski trips and camping trips taking off as scheduled, and the ninth graders finding the split among the adults just one more in a series of amazing experiences that Putney had dealt them since their arrival in early September.

Early in February, Mrs. H. said that she was ready to consider any new, representative school governing body that Putney's adults could agree on, claiming that even her own responsibility "for deciding who shall and who shall not teach at the school" might eventually be modified. She promised that Association membership would not be a factor in the ongoing individual teacher contract discussions that she was holding.

And so it was the dissident teachers' turn to respond hastily. Apparently they did not believe her—their fears for their jobs having overcome all other considerations. Heartened by news of the achievements of militant new teacher unions in urban public schools and urged along by Ed Smith, Association members voted to become CIO Public Workers Local #808 and to give over their negotiations to professional union lawyers, reserving the right to call a strike if no agreement could be reached by the time spring vacation began.

With the lines clearly drawn, all of the remaining 33 nonacademic staff members decided that they would have nothing to do with the union. A Spanish refugee teacher, who had been persecuted by Franco's goons for his leadership of a Spanish professors' union, quit Local #808, decrying the stubbornness of its leadership. The arguments were both sharpened and distorted by the national news media having entered the fray: Local #808 was the first significant teachers' union ever to be organized in a private school, and one reporter wrote darkly of the likelihood of covert Communist influence behind its respectable elected officers.

Hope returned briefly as parents began to get involved in the process of reconciliation, and an experienced, humane CIO negotiator took over the task of presenting the union's position to the equally thoughtful lawyer the school administration had hired. Both could understand why teachers would want a "clear definition of teaching loads," with salaries geared to years of service and out-of-class duties, and a due process system by which long-term "tenured" teachers could appeal the nonrenewal of a contract. The "closed shop" demand was harder to handle, yet on March 22, the two men were feeling pleased that they had reached an agreement that seemed acceptable to both sides. This was the day they heard the news that the 24 Union teachers had declared a strike, read their reasons (often speaking through tears) to their students, and walked out of their classes.

The union's president explained to Mrs. H. that it was her "uncompromising hostility" since mid-November that had brought about the strike; that "because of the recent unethical rehiring procedure on (her) part, "we repudiate all antecedent agreements as to rehiring and demand that letters of rehiring be tendered to all Union members by March 25."

Several secretaries dusted off their French, history, and math to join nonunion teachers in keeping classes going through the rest of the week until the students left for vacation. Only the oldest students were aware of the danger that no student would be able to return in April—that Putney School might well close down for good.

It took days of thought on both sides and the parents' irresistible faith in Putney's future to reopen the school after all. Groups of parents met in Boston, Chicago, and New

York; and although they could not agree on how the reopening was to take place, or even who was right and who was wrong, enough evidence accumulated about Ed Smith's career to persuade the other Union teachers that his release was not an unreasonable demand. This agreed on, Mrs. H. and the Union together appointed one parent as a mediator—an experienced school principal and college dean who brought Mrs. H. to the point where she could recognize the Union, as well as accept a due process and governance system involving teacher-elected representatives not unlike the one the teachers were requesting, and an independent board of trustees.

Given that the Union teachers had gotten virtually everything they had originally asked for, it was puzzling and sad to watch 22 of them leave by July, and the last 2 the following year. As their struggle had grown increasingly bitter and their admiration of Mrs. H. faded, the warmth of both idealism and friendship had gone out of them. Yet, remarkably, the school survived—perhaps the most powerful evidence imaginable of the strength inherent in Carmelita Hinton's original conception. Parents and alumni sent several experienced teachers her way as late as June. Through the summer, colorful neophytes showed up, attracted by Putney's 15-year reputation in a period of political retrenchment and gathering disparagement of progressive education as an indulgence in "life adjustment."

Students were ready to welcome all but the manifest "lemons." The number of new applications fell, but rose again the following year. The rationalized salary scale proved more a help than a hindrance in hiring teachers and treating them justly. The new teacher/staff/principal governing council did its job. Even Mrs. H. began to welcome the support and help she found in the Council and among the trustees, none of whom fit her fearful image of conservative bankers out to ruin her school. Within a year, neither new nor remaining faculty found Local #808 necessary and quietly voted to disband it.

In the years that followed, tears sprang to Carmelita Hinton's eyes whenever she spoke of the strike; it is hard to tell whether the conflict—the first ever that would not yield to her hard work, energy, and plain stubbornness—wounded her permanently. It seemed to students and teachers that her courage and her joy in her work quickly returned; certainly her zeal for new projects did not flag for long. She bought land on Cape Breton island and launched a summer camp. She persuaded the Putney School trustees to back the Graduate School of Education; led by one of her "finds," a radical World Federalist, the graduate school soon attracted just enough like-minded adventurers to keep it this side of bankruptcy. Against sober advice, she initiated an elementary school for faculty children on her own land near the school and recruited two Reichian therapist-educators with long-standing and successful nursery school experience (and much more controversial elementary school experience) to lead it. She defended her son Bill and daughter Joan, who had each joined the Chinese Communists in rebuilding a war-torn nation as agricultural technicians, even though Joseph McCarthy was in full cry. Mrs. H. worked with the trustees to begin the Annual Giving Fund and rejoiced as a few black students enriched the polyglot student body and alumni began entrusting their children to Putney School in large numbers.

Carmelita Hinton did not stay on indefinitely, but retired as head of the school in 1955, when she reached 65, knowing that her school had abundant strength to continue without her; indeed, one of her deepest pleasures was watching the number of applications rise so fast that most would-be students had to be turned away. Quite soon, alumni and parent donations allowed the trustees to buy nearly all of the 900 acres of land that she had owned and rented to the school. Mrs. H.'s vigorous health allowed her over 20 years more of usefulness and adventure. In 1961, she tried to get a visa to visit her daughter's Chinese family on their north China model dairy commune, and was refused. She went anyway, traveling through Russia and Mongolia—and losing her passport on her return. She helped to lead the celebrations of Jane Addams' birth 100 years before. She broke her hip walking backwards while leading a peace march, recovered, and marched again. She retrieved her passport in time to become trip leader for one of the first student trips to China since the 1930s. Entering her 80s, she took her grandchildren skiing, continued her peace work, and attended every single meeting of the Putney School Board of Trustees, winning some arguments and losing others gracefully.

As her health began to fail, Carmelita Hinton found a home in Concord, Massachusetts, with her daughter Jean. Her body hung on stubbornly even as her memory gradually failed. She never seemed to lose courage, however—perhaps because she had always had so much of it. She died at age 92 in 1983.

NOTES

1. This personal recollection of Carmelita Hinton's, like most others mentioned in the text, was recorded by me during four interviews with Hinton in her ninetieth year in preparation for writing *The Putney School.* Susan Lloyd, *The Putney School: A Progressive Experiment* (New Haven: Yale University Press, 1987). Where her memories were passed on to me by others such as her oldest daughter Jean Hinton Rosner, this is indicated. In rare cases I have drawn on my own most vivid memories of Hinton as the Director of the Putney School during my attendance there from 1948 to 1952. Certain other material used in this article is also adapted from *The Putney School.*
2. Hinton's paraphrase of James, quoted in "Carmelita Hinton at 82: From Putney to Peking," *Boston Globe Magazine* (26 November, 1972): 22. She was almost certainly summing up several of the ideas in William James, "Emotion" in *Psychology,* (New York: Henry Holt, 1892). This was James' own abridgment of his *Principles of Psychology,* designed for college students and assigned for close reading by Dr. James Leuba in the course mentioned in the text. "Refuse to express a passion and it dies," James, *Psychology,* p. 382.
3. Lawrence A. Cremin, *The Transformation of the School: Progressivism in American Education 1876–1957* (New York, Knopf, 1961), p 88.
4. From an interview with Carmelita Hinton by A. Katie Geer (hereafter referred to as AKG), a Putney School student (class of 1979) whose senior project under the guidance of history teacher Sven Huseby focused on interviewing some 20 of Putney's founders and founding teachers. The transcripts of these interviews proved to be invaluable, as several people she spoke with had died or were unreachable by 1981.
5. Neva Boyd, "Play and Game Theories in Group Work," "The Progressive Origins of the Putney School," Senior Project, 1979 (quoted in AKG).
6. Liebe Coolidge Winship, whose mother taught third grade at Shady Hill.
7. Carmelita Hinton, "History of the Putney School," from *Putney School Needs $250,000 for New Housing* (Putney School, 1946).
8. Quoted in Lloyd, *The Putney School,* p. 29.
9. Carmelita Hinton, "My Education for Teaching," in *Bryn Mawr Alumni Bulletin* (Spring 1951).
10. From Dan Morris, teacher of science, mathematics and photography, 1935–1943, quoted in Lloyd, *The Putney School,* p. 131.

CHAPTER EIGHT

FLORA J. COOKE AND THE FRANCIS W. PARKER SCHOOL

GAIL L. KROEPEL

Flora J. Cooke (1864–1953) was a disciple of the progressive educator Colonel Francis W. Parker (1837–1902). Her work was informed by a large and inclusive vision of the school, of the teacher, and of the student. To understand the impact that Flora J. Cooke had on progressive education, and secondary education in particular, and to appreciate the stature that she achieved in the educational community, it is helpful to know her history as a progressive educator.

EARLY EDUCATION AND THE INFLUENCE OF COLONEL FRANCIS PARKER

Flora Juliette Cooke did not have an extensive academic background by today's standards. Her formal training began in an elementary school in Youngstown, Ohio, and she graduated from Rayen High School in 1884 at age 19.[1] She began teaching in the fall of that year in a one-room rural schoolhouse in Auburn, Ohio, and continued the following spring in Bainbridge, Ohio.[2] Not until 1889, when she taught at the Normal School in Chicago, was she able to accumulate three years' worth of college credit in education, science, and literature through University of Chicago extension courses.[3] Her only period of protracted study in higher education was in 1900, when she studied elementary and applied chemistry at the Armour Institute of Technology, renamed Illinois Institute of Technology in the 1940s.[4] She never completed a bachelor's degree, but in 1931, Lake Forest College awarded her an honorary M.A. degree.[5]

The touch of fate that brought her into contact with Colonel Parker's educational methodology occurred in the fall of 1885, when she accepted an assignment at Hillman Street School in Youngstown, Ohio.[6] There, she came under the tutelage of Hillman's new principal, Zonia Baber, who had recently graduated from Normal School in Chicago, where Colonel Francis W. Parker was director. In a speech honoring Baber in 1944, Cooke said, "Indeed, I owe to her the greatest opportunity of my life—that of becoming a member of Col. Parker's faculty. . . . Fortunately for me that Youngstown school was at the top of a high clay hill. Miss Baber lived near the school, and I far away. Therefore, in bad weather, I spent many nights with her and received a life-long inspiration and enthusiasm for teaching children. Indeed, I had two years of intensive professional training (with much of it given after midnight) by this zealot—Zonia Baber."[7] Baber and Cooke became lifelong friends and professional colleagues.

In 1887, Zonia Baber returned to Chicago to head the Geography Department at the Normal School. Flora Cooke replaced Zonia Baber as principal of Hillman Street School and continued in that capacity until she accepted a teaching assignment at the Normal School in 1889.[8]

Francis W. Parker was the head of Normal School from 1883 to 1899.[9] It was Baber who convinced Colonel Parker to invite Cooke to teach at the Normal School. He agreed, but only if Zonia Baber would take full responsibility for Cooke. Cooke overheard Parker say, "She is yours, Zonia. I wash my hands of her. See what you can make of her."[10] Cooke took an assignment at the Normal school in 1889 and taught there for ten years, honing her teaching skills and absorbing Colonel Parker's philosophy and methodology. It was ironic that he was not impressed with her at first, because she would spend her career using his precepts as the foundation of her work.

Cooke tried to develop policies that exemplified the philosophy and educational practices that she had learned from Colonel Parker, who was to be referred to by John Dewey as "the Father of Progressive Education."[11] Colonel Parker's philosophy and practices focused on developing students as whole persons and responsible citizens. Development of teachers was also central to Parker's philosophy, and he fostered their leadership. His teachers were expected to take part in parent meetings, participate in the activities of the school community life, attend conferences, speak at institutions, and write articles for publication. Cooke wrote, "Colonel Parker expected all of his faculty to attend local education conferences and delegates were sent from the school to the National Education Association (NEA) meetings. They were expected to participate and to report back to the school any important findings or decisions."[12]

He also urged his teachers to keep developing their writing skills, as he believed that writing clarified thinking. As Cooke noted in 1941, " . . . every teacher was expected to contribute something to the 'Chicago Leaflet Envelope,'[13] a monthly publication of the school. . . . When the teachers complained that the work they did was not 'good enough to print,' or 'not ready for publication,' Colonel Parker's fixed reply was, ' . . . You need the courage to be crude. It is understood that what you write is your best only for the day. Writing will help you to find the lacks and gaps in your knowledge and equipment needed for the best work of your children.'"[14] Because of the writing discipline that she acquired at the Normal School, Cooke was able to articulate the principles that she employed at the Parker School in later documents.

Parker expected teachers to speak often before their colleagues at the Normal School, a requirement that Cooke found daunting at first. As she said to Colonel Parker, "If you ever ask me to speak before this galaxy of great teachers, I shall die or resign." He replied, "Decide quickly what you will do, for if you stay I shall try to call upon you every week. What you say will not be worth much at first, but you must be willing to 'pull your oar'—that is the only way you can learn and grow, and that's what we're all here for."[15]

The practice she received in speaking and writing while teaching at the Normal School, and the exposure to the broader educational community encouraged by Colonel Parker, proved valuable to Cooke; she became an accomplished and popular speaker at teachers' institutions. She conducted a six-week summer institute in Hawaii, and she taught Mormon teachers for one month in Salt Lake City. Throughout her career, she addressed many audiences, lecturing to teachers in 28 states about progressive education.[16]

Her position at the Normal School under Colonel Parker exposed Cooke to influential people who came to play important roles in her career. Many prominent families enrolled their children in the school. Fred Dewey, the son of the well-known educator, John Dewey, was a student in Flora Cooke's first-grade class during 1894 and 1895, and Dewey's daughter, Evelyn, attended the class the following year.[17] Anita McCormick Blaine, who later became the benefactor of the Francis W. Parker School, enrolled her son Emmons in Flora Cooke's first-grade class.[18] Carleton Washburne and Perry Dunlap Smith, who later became principals of progressive schools, were also students in Cooke's first-grade class.[19]

In 1899, Parker established the Chicago Institute, Academic and Pedagogic, on the north side of Chicago, taking with him 30 of the teachers from the Normal School, including Cooke.[20] While the Institute was being organized, the teachers were given one year to pursue advanced studies.[21] During this time, Cooke studied at the Armour Institute.[22] When the Chicago Institute opened, Cooke worked there for one year, at which time the new school became affiliated with the university of Chicago.[23] Because the Chicago Institute was transferred to the University campus on the south side, a second school, funded by Anita McCormick Blaine, was opened on the north side in 1901, and Cooke was chosen by Colonel Parker and Mrs. Blaine to be the principal.[24] The school was named the Francis W. Parker School.

EDUCATIONAL PRINCIPLES AND PRACTICES

During Cooke's first teaching assignment, in 1884, at Auburn, Ohio, she had eight students in her classroom; at Bainbridge she had 20. One year later, when she went to Hillman Street School in Youngstown, Ohio, she was confronted with 125 children in her classroom. Faced with such a large number of students, Cooke sought an activity-oriented and creative approach to teaching. She provided students with pictures, puzzles, games, modeling clay, and other activities so that children could select something of interest to do while she worked with small groups of children on their three "R's." Through Baber's influence, poems and songs were also adapted to six-year-olds, a practice that came directly from the Normal School's first-grade course of study.[25] To provide children with additional interesting reading material, Cooke published a book entitled *Nature, Myth, and Stories*,[26] which included stories that appealed to young students and encouraged them to read. Cooke considered it important to focus on students' particular interests so that they would want to learn to read, and to whet their appetite for traditional subjects.

Thirty-five years later, in 1929, Cooke responded to a colleague who wrote, asking for her advice about teaching reading, "Kindergarten . . . is a period of great sense-hunger and I believe the children should be getting first-hand experience in many directions. They should be planting, painting, modelling [*sic*], constructing, singing, and making all kinds of play things for their own use and satisfaction. At the same time there should be lovely books all about them with alluring pictures in the field of child literature. They should see their teachers using books and going to them for all kinds of lovely stories and poems for them. . . . When a child really wants to learn to read there is no difficulty whatsoever about teaching him."[27] Cooke felt that reading should be as natural as talking, and that reading and writing should draw on the natural sociability of children.

One technique that she implemented engaged children in writing “reading leaflets” about an experience. This was much like Colonel Parker’s “Chicago Leaflet Envelope.” Each child, with help from other children, dictated sentences to be written by the teacher on the board. The teacher edited the sentences into a form parents and other children would comprehend. Sharing leaflets with other first graders awakened the students’ desire to read.[28]

Cooke recounted an incident involving John Dewey’s son Fred, while the boy was enrolled in the Normal School. She used his interest in the school’s wood shop to motivate him to read. She said, “Fred loved that shop. He wanted to do things and when he was not doing things, he wanted to dream, think his own thoughts. He did not care a rap about reading and writing and those things. They seemed utterly useless things to him. . . .”[29] Because the shop did not have enough equipment for the whole class, she divided the students into groups and told them that they could take turns working in the shop if they knew what to do. She wrote directions on the board for them to follow while they were in the shop and had the students copy them. When it was time for Fred’s group to work in the shop, Cooke required them to read back the directions. Fred could not.

She concluded, “Now that was the beginning of Fred Dewey’s seeing that reading had some relation with the big group. Many directions had to be given that way. It had some relation to something that he wanted to do, and it took a good many experiences, but by the end of that year he could read almost anything.”[30]

During a conference in 1930, Cooke described how she used a student’s own interests to motivate him to learn. The student’s mother had enrolled him in the eighth grade of the Parker School. He hated school and was accepted into the Parker School on a trial basis. Cooke reported that the student did nothing for the first three months, but then he became interested in *The Parker Weekly,* the school newspaper. *The Weekly* was run by the Parker students and was not critiqued by adults until after it was distributed. This interested the new student very much, but the students on the newspaper’s staff did not accept him, because he did not have any skills to offer them. He made it his business to approach the printing teacher and learn how to set type. He brought up his other schoolwork to an acceptable standard to participate in this extracurricular activity, and by his senior year, he had become editor in chief. Cooke’s point was that something “has to be related in the child’s own mind to something that he considers worthwhile, if we are to get the whole child going.”[31]

Cooke emphasized that it was up to the school “to take these things that are in the world today that make our body of life go forward and to present in the schools problems and situations that give children opportunities to hitch up with the world and understand it in its progress.”[32] By discovering those interests that individual students held and using those interests to motivate learning, Cooke helped the students develop their abilities and experience the satisfaction that comes with the use of their own powers. She believed that this technique not only promoted a person’s individual development, but also developed a person as a useful member of the community. Through a program that created incentives, built on students’ strengths, and provided opportunities to make individual contributions in the small community of the school, the students developed the skills that eventually would allow them to make contributions to the larger world community.[33]

This approach helped to shape the policies that she was to develop as she embarked on the principalship of the Francis W. Parker School.

COOKE AS A LEADER

Flora Cooke was the principal of Francis W. Parker School in Chicago, Illinois for 33 years, from 1901 to 1934.[34] During this time, she fought tirelessly for educational innovations, traveling in the United States and abroad to carry the concepts of progressive education to many audiences. She also took on the formidable foe of progressive education—the college entrance requirements—fighting uncompromisingly for the education of the whole student, whether that student was bound for Yale or would not go on to college.

When the Parker School opened in 1901, it had approximately 60 pupils.[35] The trustees engaged Cooke as the principal at $2,500 per year.[36] Anita McCormick Blaine, one of the trustees, provided $1 million to establish the school, and she funded it heavily up to the time Cooke retired.[37] Cooke developed the educational policy of the school, following the death of Colonel Parker in 1902,[38] but she was strongly influenced by his teachings and writings. As she told a graduating class in 1935, reiterating one of Colonel Parker's primary convictions, "The needs of society should determine the works of the school. The chief need of society is good citizenship. The function of the school is, therefore, to provide conditions for the steady growth of all children towards ideal citizenship."[39]

In an article for *Coronet Magazine,* Carol Lynn Gilmer reported that Cooke "believed that Parker [School] had no validity as a private school unless its findings could be applied to public schools,"[40] because public schools were necessary to ensure a democratic citizenry. It was also important, as Cooke wrote elsewhere, that an experimental school "represent America's melting pot and have a diversified group of children differing in age, ability, and background,—economic, cultural, and religious,—learning to understand each other as growing citizens who must share in the future welfare of the nation . . ."[41] Diversity in the student population of the Parker School was possible because Anita McCormick Blaine subsidized the school annually, paying half of the entire cost of the school throughout Flora Cooke's principalship,[42] thus allowing all students to attend at a reduced cost or, in some cases, to pay no tuition.[43]

As the principal at the Parker School, Cooke sought innovative methodologies to enhance the educational experience. In 1904, she traveled east to look at other schools and interview their teachers.[44] In Cleveland, Ohio, Washington, D.C., Baltimore, Maryland, New York City, and Brooklyn, New York, she visited many types of schools, including Bryn Mawr School and Girls' Latin School, the Manual Training School, the Ethical Culture School, the Pratt Institute, Columbia Teachers College, Horace Mann School, and the Speyer School. She took detailed notes on elementary and secondary-level courses, students' responses to classwork, student-teacher ratio, types of buildings, equipment used in classes, and the amount and content of homework assigned. She discovered that students were admitted to Harvard and all other colleges, except Yale, without Latin credits.

Cooke promoted creative innovation for both students and teachers. She believed that students should have "free and spontaneous all-round growth," and teachers "must have the same kind of freedom for creative cooperation and the same responsibility for good results."[45]

Cooke demanded constant revision of teaching methods and encouraged innovation from her teachers. Under Cooke's leadership, teachers tried new teaching techniques and wrote articles about their approaches to education at Parker School, contributing to a publication called *The Francis W. Parker Studies in Education,*[46] often referred to by

Cooke as the Yearbook. This became a ten-volume series, which was published from 1912 to 1934. The publication mirrored the concrete practices of the school by describing features that were central to its philosophy. The practices that were particularly emphasized were the (1) Social Motive, which emphasized the responsibilities that the student had to the school community; (2) Morning Exercises, which helped to form a school community by bringing all of the classes together daily; (3) Social Sciences, which were used to integrate the other academic endeavors; and (4) Creative Effort, which was aimed at developing the artistic abilities of the students.

Cooke maintained a democratic relationship with her teachers, except when she believed they were forgetting their obligation to students. On one occasion, when the school was becoming overcrowded and a teacher proposed expelling some of the more difficult children, Cooke said, "If their homes are as sordid as you say and their parents as blind to our vision of education and the children as unpromising certainly their only hope of glimpsing something better is this school; and *you* have not seen child after child whom this faculty has thought impossible rise to fineness before graduation."[47]

Cooke cited Colonel Parker's insistence " . . . that every child, not subnormal, could do something well and that it was the business of a school to discover what each child's aptitudes, strengths, and gifts are, and to use and develop these until each felt some measure of success—and the confidence thus gained was to be used as a lever to help him to overcome his weaknesses and handicaps."[48]

Another principle that Cooke shared with Colonel Parker was the importance of developing the innate aptitude of each child. Cooke recalled, "Parents of unadjusted children who had been turned away from other schools wore a path to the door of his school, and unless feeble-minded they were taken in. All his faculty group were expected to find some way to help each child to develop his powers to fullest capacity and to find some aptitude which could give him a measure of success, courage, and self-respect. Colonel Parker insisted that every child possessed such aptitudes or gifts which might be developed."[49] Cooke found it challenging to continue the practice of administering to the educational needs of every student at the Parker School, however, because the trustees wanted to concentrate on the college-bound students.

Colonel Parker demanded commitment to each student, believing that, as Cooke recalled, " . . . there never was a bad child, never. There was an unadjusted child, the child that was out in opportunity [*sic*], a child who had very little power, all those things, but he said the unadjusted child was what we wanted, to try to help him in this place; . . ."[50] At the same time, nothing was allowed that bordered on license, nothing intimating an impertinence to adult or child, no "uncalled-for-liberties" were taken;[51] the child-centered approach did not mean that students were allowed to misbehave.

As a leader, Cooke modeled the values and behaviors that she expected of the faculty. Elsie Wygant, a teacher who worked for Cooke for many years, wrote of Cooke's devotion to children, "We who have worked with her year after year have learned that no individual's interest is too insignificant for her intense co-operation; no child too shiftless or crude or hopeless for Miss Cooke to perceive possibilities that would dazzle even a fond mother's hopes."[52]

Cooke's commitment to students was recalled by Perry Dunlap Smith at a luncheon in honor of Cooke in 1934. Smith, who taught algebra at the Parker School, had a student who cheated several times while in his class. Smith reported the matter, at last, to Cooke, expecting her to expel the student. Instead, Cooke admonished Smith, saying, "This boy has been expelled for cheating from four schools. . . . He is not going to get expelled from this school for cheating. If you think you have the making of a teacher in you, I should think you could probably straighten this lad out by the end of the year. If by the end of that time you haven't been able to straighten him out, I should advise you to sell bonds, or take up some other equally useful occupation."[53]

Her devotion to her students and her many successes as a principal notwithstanding, Flora Cooke remained a controversial figure throughout her long career, and withstood numerous challenges to her authority and leadership and the educational principles that she espoused.

FLORA COOKE VERSUS THE EDUCATIONAL ESTABLISHMENT

Francis W. Parker School was established in 1901 as an elementary school with a kindergarten.[54] The next year, Flora Cooke started a high school program, but almost immediately, she had difficulty applying the progressive approach to this new program, because desirable colleges would not accept high school students unless they met rigid entrance requirements that were not easily reconcilable with a progressive school curriculum. As early as 1905, Cooke was concerned because parents were withdrawing their high school age children from the Parker School and enrolling them in college preparatory schools. She tried to convince the parents that the Parker School curriculum would prepare their children for college.[55] Cooke felt strongly that the elementary school had no right to exist without the high school to test its effectiveness.[56]

In January 1906, during a trustees meeting, Cyrus Bentley, a lawyer and chairman of the Board of Trustees, raised questions about Cooke's administration of the Parker secondary school. He also expressed concern about improving the responsibility of students and raising their level of class work.[57] He questioned whether the high school should continue at all. A major conflict centered on Cooke's desire to develop the Parker School according to Colonel Parker's methodology and the trustees' desires to develop a college preparatory school. Bentley sought to put the high school into the hands of someone other than Cooke. To this end, Robert Nason was hired to be an associate principal and to head the high school.[58] Cooke was no longer allowed to make policies concerning the high school and was overruled on many high school issues.[59]

In the fall of 1907, the trustees hired Frank A. Manny, from the Teachers College at Columbia University, to do an evaluation of the school. Manny suggested that Cooke was not "fitted"[60] to supervise the high school; the school was overstaffed; several male members of the faculty should be replaced; and there were too many "subnormal" students. Furthermore, too much time was being devoted to individual students and small groups.[61]

In 1908, seven members of the faculty complained that the high school curriculum was impacting the elementary school unfavorably, stating that "the amount of time and strength taken up by the regular prescribed courses effectually prohibits the organization of the high school as a social community; . . . [that] it is not flexible enough to permit the teachers to make the necessary experiments in order to suit the work to individuals, or even roughly classes; . . . that it must inevitably influence the elementary school toward retrogression; and that it is distinctly reactionary in that it tends to put the emphasis even more strongly and unavoidably in the minds of both teachers and students upon an effort to cover ground."[62]

At the end of the 1907–1908 school year, Cooke said that it was impossible for her to work with Nason; several teachers had resigned, owing to Nason's treatment, and her authority within the school had been so eroded that two separate schools were developing. This situation was resolved when Nason left the Parker School at the end of 1908. Cooke referred to the Nason years as "disintegrating ones, but I am not without hope for the future."[63]

The challenges to Cooke continued, however. In January of 1909, Bentley hired Arthur Detmers, a former head of a large high school in Buffalo,[64] to evaluate the quality of education at the Parker School and prepare a report. Completed by January 1910,[65] Detmers' report stated that the elementary school headed by Flora Cooke was very good, but the high school was unsatisfactory. Detmers recommended that the high school "be entrusted to a man, . . . Cooke should be freed of the administrative work that she is doing and left free to supervise the educational processes of the elementary school and the kindergarten. [66] However, this recommendation was not implemented, Detmers was hired as a faculty member, and Cooke remained the head of the whole school.[67]

The difficulties of maintaining progressive qualities in the secondary school continued, but were soon to be played out in a larger arena. In 1920, Cyrus Bentley wrote to Cooke that it was becoming ever more difficult for students to gain entrance to the best colleges.[68] The Parker School had developed and restructured its curriculum several times, and some nonacademic requirements had already been sacrificed for students to be admitted to college. Another restructuring of the high school in the spring of 1921 included hiring a psychologist and organizing a standing committee of the department heads to integrate into the high school curriculum clearly defined educational policies, such as experimentation, curricula to furnish rich intellectual life, community life of mutual service and responsibility, and training in the ideals of democracy.[69]

After World War I, the use of achievement tests became important in the educational community, especially to identify college-bound students.[70] Cooke was open to using these new testing techniques and was not limited to the progressive educational community in her search for innovative teaching techniques. As she noted in a speech at a Progressive Education Association (PEA) symposium, "In 1925, after five years of experimentation with the Binet and certain group intelligence and achievement tests, the Francis W. Parker School established a Department of Measurement and Diagnosis, with a trained head, giving full time to this work, with clerical assistance."[71] Cooke considered it important not to use the tests to channel students into college or noncollege tracks but as a tool to help students uncover their potential.

At another PEA symposium, Cooke discussed the wide range of extracurricular activities that students could engage in at Parker School, if they did not neglect their academic obligations. "Some of these activities, such as the School Publication, the Student Government organization, the Forum, the School Orchestra, and divers other student-initiated clubs, are now school traditions." She discussed the Parker School requirements in music, art, and manual and physical training, emphasizing that students were expected to have a broad cultural experience. She made the case that colleges were needed to ensure that students could continue that experience without a narrowing of scope.[72] Eight years after the Parker School curriculum was challenged by Cyrus Bentley, Cooke still had not removed nonacademic subjects from the curriculum.

Cooke insisted upon linking college requirements such as Latin to literature, social studies, and drama, thus broadening the students' cultural awareness. In 1929, Cooke wrote, "We have now gotten our Latin so that it is something much more than just Latin, it is a structural basis for all the structural languages and [a] wonderful help and a nicety in the choice of words and in an increase of vocabulary. In that sense it is an experience which fits into all of their creative writing and understanding of languages."[73]

Most college preparatory schools focused on college entrance requirements for the entire four years of secondary school, but at the Parker School, the curriculum contained a diversity of course work, and an intensive review occurred at the end of the senior year. Cooke wrote, "I would rather have our pupils have three and one-half years of high school freedom and four months of intensive review and have them jump the 'fool hurdles' than for them to stay away from college, which is the alternative at present. Furthermore, if we were not willing to do this we could have no pupils at all in this school where 98 percent of them go to college."[74]

Though Cooke had made many adaptations at the Parker School to achieve accommodation between secondary school progressive curriculum and college entrance requirements, the PEA would eventually provide a resolution to the conflict.

THE RESOLUTION OF THE COLLEGE PREPARATION/PROGRESSIVE CURRICULUM CONFLICT

Cooke was an active member of the PEA, which was formed in 1919 by a small group of private school administrators and teachers to provide a forum for the many versions of progressive education being tried.[75] The chief concern of many members of the PEA was reform around the issue of the college preparatory curriculum, because college entrance preparation drained time and energy from a student's individual and social development.[76] In 1930, Cooke and the other members of the Executive Board of the PEA undertook a research project to evaluate college entrance requirements.[77] Generous grants from the Carnegie Corporation Foundation for the Advancement of Teaching and the General Education Board subsequently made possible a research project on a larger scale.[78] The group, named the Commission on the Relation of School and College, determined that a basic reorganization of the secondary school curriculum was needed.[79] The PEA initiated an experimental program, which came to be known as the Eight Year Study, and the Francis W. Parker School became one of the high schools to participate in it.

THE EIGHT YEAR STUDY

The Eight Year Study was a limited experiment in the relaxation of college entrance requirements for a small number of high schools. Selected high schools were to be released from the course work and unit requirements then in force, as well as from the entrance exams of those colleges. By 1932, more than 300 colleges and universities had agreed that selection of candidates from participating secondary schools during an experimental period would be based only on a recommendation from the principal of the school and a detailed history of the student, including activities and interests, evidence of the quality and quantity of work, the results of various examinations, and scores on scholastic aptitude, achievement, and other diagnostic tests. The commission appointed a special committee on records to determine how these nontraditional assessments of candidates should be collected and presented to colleges.[80]

After the colleges responded in 1933, the Commission and the Directing Committee chose a diverse group of secondary schools.[81] The first students entered college in 1936, and their performance was evaluated to determine whether they did better or worse than students trained with the traditional methods of college preparation. Each student in the experiment was matched with another student in the same college who was not in the study and who had met the usual entrance requirements. The performance of the matched pair was tracked during the study.[82] The Committee on Evaluation and Recording of the Eight Year Study developed new records and tests to evaluate and report the knowledge and skills of the students participating in the Eight Year Study.[83] Each school was given complete freedom to work out content and method in accordance with their best thought and experience as to what high school students should have in their last four years. Thus, a student would have a progressive secondary education, followed by four years in a traditional college or university.[84]

Before the Eight Year Study, Parker School students were required to complete the courses for college entrance requirements, plus such nonacademic course work as music and art, to earn their graduation certificate, although only courses accepted by colleges could be used for college credit. After the study, all of the courses taught at the Parker School could be considered academic courses.[85] Unless a student had an obvious talent, these courses took the form of art or music history and appreciation. Nonacademic activities involved individual student, group, and community projects, which were no longer considered extra curricular activities but were incorporated into the students' formal education.[86]

Although Cooke expressed enthusiasm for the tools that the evaluation committee had begun to develop the previous year, she spoke out against the "avalanche of tests, and particularly against the nature of the tests, which came from the commission's Testing Committee. . . ."[87] Cooke preferred to send the committee full cumulative and anecdotal records of the students, saying, "I believe the school should resist every attempt which tends to make the work uniform, rather than experimental, in character."[88]

In July 1936, to provide a thorough evaluation of the Eight Year Study, the Commission set up a subcommittee called the College Evaluation Staff, made up primarily of university deans. The College Evaluation staff judged students on intellectual competence, cultural development, practical competence, philosophy of life, character traits, emotional balance, social fitness, sensitivity to social problems, and physical fitness. These criteria were subdivided further, and "sources of evidence" were agreed upon. Evaluation tools were developed to keep track of all of the data collected.[89]

The Eight Year Study showed that progressive secondary schools could effectively educate college-bound students, and graduates of Francis W. Parker School compared very favorably to the students with whom they were matched; the Parker students' grade point average was higher than that of the control group.[90]

COOKE AS AN INNOVATOR

At a conference in 1930, Cooke stated how her policies had evolved since her years with Colonel Parker, to reflect her experience and modern ideas in education. She spoke of three new principles, in particular. One was the recognition that a child must have certain minimum skills and knowledge by the end of each grade to do his or her work creditably. The child should not be held back from doing more, but he or she must reach a minimum knowledge level.[91]

The second involved the policy of having two teachers in every classroom—an experienced teacher, along with a teacher having only a year or two of experience. The experienced teacher worked individually with those students who were not achieving the minimum knowledge level until they had developed their skills sufficiently to join the larger class.[92]

The third principle dealt with the application of the "new science" of developing standardized achievement tests. It was Cooke's practice to give students additional work when they tested high, at the same time noting their social and physical adjustment to the group. When students tested low, Cooke would continue testing until she could determine where they were strong, promoting experiences of success.[93] Though she incorporated new methodologies to enhance the educational experience, the policies that she developed remained harmonious with Colonel Parker's educational philosophy.

Cooke also mentioned her long struggle to develop a secondary school at Parker, noting three major difficulties that she had faced in the process: (1) The high school curriculum was bound by college entrance requirements; (2) there were no experimental colleges functioning on progressive principles to which graduates of progressive high schools could aspire; and (3) high school teachers were not trained in pedagogy and did not understand adolescent psychology.[94]

In addressing the first two challenges, Cooke described how she sought to provide students with experiences that she believed were necessary for full human development. The Parker School required students to work in the arts throughout high school and included such extracurricular activities as writing, debate, dramatics, and music, as well as charitable efforts that were, as she said, "considered by faculty to be the most truly educational features of the school."[95] Parker School also used three publications, clubs, and student government to develop character.[96]

As for the third challenge—the training of teachers who had the special pedagogical skills and the understanding of adolescent psychology that would make them effective in a progressive school such as Parker—Cooke was less successful. Although she did teacher training in other settings, achieving a comprehensive progressive teacher-training program took 30 years.[97]

COOKE AND THE TRAINING OF TEACHERS

In the ten years that Cooke taught at the Normal School, one of her responsibilities was the training of teachers. In a biography of Colonel Parker, Ida Cassa Heffron, describes the Normal School as a combined professional training class and a school for practice.[98] The focus of the school was pedagogy, but student teachers had to write "knowledge papers" showing their mastery of content, as well as preparing lesson plans for teaching. Cooke's duties included teaching 40 children at the primary grade level and serving as a "critic teacher" to 20 student teachers, who observed the classroom for one hour each day and performed one hour of teaching under the supervision by the "critic teacher." Student teachers were not allowed to teach until the head of the department and the "critic teacher" approved them.[99]

Cooke wanted to develop teachers who could teach in progressive schools and repeatedly submitted to the trustees plans for a teacher training component at the Parker School, but without success. She received discouragement from other quarters as well. Patty Smith Hill, the Director of the Department of Lower Primary Education in the Teachers College at Columbia University, also advised Cooke against training teachers at the Parker School. Hill, noting that the Teachers College was "too burdened with numberless teachers to get the very best results," did not believe that Cooke could effectively train teachers while teaching students.[100]

There is evidence in Cooke's correspondence, however, that she was doing teacher training in other educational settings. For example, up to 1906, Cooke had been teaching at the College of Education at the University of Chicago, as evidenced by a letter from the dean requesting that she not withdraw her courses. The dean states, "The withdrawal of your courses from the second term of the present quarter would be a most serious and irreparable damage to us. Your courses are most attractive, and they materially meet the needs of a very large proportion of the men and women who are here for special study. . . . [F]rom our point of view the withdrawal of your courses would mean a very material injury to the college of education from which we could not well recover."[101]

In 1908, Cooke was invited by the North Texas State Normal College to give a series of lectures in their summer school.[102]

In February 1910, she was invited to speak at the University of Wisconsin to talk to their teachers in their training program, as well as the Collegiate Alumnae Association.[103]

In March of 1910, the following request came to Cooke from Peru State Normal School in Nebraska: " . . . [Y]ou have a host of disciples in York County of this state who want you above every one else to come to the County Teachers' Institute for the week beginning June 13. . . . I am sure you can command the highest price ever paid a Primary instructor in Nebraska, because we really want you."[104] As this letter shows, Cooke had developed a following in the educational community that regarded her as an authority.

In November 1911, she was invited to offer a summer course in Methods of Teaching Reading at the School of Education of the University of Chicago.[105] In May 1912, she was appointed to give two courses of instruction in the Department of Education at the University of Chicago.[106]

In May 1914, Cooke was invited by the Kansas State Teachers' Association to speak before the primary and kindergarten teachers.[107] In November of that same year, Cooke was a speaker at the Kansas State Teachers' Association. She spoke on the "Social Motive in Education,"[108] a topic that she had long advocated.

In 1932, Cooke founded the Graduate Teachers College of Winnetka,[109] with cofounders Carleton Washburne, the superintendent of the Winnetka Public Schools, and Perry Dunlap Smith, the principal of the North Shore Country Day School,[110] both of whom had received their primary and secondary education under Cooke at the Parker School.[111] The Graduate Teachers College was a laboratory school offering observation, experience, and research opportunities to graduate students. In a letter to Dr. Hilman Sieving, Cooke described the Teachers College as " . . . a small graduate college for well-qualified college graduates, who wished to prepare themselves for teaching the type of education which the founding schools represented. . . . The teacher training is given through a system of seminars and apprenticing, and each individual student is assigned for various periods of time to the school in which he can receive the preparation best suited to his interests, aptitudes, and the post-graduate field of work."[112]

It is likely that Cooke intended that the Teachers College provide a steady supply of teachers for the Parker School. Although Cooke never achieved a formal training program for teachers at Parker School, these teachers from the Teachers College did serve as "interns" at Parker. More important, however, was the opportunity to shape the way teachers were trained, and informal teacher training was a lifetime practice for Cooke.

Cooke was also involved with the founding of North Shore Country Day School, and Roosevelt University.[113]

COOKE'S INFLUENCE ON THE EDUCATIONAL COMMUNITY

Cooke was sought after as a speaker and authority by many educational institutions. Her appearances in schools around the country enabled her to become widely known as an educator and to have a strong influence on the preparation of teachers. By 1935, six progressive schools like the Parker School were scattered around the United States, headed by graduates of the Francis W. Parker School.[114]

On one occasion, she was consulted on the evaluation of an entire school system. In March 1914, a Leavenworth, Kansas, newspaper ran the following article. "Miss Flora J. Cooke, . . . a woman of national reputation in educational work, is in Leavenworth, assisting in the survey of the school system. Miss Cooke will be here for the balance of the week and will devote the greater part of the time in the primary grades in which work she is particularly well known. Miss Cooke received her training under the late Colonel Parker and one of the great educators of the country. In addition to her observations she will hold roundtable discussions with the primary teachers and give model lessons. . . . This city is one of the first to have its school system tested."[115]

By 1914, she was becoming increasingly active in professional associations. In April of that year, Cooke was invited by the commissioner of the U.S. Department of the Interior in the Bureau of Education to participate in an educational conference.[116] At that same time, Cooke became a member of the Advisory Council of the Chicago Kindergarten Institute Society, which was raising an endowment for a Kindergarten Training School,[117] as well as the National Kindergarten Association, the Chicago Teachers' Association, and the Northern Illinois Teachers' Association.

In 1923, Cooke became the Chairman of the Publicity Committee of Deans of the Women Association [*sic*],[118] and in 1925, she became President of the Superintendents and Principals Association of Northern Illinois. She also served as a trustee of the Chicago Teachers' College[119] and was an active member of the NEA.

In January 1927, Cooke received a request from the superintendent of the Ethical Culture School in New York City to make a thorough evaluation of their Elementary Department through the seventh grade. The evaluation was to cover the content of their course of study, the uniformity of work in parallel divisions, the sequence of formal subjects and the methods employed, class and school government, and the quality of their teachers. The superintendent said, "We estimate that about a month's residence in our School would be necessary for this task."[120] In May 1927, Cyrus Bentley, one of the trustees of the Parker School, asked Cooke to decline the invitation from the Ethical Culture School, so as to remain available to the trustees of the Parker School.[121]

Also in 1927, Cooke was a delegate to the National Educational Foundation conference on the "Meaning of the New Education" at Locarno, Switzerland. In conjunction with the trip to Locarno, Cooke received a letter written on behalf of the head of the National Education Foundation that stated, "Mrs. Ensor begs you to come to England as she would so much like to have you at her school, . . . to see the work that she and her co-principal, . . . are doing there. Also we could arrange for you to see . . . other very interesting experimental schools. . . . [122] Sometime during the Conference we shall call upon you to speak of the work of the Francis [W.] Parker School to one of our study discussion groups!"[123]

Describing to her constituency the Locarno Conference, which had attracted 1,200 people from 52 nations, many of them educators representing experimental schools,

Cooke expressed concern that "some of the experiments seemed to go too far in the direction of following the child's instinctive powers and interests *only.* In fact, quite a group of us agreed that Colonel Parker and Dr. Dewey had the vision for all-round education which has not yet been fully realized in any European school, . . . [we] found that Colonel Parker's idea of 'social education' and Dr. Dewey's as expressed in 'School and Society,' were not being emphasized. . . ."[124] Cooke applauded the European emphasis on the development of the individual child but was concerned about the lack of emphasis on social integration. Here, also, we see that she viewed both Parker and Dewey as important exponents of the concept of balancing individual and social development.

Cooke described how, prior to the conference, she and some colleagues spent a week in Paris visiting Versailles, the Louvre, Notre Dame, and the Cluny Museum. In Geneva, Switzerland, they witnessed the Labor Bureau and the League of Nations activities. After the conference, Cooke and her colleagues went to Florence, Italy, for ten days, and finally visited Venice and Holland.[125] This trip, along with a previous summer spent in Europe, as well as her many trips throughout the United States, broadened her awareness of educational issues far beyond those of her school in Illinois.

In 1929, Cooke attended the National Educational Foundation conference at Elsinore, Denmark,[126] at the time when the Foundation was promoting a merger with the PEA.[127] Cooke was a long-time member of the PEA and often participated in their activities, speaking at conferences,[128] serving as committee chairman,[129] and acting as presiding officer at sectional meetings.[130] In February of 1933, the PEA invited her to a closed meeting on the training of progressive teachers.[131]Her major contribution to the PEA was her involvement in the Eight Year Study.

In 1933, the PEA held their annual conference in Chicago, and the Parker School was included in a tour of the outstanding schools of the Chicago area.[132] The school set up an extensive series of exhibits for the visitors,[133] thus giving students the opportunity to show their work.

Cooke maintained a relationship with the PEA, and, in April 1933, the vice president notified her that her name had been added to its list of honorary vice presidents. The president wrote of her, "Your early leadership in the progressive movement, your loyalty to its ideals, and the wise advice and guidance which you have placed at the service of the Association in the past is appreciated. . . . [This is] also an invitation to you to retain your active participation in the councils of the Association, . . . [and] attend the meetings of the Executive Board, . . . that we may continue to profit by the wisdom which we have so long respected."[134]

COOKE'S ENDURING CONTRIBUTION

In the spring of 1934, at the age of 70, Flora J. Cooke retired as principal of the Francis W. Parker School. Retirement did not end her involvement with the Parker School, the PEA, or many national and local organizations, however. Cooke was named a member of the Parker School Board of Trustees shortly before she retired, and she served in this capacity until 1948. She also served as a member of the Educational Council of the Parker School from 1932 until 1953.[135] Cooke remained active in the educational life of 23 city and national organizations, expanding social awareness of progressive education.[136]

During the summer immediately after her retirement, she visited 23 of the high schools that participated in the PEA's Eight Year Study.[137] She visited schools in California, as well as the cities of Denver, Des Moines, and St. Louis.[138] During the winter months of that year, she taught a "seminar course with the students of the Winnetka Graduate Teachers College who will be working in the Francis W. Parker School that term."[139] Practice teaching could at last take place at the Parker School, realizing Colonel Parker's model of training teachers in progressive principles and providing a practice school for the application of these principles.

The following spring, she visited schools in the east. "Gathering data from these schools as to how each meets our common problems," she wrote to an administrator of one of the schools in the study, "seems most interesting, and I shall give the data obtained to the Francis W. Parker School and shall be glad to offer it without expense to any other schools who could use it."[140]

In 1933, more than 30 years after Cooke taught a summer institute in Hawaii, Henry S. Townsend, Professor Emeritus of Philosophy at the University of the Philippines, expressed his desire that one day she would conduct an institute in Honolulu, sponsored by the PEA. Townsend said, "No one living could arouse more enthusiasm among the teachers who know than Flora J. Cooke. . . ."[141] He also said, "It remained for Dewey to give us the classical statement of the Philosophy underlying activity-centered schools . . . and for Miss Flora J. Cooke, in our very midst, to give us the well-digested illustration of it in actual practice, in 1899."[142] Cooke, herself, often noted that Dewey was the theoretician and Colonel Parker was the practitioner.[143]

In 1937, Cooke edited and published[144] the only lengthy work by Colonel Parker, *Talks on Pedagogics.*[145] In 1920, she had bought the plates and publishing rights[146] to his work. She previously had edited and written the preface for a book by Ida Cassa Heffron about Colonel Parker and the Normal School. Heffron had been an art teacher at the Normal School for many years and had known Colonel Parker well. Cooke not only edited the book, but also raised the money to have it published.[147] She campaigned among friends and educators to promote its sale and often referred to its accuracy in portraying Colonel Parker and his school. Cooke remained intensely loyal to the ideas of Francis W. Parker and sought to keep his philosophy alive. She wrote many articles detailing his influence on her and the Parker School, and when she died, she left an unfinished manuscript about his life's work. Flora Cooke was active until the end of her life. In 1948, Margaret Rush noted in a newspaper article, "She receives and answers over 500 letters per month. She attends club meeting[s] and conferences. She lectures and makes speeches—and she writes them herself!—before civic and educational groups."[148] Even in her last years, Cooke lectured extensively in the United States and abroad in Honolulu, Switzerland, England, and Denmark.

Flora Cooke died of a heart attack in her Chicago home on February 21, 1953, at 88 years of age.[149]

COOKE IN RETROSPECT

Traditionally, contributions of women in education have been overlooked. But here was a woman who rose to a leadership position in education at the turn of the twentieth century, and in a curious way, innovation and creativity in education in Flora Cooke's day seemed more possible in the primary grades than in the upper grades, where the pressure to prepare students for college began narrowing the focus away from the development of whole persons and their social awareness.

Although Cooke rarely commented directly on the issue of women's rights, at the Chicago's World Fair of 1933, she gave a speech entitled, "Childhood Education

1833–1933," extolling the contributions that women, largely unnoticed by the world, have made to progressive education. She cited the strides in the liberalization of secondary education achieved because of "principles learned in the earlier grades,"[150] where teachers were, for the most part, women. Progressive practices in elementary grades, she noted, had escaped notice because boards of education did not consider the education of young children important.

Cooke continued, " . . . in this century, all authorities seem to agree that in Childhood Education, the most rapid and satisfying advance has been made with children of primary and pre-school age. Gradually, but steadily, the higher grades have been liberalized by teachers influenced largely by the principles practiced successfully in the lower school."[151]

An assessment of Flora Cooke's contributions as an educator reveals several noteworthy achievements:

1. She created a school that modeled progressive education at its best. The progressive approach to education sought to free the individual student from the tyranny of rote learning without meaning and discipline imposed by outside authorities, which characterized many of the educational methods of the nineteenth century. In Cooke's educational environment, students developed self-discipline. Learning was informed by a student's interests, with the teacher providing guidance for the broadening of those interests through discipline, practice, and creative expression.
2. She extended Colonel Parker's theories from elementary education to secondary education and participated in the development and implementation of the Eight Year Study of the PEA, which proved that progressive education at the secondary level was valid. Her students not only succeeded but, in many instances, excelled in a traditional educational environment.
3. She created a school that served as a laboratory for experimentation, with policies and procedures to enrich public education. For Cooke, the Parker School had no justification to exist unless it could provide a model for public education and do it so well that the public would demand similar conditions for all students.
4. She modeled congruence between principles and practices for the teachers in her school and for those she taught in other settings. She insisted on adherence to her expressed principles, even in the face of opposition.
5. She created an educational environment based on developing whole, creative individuals who could be of service as citizens in a democratic society. She consistently found ways to balance the tensions between two main themes of Colonel Parker's pedagogical methodology: the social motive and individual development.
6. She saw the school as a microcosm of society; she constantly strove to make her school responsive to the broader cultural and social context, incorporating and integrating the cultural riches of Western civilization while taking into account the many transformations that were occurring in American society. Topics like poverty in urban living, World War I, the depression, scientific methods of testing and measurement; the psychological stages of growth, the impact of technology, and racial conflict were all part of the dialog at the Parker School.

It is ironic that Flora Cook spent almost her entire career making the secondary school progressive, and then World War II took the nation's attention away from educational endeavors that demonstrated the merit of progressive education. The returning veterans demanded a more pragmatic approach to education. By the mid-1940s progressive education had fallen out of favor.[152]

At the beginning of the twenty-first century, as at the beginning of the twentieth, education is in transition, and many of the issues Flora Cooke struggled so valiantly to address remain unresolved. The Francis W. Parker School continues to thrive, and many of the traditions begun by Flora Cooke are still practiced, yet policies of which Cooke disapproved, such as formal grades and the use of College Board Examinations, have been instituted; and in some ways, the Parker School has became more traditional than it was under her direction. The pressures on high school students to concentrate their efforts on preparation for college entrance exams is greater than ever before, as they prepare to take their place as citizens of a new era.

It is not unreasonable to hope that a twenty-first century Flora Cooke will come forward who can champion a grand vision of education such as the one developed by Colonel Francis Parker and other progressive educators—a vision that was served in the early twentieth century with such dedication by the remarkable woman, Flora J. Cooke, who came to be known as "The Grand Old Lady of Education."[153]

NOTES

1. Flora J. Cooke, "Opportunities and episodes of a teacher's life in America during the last half century—born 1864—teaching life 1884–1934—present date 1941," box 17, folder 100, p. 2, Francis W. Parker School records, 1884–1960, Manuscript Collection in the Chicago Historical Society (hereafter cited as Parker MSS).
2. Cooke, "Opportunities," p. 2.
3. Ibid., p. 3.
4. Ibid., p. 7.
5. Herbert McComb Moore (Address by the President of Lake Forest College), box 9, folder 52, p. 1, Parker MSS.
6. Cooke, "Opportunities," p. 2.
7. Flora J. Cooke, "Testimonial Luncheon in Honor of Miss Zonia Baber," 21 October 1944, box 19, folder 111, p. 1, Parker MSS.
8. Cooke, "Opportunities," p. 3.
9. Jack K. Campbell, *Colonel Francis W. Parker: The Children's Crusader* (New York, Teachers College Press, 1967), p. 113.
10. Cooke, "Testimonial Luncheon," p. 1.
11. Flora J. Cooke, "Childhood Education 1833–1933" (address given at the Century of Progress, World's Fair), Chicago, Ill., 1933, box 10, folder 58, p. 23
12. Cooke, "Opportunities," p. 5.
13. Ibid., p. 4. The "Leaflet" was a collection of both students' and teachers' writing. There were other names for this, such as the "Chicago Normal School Envelope." During the last years at the Cook County Normal School, Colonel Parker secured a printer to print "reading lessons," the children's own observations and expressions.
14. Cooke, "Opportunities," p. 4.
15. Flora J. Cooke, "A Brief Sketch of a Chapter in the History of the Chicago Teachers College," 8 February 1945, box 19, folder 112, p. 2, Parker MSS.
16. Cooke, "Opportunities," p. 5.
17. Ibid., p. 7.
18. Ibid.
19. Ibid.
20. Ibid.

21. Campbell, *The Children's Cursader*, pp. 214–215.
22. Cooke, "Opportunities," p. 7.
23. Ibid.
24. Memorandum of Agreement, 11 June 1901, box 1, folder 3, p. 1, Parker mss.
25. Cooke, "Opportunities," p. 3.
26. Flora J. Cooke, *Nature, Myth, and Stories* (Chicago, A. Flanagan Co., 1922).
27. Flora J. Cooke to Miss Steln, 14 February 1929, box 6, folder 41, Parker MSS.
28. Beryl Parker to Francis W. Parker School, 23 November 1916, box 4, folder 25, Parker MSS.
29. Conference on School Methods, box 7, bound typescript, pp. 169–171, Parker MSS.
30. Ibid., pp. 169–171.
31. Ibid., pp. 161–162.
32. Ibid., p. 193.
33. Ibid., p. 172.
34. Cooke, "Opportunities," p. 8.
35. Ibid.
36. Memorandum of Agreement, p. 1.
37. Cooke, "Opportunities," p. 8.
38. Carol Lynn Gilmer, "Flora Cooke: Grand Old Lady of Education," *Coronet Magazine* (October 1947): 80, box 20, folder 123, Parker MSS.
39. Flora J. Cooke (Address to the class of 1935), box 11, folder 63, pp. 8–9, Parker MSS.
40. Gilmer, "Flora Cooke," p. 81.
41. Cooke, "Opportunities," p. 8.
42. Flora J. Cooke to Helen E. Wells, 15 March 1935, box 11, folder 65, p. 1, Parker MSS.
43. Cooke, "Opportunities," p. 8.
44. Flora J. Cooke, "Record of Trip up to Date," 1904, box 1, folder 4, Parker MSS.
45. Elsie A. Wygant, "Flora J. Cooke," *Bulletin National Council of Primary Education* 13, no. 3 (1930): 1–2.
46. *Francis W. Parker School Faculty, Francis W. Parker School Studies in Education,* vols. 1–10 (Chicago: Francis W. Parker School, 1912–1934). Volumes 1–5 were entitled *Francis W. Parker Yearbooks.*
47. Wygant, "Flora J. Cooke," p. 1.
48. Cooke, "A Brief Sketch," p. 6.
49. Cooke, "Opportunities," p. 3.
50. Conference on School Methods, pp. 164–165.
51. Flora J. Cooke, "Lectures on Francis W. Parker School," lecture 3 (delivered at the Webster Hotel), box 12, (bound typescript), pp. 6–7, Parker MSS.
52. Wygant, "Flora J. Cooke," pp. 1–2.
53. Perry Dunlap Smith, "Luncheon in Honor of Miss Flora J. Cooke," 7 April 1934, bound typescript, box 10, folder 61, p. 9, Parker MSS.
54. Flora J. Cooke, "Making the Secondary School Progressive," June 1930, box 8, folder 45, p. 2, Parker MSS.
55. Open Meeting, Francis W. Parker School, 27 January 1905, box 1, folder 5, pp. 1–2, Parker MSS.
56. Flora J. Cooke to the Trustees, 25 January 1909, box 2, folder 13, p. 3, Parker MSS.
57. Flora J. Cooke and Carley, Ira to Cyrus Bentley, Esq. and Mrs. Emmons Blaine, 10 January 1906, box 1a, folder 6, p. 9, Parker MSS.
58. Marie Kirchner Stone, *Between Home and Community* (Chicago, Francis W. Parker School, 1976), p. 39; see also idem. *The Progressive Legacy: The Francis W. Parker School, 1898–2000* (New York: Peter Lang, 2001).
59. Flora J. Cooke and Robert Nason to the Trustees, 23 January 1908, box 2, folder 9, pp. 1–3, Parker MSS.

60. Frank A. Manny, "Memoranda for the Trustees of the Francis W. Parker School," 22 January 1906, box 2, folder 9, p. 2, Parker MSS.
61. Ibid., p. 4.
62. Faculty Committee, "What is to be . . . ," box 2, folder 12, pp. 1–2, Parker MSS.
63. Flora J. Cooke to Anita McCormick Blaine, 4 December 1909, box 2, folder 14, p. 5, Parker MSS.
64. Cyrus Bentley to Flora J. Cooke, 14 January 1909, box 2, folder 13, pp. 1–2, Parker MSS.
65. Arthur Detmers, "Preliminary Report Made to the Trustees of the Francis W. Parker School by the Educational Adviser to the Trustees and Faculty," 4 January 1910, box 3, folder 15, Parker MSS.
66. Ibid., p. 14.
67. Stone, *Between Home and Community,* p. 88.
68. Cyrus Bentley to Flora J. Cooke, 11 October 1920, box 4, folder 27, p. 2. Parker MSS.
69. Parker Committee, "The School," 21 June 1921, box 4, folder 28, p. 1, Parker MSS.
70. Lawrence A. Cremin, *The Transformation of the School: Progressivism in American Education 1876–1957* (New York: Vintage Books, 1964), pp. 186–190.
71. Flora J. Cooke, "Use of Tests and Measurements, the Three R's," *Progressive Education* 5 (1928): 147.
72. Flora J. Cooke, "Problems of the Progressive Secondary School," *Progressive Education* 5 (1928): 317.
73. Flora J. Cooke to Mr. Storm, 8 February 1929, box 6, folder 41, Parker MSS.
74. Ibid.
75. Patricia Albjerg Graham, *Progressive Education: From Arcady* to Academe (New York: Columbia University, 1967), p. 1.
76. Ibid., p. 89.
77. "Report of Executive Board Conference: Vassar College," 18–19 April 1930, box 7, folder 44, pp. 1–2, Parker MSS.
78. Graham, *Progressive Education,* pp. 89–90.
79. Wilford M. Aikin, "Report of the Committee on College Entrance and Secondary Schools," *Progressive Education* (1931): 319.
80. Wilford M. Aikin, *Story of the Eight Year Study,* (New York and London: Harper & Brothers, 1942), pp. 12–13.
81. Ibid., pp. 13–15. Here is a list of the original high schools chosen.
82. Ibid., pp. 109–110.
83. Eugene R. Smith and Ralph W. Tyler, *Appraising and Recording Student Progress* (New York and London: Harper & Brothers, 1942), pp. 3- 4.
84. Robert D. Leigh, "Twenty-Seven Senior High School Plans," *Progressive Education* 10 (1933): 373.
85. Flora J. Cooke, "Speech on the Eight-Year Experiment," 19 April 1934, box 10, folder 61, p. 2, Parker MSS.
86. Flora J. Cooke to the Directing Committee, 18 March 1933, box 10, folder 56, p. 2, Parker MSS.
87. Flora J. Cooke, "North Central Association," 19 April, 1934, box 10, folder 61, p. 3, Parker MSS.
88. Ibid., p. 4.
89. Smith and Tyler, *Appraising and Recording Student Progress,* p. 18.
90. *Thirty Schools Tell Their Story* (New York and London, Harper & Brothers, 1943), p. 318.
91. Conference on School Methods, p. 167.
92. Ibid., pp. 167–168.
93. Ibid., p. 168.
94. Cooke, "Making the Secondary School Progressive," p. 1.
95. Ibid., p. 3.

96. Ibid., p. 4.
97. Carleton Washburne, "A New Venture in Teacher Training," *The Nation's Schools* 14, no. 2 (1934): 2, box 11, folder 62, p. 1, Parker MSS.
98. Ida Cassa Heffron, *Francis Wayland Parker: An Interpretive Biography* (Los Angeles: Ivan Deach, Jr., 1934).
99. Cooke, "Opportunities," p. 4.
100. Patty Smith Hill to Miss Flora J. Cooke, 24 May 1920, box 4, folder 27, Parker MSS.
101. Nathaniel Butler to Flora J. Cooke, 21 June, 1906, box 1a, folder 6, p. 1, Parker MSS.
102. W. H. Bruce to Flora J. Cooke, 23 January 1908, box 2, folder 9, Parker MSS. Cooke did not accept this invitation.
103. Director of the Course for the Training of Teachers to Flora J. Cooke, The University of Wisconsin, box 3, folder 15, Parker MSS.
104. J. W. Gearson to Miss Flora Cooke, Peru State Normal School, 24 March 1910, box 3, folder 15, Parker MSS.
105. Charles H. Judd to Miss F. J. Cooke, The School of Education of The University of Chicago, 9 November, 1911, box 3, folder 19, Parker MSS.
106. H. Judsom to Miss F. J. Cooke, The University of Chicago, 9 November, 1911, box 3, folder 20, Parker MSS.
107. Mary Pfefferkorn to Miss Flora J. Cooke, Kansas State Teachers Association, 16 May 1914, box 4, folder 23, Parker MSS.
108. Flora J. Cooke, "Social Motive in Education" (Address delivered at the Kansas State Teachers' Association), box 4, folder 24, p. 2, Parker MSS.
109. Washburne, "A New Venture," p. 2.
110. In 1933, Cooke, Carleton Washburne, and Perry Dunlap Smith served as Educational Directors. Frances L. Murray was the Dean. Carleton Washburne, to Flora J. Cooke, 18 February 1933, box 10, folder 56, Parker MSS.
111. Cooke, "Opportunities," p. 7.
112. Flora J. Cooke to Dr. Hilman Sieving, 2 November 1951, box 1, folder Provenance, p. 2, Parker MSS.
113. "Flora J. Cooke, 84, Heads Roosevelt Dinner Plans," 12 May 1949, box 30, folder (F.J.C. clippings), Parker MSS.
114. Cooke to Wells, p.1.
115. "Miss Cooke Here to Take Part in School Survey" (unidentified newspaper clipping from Leavenworth, Kansas), March 1914, box 4, folder 23, Parker MSS.
116. Commissioner P. P. Claxton to Miss Flora J. Cook[e], Department of the Interior, 8 April 1914, box 4, folder 23, Parker MSS.
117. Ethel Lindgren and Mary Page to Miss F. J. Cook[e], Chicago Kindergarten Institute, 1 April 1914, box 4, folder 23, Parker MSS.
118. Historical Encyclopedia of Illinois, box 6, folder 38, Parker MSS.
119. The Chicago Teachers' College evolved out of the Chicago Normal School, where Cooke had worked under Colonel Parker.
120. Franklin C. Lewis to Miss Cooke, 12 January 1927, box 6, folder 36, Parker MSS.
121. Cyrus Bentley to Miss Cooke, 14 May 1927, box 6, folder 36, Parker MSS.
122. Clare Soper to Miss Cooke, 22 June 1927, box 6, folder 36, Parker MSS. Mrs. Ensor was the executive director of the New Education Fellowship. Apparently, Cooke did not accept Mrs. Ensor's invitation to visit her in England, as indicated from a letter that she wrote after returning home.
123. Ibid.
124. Flora J. Cooke to supporters, n.d., box 6, folder 37, p. 2, Parker MSS.
125. Flora J. Cooke, "personal journal about her trip through Europe," (It is very difficult to read the handwriting because it is so small), box 5, folder 30, Parker MSS.
126. Moore (Address by the President of Lake Forest College).

127. Graham, *Progressive Education,* p. 43.
128. Flora J. Cooke (Address at the Ninth Annual Conference of the Progressive Education Association [PEA]), box 6, folder 41, Parker MSS.
129. PEA to Flora J. Cooke, 9 January 1933, box 10, folder 56, Parker MSS.
130. PEA Annual Conference Program, 3 March 1933, box 10, folder 56, p. 2, Parker MSS.
131. PEA to Flora J. Cooke, 20 February 1933, box 10, folder 56, Parker MSS.
132. PEA, Annual Conference, box 10, folder 56, Parker MSS.
133. Exhibits: Francis Parker School, box 10, folder 56, Parker MSS.
134. Willard W. Beatty to Flora J. Cooke, 3 April 1933, box 10, folder 57, Parker MSS.
135. Stone, *Between Home and Community,* pp. 35, 52.
136. Margaret Rush, "Flora Cooke: Progressive Teacher," *Oregonian* (1948): 2, box 30, folder (F.J.C. clippings), Parker MSS.
137. Cooke (Address to a class of 1935), p. 1.
138. Flora J. Cooke to Mr. George A. Walton, 8 August 1934, box 11 folder 62, p. 1, Parker MSS.
139. Ibid., p. 1.
140. Ibid.
141. Henry S. Townsend to Miss Cooke, 26 August 1933, box 10, folder 58, p. 1, Parker MSS.
142. Ibid., p. 2.
143. Cooke, "Childhood Education 1833–1933," p. 23.
144. Cooke, "Lectures," lecture 2, p. 4.
145. Francis W. Parker, *Talks on Pedagogics* (New York and Chicago: E. L. Kellogg & Co., 1894).
146. John Laidlaw to Miss Flora J. Cook[e], 28 April 1920, The A. S. Barnes Company: Educational Publishers, box 4, folder 27, Parker MSS.
147. Flora J. Cooke to two hundred people, 15 July 1933, box 54, folder "Cooke Special, 1933–34," Parker MSS.
148. Rush, "Flora Cooke: Progressive Teacher," p. 1.
149. "Miss Flora J. Cooke," (Obituary), *New York Times,* 22 February 1953, p. 63.
150. Cooke, "Childhood Education 1833–1933," p. 2–3.
151. Ibid.
152. Cremin, *The Transformation,* pp. 324–327.
153. Gilmer, "Flora Cooke," p. 80.

CHAPTER NINE

MARGARET HALEY: PROGRESSIVE EDUCATION AND THE TEACHER

KATE ROUSMANIERE

The cause of the teacher is the cause of the people
and vice versa, and their common cause is that of the children.

—Margaret Haley, 1901[1]

This chapter is a study of the way in which Margaret Haley, the early-twentieth-century teacher union leader, saw herself as a progressive educator. My task is complicated by the fact that Haley never described herself as a "progressive educator" and that she often worked *against* what some historians have identified as progressive educational practices. Nor did she work in or with the wide community of educators who are commonly referred to as progressives. Indeed, as a teacher union leader concerned primarily with administrative change and teachers' working conditions, Haley was not always in a position to promote child-centered classroom practices, alternative classroom structures, or other curricular innovations. In her 30-year career with the Chicago Teachers' Federation, Haley was more apt to describe herself as a political activist than as a progressive educator. Yet, in fact, she saw herself as both. We too, can see her as both, but only if we understand the complexities of her role as a teacher union leader and her own identity as an Irish Catholic woman in early–twentieth century Chicago.

In the myriad historical discussions about the Progressive Education movement at the turn of the century, a distinction is usually made between pedagogical progressives and administrative progressives. Pedagogical progressives, it is generally argued, promoted child-centered curriculum and classroom activities and the creation of a more democratic climate in schools. Pedagogical progressives thus focused on students and teachers in the classroom and on the school as a community. In contrast, administrative progressives are described as reformers who focused on administrative changes in schools and school systems. Their attention usually centered on the centralization and systematizing of school bureaucracies, the creation of standardized curriculum and systems of delivery, and the development of school administration as a scientific and professional field.[2]

The distinction between pedagogical and administrative progressives is helpful but it can also be misleading because many progressive educators saw that the two phenomenon were closely linked. Most notably, John Dewey argued that it was difficult to develop cooperative democratic schooling within an oppressive hierarchical bureaucracy. He realized that many educational reformers might prefer to focus simply on the classroom, believing that changing the curriculum and classroom practice would, in and of itself, change the school. But, as he wrote in 1902, pedagogical reformers should take care to not see the "mechanics of school organization and administration as something comparatively external and indifferent to educational purposes and ideals." It was the ways in which administrative decisions were made, and the curriculum and instruction were developed, that "really control the whole system."[3] This was precisely the reason why pedagogical change was so hard to accomplish in large public urban school systems. As David Tyack observed, it was "no accident" that when Dewey and his daughter Evelyn described teachers who exemplified his ideals of progressive education in *Schools of To-morrow,* they found them in small private schools rather than large public schools.[4] Pedagogical reform faced great barriers in large bureaucratic school systems.

For Margaret Haley, who worked in the large urban school district of Chicago, the necessity for administrative change to develop pedagogical change was obvious. Haley took Dewey's notion of democracy in education and applied it first to the school organization, with the understanding that only then could democracy happen in the classroom. She paid less attention to pedagogy and classroom practice, not because she did not believe in the importance of such changes—she was, in fact, well-educated in progressive classroom practice—but because she inherently believed that in a truly democratic school organization, democratic classrooms would naturally evolve. Her self-defined job as a progressive educator, then, was not to focus on students' education per se, but on the teacher and the school organization.

How did Haley shape her particular understanding of progressive education and how effective was she at promoting her vision? This essay explores how Margaret Haley learned about progressive education and how she developed her own interest in it. I also examine the ways in which Haley's vision and her identity as a teacher union leader did and did not match with those of other progressive educators. I argue here that Margaret Haley promoted a particular brand of progressivism, and in this she was not always understood or welcomed.[5]

MARGARET HALEY: PROGRESSIVE EDUCATOR

Margaret Haley is one of the most popular and popularized icons of American teachers' history, heralded as a radical school reformer, labor activist, and feminist. In the first three decades of the twentieth century, Haley led the Chicago Teachers' Federation to be the first teachers' labor union in American history, and in so doing, she changed the shape of American educational politics forever. Haley's campaigns for increased teacher salaries, pensions, and tenure laws attracted over half of all Chicago's teachers, most of them women, into the Federation. Haley also led teachers into municipal and state reform activities, challenging business interests in school management and promoting municipal and electoral reforms, including women's suffrage. In 1902, she negotiated an unprecedented affiliation between teachers and the Chicago Federation of Labor, and in

1916 she led her organization to become Local #1 of the newly formed American Federation of Teachers. Haley worked at the national level too, traversing the country to promote political activism among the nation's predominately female teaching force.

Born in 1861 in Joliet Illinois, Margaret Haley was the eldest child of Irish Catholic immigrants. Her father had been a laborer on the Illinois-Michigan canal who had honed his skills to become a stonecutter and engineer. A public spirited and politically minded man, Haley's father taught his children about the passions and intrigue of working class politics. As an older woman, Haley recalled that a defining moment in her childhood was at age 12, when her father received a pamphlet on labor rights and currency reform from the Greenback party. Young Margaret Haley read the pamphlet with intense interest and later credited that experience with her recognition "of the association between a governmental currency system and the daily lives of the people of the country."[6] Raised on the image of the Knights of Labor as an inclusive force for all laborers and educated in the economic politics of farmers, canal workers, and Irish immigrants, Margaret Haley learned about both the potential of liberal capitalism and its dangers, especially in a world where the interests of monopoly were not controlled.[7] Her father's own erratic career confirmed these lessons: at age 16, Haley began teaching school to support her family after her father's third business disaster.

As a child, Haley loved school, in part because of the luck of her earliest education. In a small schoolhouse in rural Illinois, she had a teacher whose untrained pedagogical practices rejected formalized structures, including textbooks. This itinerant male teacher was Cyrus Winthrop Brown, a Civil War veteran, who told stories that engaged his students, inspired young Haley and became an early model of what she would call, throughout her life, "wide awake education."[8]

When Haley herself became a teacher in a local village school, she spent her weekends and summers attending classes at a new normal school that opened up in the local town of Morris, Illinois. Like other normal schools, the Morris school served as both a general high school and a teacher training institution. Morris was special in that its instructors came from the Illinois State University, a well-funded and expansive institution that delivered up-to-date educational methods and theories to the state's teachers.[9] The Morris Normal School instructors rejected old-fashioned rote memory learning and instead promoted problem solving and close analytical work. Science classes took frequent trips to the field to study nature; in mathematics class, students explored the structure of mathematical problems; history classes studied the nature of law and citizenship, as well as historical events; and students studied debating and composition to encourage active participation in society as citizens.[10] Haley was inspired by the teaching, recalling that she learned not only about instructional methods but also about the broad commitment to social change that school teachers should hold. About this experience she recalled, "If you had any sense of responsibility to children, it would awaken it. If you had any conscience, it would arouse it."[11]

At the end of her fifth year of teaching, Haley made a further commitment to her education by registering for a four-week summer session at Illinois State University.[12] On campus, she heard some of the leading proponents of Herbartian curriculum theory who worked there. The Herbartians were part of a trans-Atlantic educational movement that emphasized the interrelation and connection between subjects over the scientific training of skills. The Herbartians also delved into early psychological concepts related to student interest, for example, by promoting curriculum in stages that paralleled the stage of the development of the child. In this, the Herbartians were an influence on John Dewey's later understanding of child-centered and experiential education.[13]

However, indicative of Haley's emerging interest in school politics, her favorite class at Illinois was not about pedagogy but about political economy. Her favorite teacher was Edmund Janes James, a scholar of education and economics whose research interests included the labor movement, tax reform, and school finance.[14] In James' economics class, Haley read Henry George's recently published *Progress and Poverty,* a book that was revolutionizing liberal American economic theory with its proposal for a single-tax system that would allow a more egalitarian and benevolent operation of the capitalist system. George's theory, Haley recalled, opened up to her "a wide world" of economic restructuring for social improvement.[15]

In 1883, Margaret Haley, her parents, and her five younger brothers and sisters moved to Chicago. Attracting them was the possibility of civil service jobs in the expanding city and the large Irish Catholic community there. Further drawing Margaret Haley to Chicago was the Cook County Normal School which, under the direction of Francis Wayland Parker, had recently become a famous center of progressive educational practice.

Parker's educational philosophy was, in some ways, a microcosm of progressive educational thought, linking together the work on pedagogy by nineteenth century Europeans and Americans. Parker was particularly influenced by Friedrich Froebel, the German originator of the kindergarten, who promoted respect for the activities, ideas, and expressions of individuality of even the youngest children. For both Froebel and Parker, the child was the subject and the center of the entire educational enterprise; both rejected the traditional structure of schooling in which the schoolmaster imposed his or her will on the child. Parker's interpretation of these theories reached outside of the classroom into society. He had a simple educational credo: The child should be the center and inspiration of the school's activities, and the school should pursue two objectives—character development and a democratic social order. Parker argued that traditional teaching methods squelched the child's natural curiosity of the world and that traditional education produced passive citizens. The new education, he promised, would create "free people."[16]

Parker's ideas about the role of teachers were as revolutionary as his ideas about the role of children in schools. Given the educational importance of classroom relations, the role of the teacher was elevated to that of the shaper of children's values and ethics. Parker believed that the individual classroom teacher needed to have the authority of a policy maker level, a novel-role for female elementary schoolteachers.[17] Teachers should be free to experiment. What was important to Parker was "constant change, elimination, innovation, experiment, tentative conclusions—this was the manner of progress."[18]

Haley spent only a few months at Cook County Normal School, but she was profoundly influenced by Parker's personality and teachings. She described him as someone who broke old traditions of education, freeing women teachers to use their minds in the same way that modern fashions had freed women's bodies from the corset. She was particularly inspired by Parker's lectures on the school as a social community and his insistence on academic freedom. In Parker, she saw the importance of new methods in education and how it could lead to change in society. Francis Parker was "a great light shining on the barren fields of education in the United States."[19]

From this great light, Haley went to teach in one of the darkest educational environments in the city. Haley taught for 16 years in the Hendricks School in the heart of the poverty-stricken meat-packing district of Chicago, an experience about which she wrote very little. Given the strict public school curriculum, the crowded classroom, and the problems raised by students' poverty and linguistic diversity, it is hard to imagine how any Chicago teacher devised anything but the most mechanical rote lessons for his or her students, and at least one observer described exactly that. In his scathing 1892 survey of American city schools, muckraker Joseph Rice found Chicago classrooms full of students glued to their desks, monotonously reading, writing, ciphering, and reciting from textbooks all day long.[20]

But progressive pedagogy may have been more prevalent in Chicago schools than Rice saw, as less than a year after his observations, the Chicago Board of Education erupted in controversy over a charge that the city curriculum was driven by too many progressive "fads," including too many courses in art, music, and physical culture.[21] Margaret Haley herself believed that even at their worst, Chicago schools were less restrictive than those in many eastern cities because they were lacking "the melancholy attitude of Puritanism."[22] Indeed, the years that Margaret Haley was a teacher were the years that some of the most innovative educators worked in the Chicago public school system.

Notable among these educators was Albert G. Lane, the superintendent of Chicago schools from 1891 to 1898. A graduate of the first class of the Chicago High School teacher training program and a former school principal and county school superintendent, Lane oversaw the expansion of manual training classes, kindergartens, classes for students with special needs, night school for working people and immigrants, and a residential parental school.[23] The combined influence of Parker at the normal school and Lane in the superintendency led to the hiring and nurturing of some remarkable school administrators.

Josephine C. Locke was head of the Art Department of the Chicago public schools between 1890 and 1900. Prior to that, she taught in the Cook County Normal School with Parker. The promotion of art in public schools, particularly in working class schools, was a central part of school reformers' work, particularly in urban districts where the lives of city children were so bereft of creative influences. According to Locke, a main purpose of art instruction was to restore the life of nature to city dwellers, and she fought against cuts in art teachers' salaries in working class schools.[24] Haley called Locke, "a bubbling fountain of joy," who came to Chicago "to teach the art of life as well as of technic."[25]

Another progressive leader was District Superintendent William Speer. Speer was a former superintendent of schools in Marshall County, Iowa, where he had introduced manual training and modern methods of teaching writing, science, and arithmetic. He was most famous for developing a new method of teaching arithmetic that emphasized the "visualization" of the entire concept of the mathematical problem. Speer had also worked with Parker, originally at a summer school in Martha's Vineyard and, later, as vice principal in the Department of Mathematics at Cook County Normal School.[26] Between 1896 and 1902, Speer was district superintendent in one of the poorer districts of Chicago. Common to both Speer and Locke's work was lessening the influence on textbooks and increasing students' use of their own skills in inventive exercises. To Haley, this group of educators promised to "leaven the lump" of the city's school system.[27]

The most inspiring school administrator for Chicago's women teachers was, however, District Superintendent Ella Flagg Young. Young had been a rising star in Chicago's school system since 1865 when, at age 20, she was appointed principal of the Model School of the Chicago Normal School. She rose through the ranks of the Chicago school administrative structure until she was appointed district Superintendent in 1887, a position that she held for the next 12 years. In those years, Young won the respect of Chicago teachers because of her commitment to teacher education and to a belief that teachers should be involved in school administrative decisions. Later, Young went on to explore these issues further in graduate work with John Dewey, but in the 1880s and early 1890s, she lived out this work as the district superintendent of a poor district on Chicago's South Side. For Young, teaching was deeply personal work that needed to be supported with sensitivity and respect.[28] For Margaret Haley and other young teachers in Chicago's schools, the influence and image of people like Ella Flagg Young, William Speer, and Josephine Locke, in the higher levels of the city school administration, showed some commitment to progressive educational principles.

Yet even such inspiration could not erase Haley's crowded classroom, the exhausted and hungry students, the absence of supplies and space to lead creative lessons, and the lack of parks or a safe neighborhood in which to take field trips. Haley herself described her school in a sweeping statement as "almost hopeless."[29] Furthermore, her later pronouncements that teachers should have the right to shape the classroom into a caring and supportive environment for children may suggest that she, herself, did not have much of that opportunity. Indeed, part of her own politicization process may have been the contrast between her own progressive teacher training and the restrictive working conditions in her school. If the school could not "bring joy to the work of the world," she would argue later, then "joy must go out of its own life, and work in the school as in the factory will become drudgery."[30] Did she know this because she had seen such drudgery in her own classroom?

Two other educational experiences shaped Haley's thinking. In 1896, Haley went to a two-week Summer Institute for teachers at the Buffalo School of Pedagogy in New York State. Opened in 1895 under the direction of Frank McMurry, a former faculty member from Illinois State University who promoted the progressive educational ideas of that institution and of Francis Parker, the school was designed for advanced study by current teachers and school administrators.[31] There, Haley heard William James deliver his "Talks to Teachers on Psychology" which she praised as a "mighty wind whirling through dark corridors, clearing out cluttered corners of the mind."[32] Haley was one of thousands of teachers who were influenced by James' "Talks," which introduced the notion that teachers' individual work could change individual students. The teacher, James argued, had a role, a responsibility, and an effect on his or her students. The teacher could prepare new citizens who could create a new world. Because of James, Haley left Buffalo "thrilled with the pride" of her profession.[33]

The following summer, Haley went to a Catholic summer school in Madison, Wisconsin, led by the Dominican Sisters from Sinsinawa, Wisconsin.[34] Catholic summer schools were part of the liberal social movement of the American Catholic church, designed to expose parochial and public school teachers to contemporary social and economic problems and secular intellectual debates.[35] The summer school in Madison was the brainchild of the Sinsinawa Mother Superior and Thomas Edwards Shields, one of the leading Catholic educators of the day. Their joint goal was to promote a reconciliation between the study of science and the practice of faith, and to introduce the application of scientific principles to education. Although Shields critiqued the philosophy of John Dewey as being "materialistic," he promoted a very Deweyan curriculum of socialized education. The elementary school classroom, Shields later wrote, should not be "a quiet, sad place, where little children fear to move lest they should disturb a nervous teacher," but it should be an active and social place where children learned "from each other more than they learn from the teacher and where they learn by doing rather than by hearing."[36] Teachers needed appropriate training in pedagogical techniques and psychology to create such an educational environment. It was from Shield's visiting lectures and other courses in literature, psychology, political science, and pedagogy at the Sinsinawa summer school that Haley recalled she learned the "widening view" about "the fundamental issue in the unending war for academic freedom: the right of the teacher to call his soul his own."[37]

By the time she joined the Chicago Teachers' Federation in 1897, Margaret Haley had a good training in progressive pedagogical principles and theories. In each of these lessons, she was particularly interested in the application of such theories to the organization of schools and the improvement of teachers' work. For Haley, the interesting thing about progressive education was the way in which it promised changes for teachers by raising issues of academic freedom, democracy in education, and the nature of the school as a community.

THE TEACHERS' FEDERATION AND PROGRESSIVE SCHOOL CHANGE

When 36-year-old Margaret Haley joined the newly founded Chicago Teachers' Federation, her lifelong interest in politics and economics took root and quickly blossomed. As the paid business representative of the Federation, she developed her strengths as a political activist, lobbyist, and public speaker for the teachers' cause. Almost immediately, she became involved in a complicated legal case in which the Federation sued a state agency for underassessing the taxes of major corporations in Chicago. The tax fight lasted for five years and led Haley into the bowels of city and state government, fighting through obscure legal proceedings, male-dominated administrative systems that were hostile to women, and the tangled underground of Chicago politics. In later battles she challenged the Chicago Board of Education's corrupt leasing of school properties, fought for teachers' pensions and salary increases, and organized against the centralization of the Chicago school system. Haley was wildly popular at the local and national level. For over two decades, she was often on the road, speaking to enthusiastic crowds of teachers and social reformers about tax reform, teachers' rights in school organizations and their responsibility in society, and broad social topics including women's suffrage, municipal ownership of public utilities, and the initiative and referendum movement. When not fighting in specific school districts, she was an active and often oppositional member of the administrator-dominated National Education Association (NEA), from where she asserted her message of democratic practices in school organizations and in the right of teachers to organize.[38]

In all of this work, Haley focused on organizational change more than pedagogical reform. Certainly her vision implied classroom change: An early critic of what she called the "factoryization" of education, Haley believed that teachers should have the authority to shape the classroom into a caring and supportive environment for children. Echoing John Dewey's faith in the school as a potential agent of social change, Haley believed that the classroom teacher could be the center of humanitarian reform. She believed that "the public school must become a more potent, conscious, and recognized factor in the civic life of the large communities" and that this could only happen when teachers secured better working conditions.[39] The free and empowered classroom teacher was her goal; her method toward achieving that was to reform school organizations.

Haley's oft-cited 1904 speech before the NEA, "Why Teachers Should Organize," lays out Haley's vision. One can read in her speech her lessons from William James—that teachers can have an effect on individuals—as well as her lessons from Parker about the school as an incubator for social change. These notions were embedded in Haley's identification of the teachers' union as the best agency for promoting progressive social change. Only organized teachers could work against financial and political interests to keep schools up to date, properly equipped, and adequately supported. Only organized teachers could keep academic ideals alive by maintaining adequate working conditions.[40] This was the teachers' responsibility to a democratic society.

This belief was Haley's underlying support for representative teacher councils in schools. In the same 1904 speech, Haley quoted Dewey, who argued that teachers required a "regular and representative way" to register judgment upon matters of educational importance, with the assurance that this judgment will somehow affect the school system. . . . [41] What does democracy mean save that the individual is to have a share in determining the conditions and the aims of his own work?" Dewey asked in an essay that was reprinted in the *Chicago Teachers' Federation Bulletin* in 1904.[42] As Haley argued, teacher councils were a "declaration of independence on the war for academic freedom."[43]

Haley also fought against the authority of textbook companies to influence city curriculum. Haley's argument here was both that democracy in education began with the right of teachers to decide on their own classroom technique *and* that organized teachers needed to moderate outside influences that used the schools for their own profit. In 1902, she organized teachers and labor unions to protest the Chicago Board of Education's removal of District Superintendent William Speer. Haley and her supporters argued that Speer had angered the textbook companies because his "visualization" system of teaching mathematics did not require textbooks.[44] John Dewey, Francis Parker, and Thomas Shields wrote letters to the board protesting Speer's removal and advocating Speer's work as a creative educator who promoted both students' freedom of thinking and parents' freedom from purchasing unnecessary textbooks.[45] Upon his removal, Speer wrote an essay for the *Chicago Teachers' Federation Bulletin,* significantly entitled "Freedom: The Condition of Individual Development."[46] Haley's other administrative role model, Josephine Locke, was also driven from the school system because she promoted a method of teaching art that did not require textbooks. As Locke told the Federation teachers in a public address in 1902, "Until the teachers are equipped sufficiently to discuss and initiate new methods, new subject matter, not only in school instruction but in school administration and organization, the schools will remain practically what they are today, an exploiting field to be manipulated by the ignorant, the selfish, and the unscrupulous."[47] The promotion of progressive educational practices, then, required defensive actions against dangerous outside influences.

Given Haley's primary concern of teachers' working conditions, those dangerous outside influences could include progressive education itself. Haley's promotion of curricular and pedagogical reform in schools was consistently tempered by her protection of teachers' working conditions. In May 1902, for example, she responded to the director of Physical Education in Columbus, Ohio, who had asked Haley to comment on the need for physical education in schools. There was no question in Haley's mind that physical education was necessary in schools because of its value to students' physical and intellectual development. Yet she cautioned the director about offering such a curriculum in schools without the revenue to support it. It was a "fatal mistake" to extend education without providing the necessary means. "Failing to do this, the burden is thrown entirely on the grade teachers who never receive any additional compensation for the additional work required of them and the additional preparation for the work." She offered as an example the situation in Chicago where kindergartens, manual training, and physical culture were offered in the schools without financial provision to accommodate those new departments. This meant that these improvements had been made at the expense of grade school teachers. Haley asserted that "aside from the injustice and lack of foresight involved in this policy, it is a question if there has been anything gained, for the standard of living of the teachers has been lowered and the efficiency of the schools must be to that extent impaired." She argued that overworked and underpaid teachers could effectively negate any of the proposed benefits of progressive education.[48] The reverse was also true: The introduction of progressive education could lead to overworked and underpaid teachers.

This thinking led Haley to take some positions that clashed with other educators. For example, she consistently opposed the introduction of new educational requirements for teachers, rejecting board recommendations for increased training for aspiring teachers. Haley's reasoning was that teachers, not administrators, should decide the qualifications of the occupation. She also rightly argued that increased requirements would make teaching less accessible to daughters of the working class. But by opposing such reforms, Haley in effect supported the continuation of poorly trained teacher applicants who were hired through local and political connections, and not by skill or training.[49]

And although Haley recognized from her own experience the great value of ongoing education for teachers, she insisted that such practice remain voluntary. She adamantly opposed the addition of higher education requirements for teachers for promotion. Haley argued that teachers, like other workers, should be entitled to higher salaries because of the good work that they performed in the classroom, and not because of extra work that they took on outside of class. In 1907, Superintendent Edwin Cooley proposed a promotion plan that linked teachers' salary increases with five extra education classes. Haley critiqued the plan because it favored teachers who could afford to take those classes, and it forced teachers to see those classes as merely hurdles to leap to gain a raise. Furthermore, because the plan effectively restricted the number of teachers who qualified for an increase, Haley saw the new promotional system—newly garbed in a plan that looked like it promoted increased academics—as a continued attempt to cut school costs.[50] So adamant was she about this principle that she devised a scheme whereby over 1,000 teachers surreptitiously enrolled in five concurrent classes at the Chicago Art Institute, thereby earning their raises cheaply, within merely one semester.

Haley was vindicated in that she showed the board that teachers would not be pushed around. But her scheming hardly spoke well for her commitment to teacher education. Jane Addams, who was on the Chicago Board of Education at the time, objected to Haley's stand, arguing that the Federation should *want to* promote an education plan that would improve teachers' classroom practice and thereby lighten their burdens at work.[51]

RELATIONS WITH OTHER PROGRESSIVES

Haley's focus on teachers' working conditions contributed to her exclusion from other progressive education activities in Chicago. Given that Haley worked for women teachers on school and social reform in Chicago in the early decades of the twentieth century, it would seem that she would have coordinated her work with the large community of women reformers in the city at that time. To some extent she did: She allied with Jane Addams on school and child labor issues, and she lobbied for the appointment of Addams, reformer Anita McCormick Blaine, and Dr. Cornelia DeBey to the Chicago Board of Education.[52] But a number of prominent women reformers who focused specifically on school reform in Chicago had very little contact with Haley. The blame for this rift is two-sided: Not only was the Federation excluded from some of these communities, but Haley specifically kept her teachers removed from these groups.

One group of school reformers that Haley refused to work with were African American club women. By the turn of the century, more than 50 black women's clubs in Chicago were organized for community reform. Education was the single most important area of activism for these women, primarily because of the poor educational conditions that the city offered African American children. Furthermore, many of these women were teachers, themselves, who were all but excluded from teaching in Chicago schools. In response, black women club members established and supported their own independent schools, kindergartens, libraries, training schools, and summer programs.[53] There is some indication that Haley may have worked with a Black women's club in her campaign to publicize the tax fight, but she generally ignored the Black community and the significant problems that Black teachers and students faced in Chicago schools.[54]

In the early years of the Federation, Haley worked somewhat closely with white middle-class women's clubs on school and social reform movements, including women's suffrage.[55] The political culture of Chicago's middle-class women reformers at the turn of the century centered around the notion of "municipal housekeeping"—a city-wide agenda that implied not just social betterment, but also the institutionalization of a publicly financed welfare structure, modeled on the ideal caring home and community. Municipal housekeeping placed the interests of the family and community as central to the interests of the state, and the school was included in the social equation of the state as a large home. School issues as they related to poor urban children—compulsory enrollment, curriculum, school building maintenance, and school health—were a critical part of Chicago women reformers' programs.[56]

Although such reform agendas paralleled Haley's vision of an improved school system, they did not speak directly to her concern for teachers' improved working conditions. Women reformers' initiatives for the introduction of kindergartens, school lunch programs, public playgrounds, school guidance, and vacation schools were valiant efforts but were not central to teachers' work, and Haley believed that they risked being introduced without adequate supports, thereby further burdening teachers. Haley once publicly lashed out at Jane Addams, after Addams had opposed the teachers on a proposed change in the teachers' pension law. Haley belittled Addams' involvement in social reform work at the expense of teachers. "You pay too dear for your forty playgrounds, or forty times forty playgrounds, when you trade them off for the fundamental rights that the teachers have."[57]

Class differences carved a great divide between Haley and her middle-class progressive colleagues. The wealthy Protestant women who made up the social reform community lived in a different world than Haley did. Native born, raised in economically comfortable families, often educated in private schools, well-read, and cultured in middle-class morals and values, these women often furthered their education at elite universities like Northwestern or the University of Chicago. Some of these women were among the vanguard of progressive education in private schools. Anita McCormick Blaine and Flora Cooke, for example, were behind the founding and early growth of the Francis Parker School.[58] Other women worked in or on behalf of settlement houses like Jane Addams' Hull House. Passionate, committed, and visionary, these women were the backbone of Chicago social and educational reform.

But as an Irish Catholic, Haley was, both by practice and social custom, excluded from the world in which these women lived. Through the turn of the century, Irish

Catholics in America were viewed as almost a distinct race: scourged and criticized by white Protestants as insolent, lazy, and unreliable; they were also seen as unpatriotic because of their allegiance to the Pope. The social distinction between Irish Catholics and Protestants in turn-of-the-century Chicago cannot be underestimated. Margaret Haley lived a life that was largely separate from Protestant people. The community in which Haley lived, the church to which she belonged, and the parish around which her social and religious life circled on the South Side of Chicago; the labor unions and fraternal organizations with which her father and brothers identified; and the community and reform activities that she engaged in were all Roman Catholic.[59]

Haley's identity as a public school teacher furthered the cultural differences between her and middle-class reformers. Women teachers in Chicago were predominantly Irish Catholic and from the rising working class, or labor aristocracy. They worked in a feminized labor force with explicit gendered and classed prescriptions for behavior, yet they also shared their days in relatively heterosocial environments in interactive relationships with men and women staff, parents, school officials, and community members from a variety of class and cultural backgrounds. Teachers' occupational identity thus shared components of the industrial proletariat, feminized social service work, and white collar professionals and they engaged in a range of alliances and conflict across the classes and genders of social spheres.[60] These women public school teachers had more in common with their brothers in the police force than with women private school teachers at Francis Parker School.

Because of their different class background, middle-class Protestant reformers held a distinctly different understanding of school politics than organized teachers did. The reformers were apt to see the Federation politics as too narrow and self-interested. They attempted to diversify the types of public school teachers with whom they worked, and they often did this by working with Haley's opponents and by ignoring the significance of occupational divisions. Reformer Ethel Sturgis Dummer, for example, worked closely with the school superintendent and with the Chicago Board of Education but had few connections with the Chicago Teachers' Federation, which represented over three-fourths of the city's elementary school teachers. As Chair of the Joint Committee on Education, a group set up by a collection of city women's clubs, Dummer contacted a member of the Teachers' League, a group opposed to the Chicago Teachers' Federation. The league member who was appointed was Lucy Laing, who was not a teacher but the principal of the Bass Elementary School.[61] In later years, this joint committee enraged the Federation by supporting school Superintendent William McAndrew, who abolished teacher councils, advocated close administrative supervision of teachers, and promoted school organization reforms like the platoon plan that the Federation adamantly opposed.[62]

Different social norms and understandings of political leadership further divided the two groups. Given the nature of Haley's work—political lobbying with elected officials for government and civil service reform—she adopted the sharp-edged uncompromising leadership style of labor leaders. In contrast, she saw club women as intellectually abstract people driven by noblesse oblige for vague social goals, ignorant of the tough office politics of legislative lobbying. Commenting on the testimony of educational reformer and club woman Helen Heffernan before a state legislative committee on schools, Haley tartly explained that Heffernan only made a general "academic speech" about what was good for schools, and "didn't come within a million miles of the subject of the amendment."[63]

Haley's intense character also may have made some middle-class women uncomfortable. Jane Addams originally told Anita McCormick Blaine that she admired Haley's "pluck and energy," but within a few years Addams may have described Haley with different adjectives.[64] In 1905, Haley helped to convince Addams to join the Chicago Board of Education under a reformist mayor. Although the same age, and sharing similar politics, the two women had vastly different personalities and leadership styles. Irish Catholic working-class Haley was a hard-boiled, uncompromising political leader and lobbyist, whereas the middle-class, college-educated Addams was known for her compromising leadership style and her philosophical meditations on justice and responsibility. As a union leader, Haley's demands for teachers' specific interests often conflicted with Addams' broader political vision and her compromising tactics. Haley disparagingly referred to her as "Gentle Jane" and mistrusted her commitment to teachers.[65]

Addams, herself, ultimately regretted taking the job on the Board of Education. Her first year on the board was "stormy and very unsatisfactory," she wrote to a friend, and she wondered if there was any way to make education more democratically self-governing.[66] Prior to her board experience, Addams had felt that all political problems could be mediated and compromised, but on the school board she learned that "all such efforts were looked upon as compromising and unworthy, by both partisans."[67] Years later, she still found it difficult to write about her experience on the Chicago Board of Education because it was "dramatized in half a dozen strong personalities."[68] Haley was certainly one of those figures, and their relationship remained strained. Addams ultimately felt that Haley had pressured her into taking stands with which she did not believe, and when Addams stood up for herself, Haley rejected her.

John Dewey, whose words Haley used liberally in the *Federation Bulletin,* recorded only one interaction with Haley, and his experience paralleled Addams'. In 1907, Haley went to Dewey to ask him to run for Superintendent of Schools. Dewey wrote his wife that not only was the suggestion absurd, but that Haley came to him with "pomp and circumstance," claiming to be secretly negotiating a deal for Dewey with friendly Board of Education members who would nominate him. Dewey read through Haley's feigned praise of him, knowing full well that the recent election of a mayor who was unfriendly to the Federation meant that the teachers were desperate for a powerful ally like the famous John Dewey. He also knew full well that his asset was his name only and not his dubious skills as an administrator.[69] Dewey's observation of Haley as overbearing and manipulative was not unique. In over 30 years of political leadership, she was often accused by friend and foe alike of vindictively setting out personal attacks and of damning anyone who disagreed with her as being wrong at best and ignorant at worst.[70]

Further distancing Haley from middle-class women reformers was Haley's notorious bold and unfeminine behavior. Not only did she speak publicly and enthusiastically before crowds of people, but she traveled alone and worked closely with male politicians and lawyers—a point that concerned some of her friends and family. She enjoyed telling stories of drunken politicians who leered or yelled at her, cursing her intrusion into their privileged world. Her fearless character was renown among teachers; one teacher penned a verse description of Haley as being "urged on to danger by a selfless will."[71] In public she humiliated public officials by mocking their actions, and in private she fiercely lectured opponents. Haley once accused a teacher of being the equivalent of a Tory in the American Revolution and that any teacher who did not take Haley's side in this issue was simply unfit to teach children American History.[72]

In a few rare moments, Haley recognized that she may have gone too far. After lecturing Jane Addams for not obeying Haley's wishes, she wrote Addams an apologetic letter, admitting that her history of "irritations" with school politics often made her lose her temper so that she expressed "ugly feelings" of which she was deeply ashamed. The problem was that her passion for democratic principles was so strong, and the work was often so discouraging, that she fell prey to small feelings of "prejudices and antagonism. . . ."[73] Forgive my crass behavior, Haley pleaded, "My only crime is that I care too much about democracy."

CONCLUSION

Margaret Haley's position as a progressive reformer provides an alternate model to the term "progressive educator" and offers an alternative perspective to what it meant to be a woman educational leader in the Progressive Era. As a union leader, Haley consistently argued that her work for teachers was not only for labor interests but "far more truly . . . for the children of the city."[74] Yet sometimes it is not easy to see this in her work, particularly because her peer progressive educators rarely mentioned her or worked with her. By keeping the political economy of schooling front and center in her work, Haley promoted a pragmatic *real politik* perspective on school reform. One might say that Haley believed that only with appropriate funding and support by the state could city schools implement child-centered education and truly make the school responsible to the needs of society.

NOTES

1. Handwritten note by Margaret Haley, Buffalo, N.Y., 3 August 1901, box 36, folder 1, Chicago Teachers' Federation Archives, Chicago Historical Society (Hereafter referred to as CTF Archives).
2. Studies of turn-of-the-century pedagogical progressives include Lawrence A. Cremin, *The Transformation of the School: Progressivism in American Education, 1876–1957* (New York: Vintage, 1964); Patricia A. Graham, *Progressive Education From Arcady to Academe: A History of the Progressive Education Association, 1919–1955,* (New York: Teachers College Press, 1967); and Susan F. Semel and Alan R. Sadovnik, eds., *"Schools of Tomorrow," Schools of Today: What Happened to Progressive Education?"* (New York: Peter Lang, 1999). On administrative progressives, see David B. Tyack, *The One Best System: A History of American Urban Education* (Cambridge: Harvard University Press, 1974).
3. John Dewey, *The Educational Situation* (Chicago: University of Chicago Press, 1902), p. 23.
4. Tyack, *One Best System,* pp. 197–198.
5. Haley is often included in general discussions about progressive education, although her position as a pedagogical or administrative progressive has not been described adequately. See Joan K. Smith, "Progressivism and the Teacher Union Movement: A Historical Note," *Educational Studies* 7 (1976): 44–61.
6. Robert Reid, ed., *Battleground: The Autobiography of Margaret A. Haley* (Urbana: University of Illinois Press, 1982), pp. 11–12.
7. Marjorie Murphy, "Progress of the Poverty of Philosophy: Two Generations of Labor Reform Politics: Margaret and Michael Haley" (Paper delivered at the Knights of Labor Centennial Symposium, Chicago, Ill., 17–19 May 1979).
8. Reid, *Battleground,* p. 12. Cyrus Winthrop Brown was born in West Behavia, N.Y. 20 July 1844 and died in Joliet, 10 January 1921; "Memorials of Deceased Companions of the

Commandery of the State of Illinois Military Order of the Loyal Legion of the United States," from 1 January, 1912 to 31 December, 1922, Chicago, Ill., 1923, pp. 629–631.

9. Laura Doctor Thornburg and Christine A. Ogren, "Normal Schools," in *Historical Dictionary of American Education,* ed. Richard Altenbaugh (Westport, Conn.: Greenwood Press, 1999), pp. 260–262; Jurgen Herbst, "Teacher Preparation in the Nineteenth Century: Institutions and Purposes," in *American Teachers: Histories of a Profession at Work,* ed. Donald Warren (New York: MacMillian, 1989), pp. 224–226.
10. Fourth Annual Announcement of the Normal and Scientific School, Morris, Grundy County, Illinois, 1881–1882.
11. Margaret Haley's autobiographical manuscript, 1934, Chicago, Ill., box 34, p. 414, CTF Archives.
12. "Names and Addresses of Persons Attending the August Term at Normal, 1882," *Illinois School Journal* (September 1882): 156.
13. Herbart M. Kliebard, "Dewey and the Herbartians: The Genesis of a Theory of Curriculum," in *Forging the American Curriculum: Essays in Curriculum Theory and History* (New York: Routledge, 1992).
14. Richard Allen Swanson, "Edmund J. James, 1855–1925: A Conservative Progressive in American Education," (Ph.D. diss., University of Illinois, 1966); "Normal News," *Illinois School Journal* (January 1883): 284; Daniel T. Rodgers, *Atlantic Crossings: Social Politics in a Progressive Age* (Cambridge: Harvard University Press, 1998), pp. 84–85, 107–108.
15. Reid, *Battleground,* p. 20.
16. David Hogan, *Class and Reform: School and Society in Chicago, 1880–1930* (Philadelphia: University of Pennsylvania Press, 1985), pp. 82–85; Robert Eugene Tostberg, "Educational Ferment in Chicago, 1883–1904," (Ph.D. diss., University of Wisconsin, 1960), pp. 53–54.
17. Tostberg, "Educational Ferment," pp. 75–76.
18. Francis W. Parker, "An Account of the Work of the Cook County and Chicago Normal School from 1883–1899," *Elementary School Teacher* 2 (June 1902): 754.
19. Reid, *Battleground,* p. 23.
20. J. M. Rice, *The Public School System of the United States* (New York: Arno Press, 1969 [originally published 1893]), pp. 166–183. See Kate Rousmaniere, "Sixteen Years in a Classroom," in *Silences and Images: The Social History of the Classroom* eds. Ian Grosvenor, Martin Lawn, and Kate Rousmaniere (New York: Peter Lang, 1999), pp. 235–255.
21. Hannah Belle Clark, *The Public Schools of Chicago: A Sociological Study* (Chicago: University of Chicago Press, 1987), pp. 31–32, 77–78.
22. Reid, *Battleground,* p. 23.
23. Mary J. Herrick, *The Chicago Schools: A Social and Political History* (Beverly Hills: Sage, 1971), pp. 72–73.
24. Josephine Locke to Ethel Sturgis Dummer, n.d. Ethel Sturgis Dummer Collection, box 20, folder 318, Schlesinger Library, Radcliffe College; "Art in the Public Schools" *Chicago Tribune* (16 December 1899); "Aroused by Cut" *Chicago Times,* (9 January, 1896).
25. Reid, *Battleground,* p. 23.
26. *A History of Marshall County, Iowa* (Chicago: Western Historical Society, 1878), p. 571; Clarence Ray Aurner, "History of Education in Iowa" (Iowa Historical Society, 1916), p. 136; W. W. Speer, "Observation and Natural Science," in Biennial Report of the Superintendent of Public Instruction State of Iowa (Des Moines, 1881), pp. 49–52.
27. Reid, *Battleground* p. 23.
28. Carolyn Terry Bashaw, "Ella Flagg Young," in *Historical Dictionary of Women's Education in the United States,* ed. Linda Eisenmann, (Wesport Conn., Greewood Press, 1998), pp. 496–498.
29. Reid, *Battleground,* p. 24.
30. Ibid., p. 286.

31. *Educational Review* 10 (June 1895): 104; Charles R. J. Collins, "The University of Buffalo School of Pedagogy, 1895–1898," *Niagara Frontier* 19 no. 2 (Summer 1972): 30–41.
32. Reid, *Battleground,* p. 27.
33. Ibid.; Educational *Review* 12 (September 1896): 194. Cremin, *Transformation of the School,* pp.105–109; Merle Curti, *The Social Ideas of American Educators* (Totowa, N.J.: Littlefield Adams, 1959), pp. 429–458.
34. "Summer School at Sinsinawa," *The Young Eagle,* St. Clara Academy, Sinsinawa, Wis., September 1897. Sinsinawa Dominican Archives, Sinsinawa, Wis.
35. Robert D. Cross, *The Emergence of Liberal Catholicism in America* (Cambridge: Harvard University Press, 1958); Henry J. Browne, *The Catholic Church and the Knights of Labor* (Washington, D.C.: Catholic University of America Press, 1949).
36. Thomas Edward Shields, *Teachers Manual of Primary Methods* (Washington, D.C.: The Catholic Education Press, 1912), p. 35; Justine Ward, *Thomas Edward Shields: Biologist, Psychologist, Educator* (New York: Scribners, 1947), pp. 125–214.
37. Reid, *Battleground,* p. 28.
38. Wayne J. Urban, *Why Teachers Organized* (Detroit: Wayne State University Press, 1982), pp. 113–117; idem, *Gender, Race and the National Education Association: Professionalism and its Limitations* (New York: Routledge Falmer, 2000), pp. 14, 26; Marjorie Murphy, *Blackboard Unions: The AFT and the NEA, 1900–1980* (Ithaca: Cornell University Press, 1990), pp. 72–79.
39. Margaret Haley to Franklin S. Edmonds, Central High School, Philadelphia, 2 June, 1903, box 37, folder January-June 1903, CTF Archives.
40. Lois Weiner, "Teachers, Unions, and School Reform: Examining Margaret Haley's Vision," *Educational Foundations* (Summer 1996): 85–96.
41. Quoted in Weiner, "Teachers, Unions, and School Reform," p. 89.
42. John Dewey, "Address," *Chicago Teachers' Federation Bulletin* 2 (26 February 1904): 1–3, 5–6.
43. Reid, *Battleground,* pp. 88–89. Originally proposed by Ella Flagg Young in 1898, teacher councils were not instituted in Chicago until 1919 and were abolished in 1925. See Herrick, *The Chicago Schools,* pp. 139–141, 150–154.
44. J. H. Bowman, "Chicago Public Schools: what the Cleveland Educators Think of the Speer System" *Union Labor Advocate* 2, no. 8, (April 1902): 13.
45. Letters regarding William W. Speer, box 37, folder January–June 1903, CTF Archives.
46. William W. Speer, "Freedom: The Condition of Individual Development," *Chicago Teachers Federation Bulletin* (3 July, 1903): 6.
47. Reid, *Battleground,* p. 87; Josephine C. Locke, "Address to the Chicago Teachers' Federation, 5 September, 1902," *Chicago Teachers' Federation Bulletin* (12 September 1902).
48. Margaret Haley to Mrs. Frances W. Leiter, 4 May 1902, box 36, folder January-May 1902, CTF Archives.
49. Robert Lowe, "Chicago Teachers' Federation and its Legacy," in *Transforming Teacher Unions: Fighting for Better Schools and Social Justice,* eds. Bob Peterson and Michael Charney, (Milwaukee: Rethinking Schools, 1999), p. 83.
50. Margaret Haley's autobiographical manuscript, 1911–1912, Seattle, Wash., box 32, p. 165, CTF Archives.
51. Jane Addams, cited in Lowe, "Chicago Teachers' Federation," p. 83.
52. Mary C. Schiltz, "Women on the Chicago Board of Education: A Collective Biography of Three Members in the Early Twentieth Century," *Journal of the Midwest History of Education Society* 21 (1994): 215–227.
53. "Notes on the Chicago Public Schools, 1880–1917, Illinois Writers' Project, box 14, folder 20, Carter Woodson Library, Chicago, Ill.; Anne Meis Knupfer, *Toward a Tenderer Humanity and a Nobler Womanhood: African American Women's Clubs in Turn-of-the-Century Chicago* (New York: New York University Press, 1996); Wanda A. Hendricks, *Gender, Race*

and Politics in the Midwest: Black Club Women in Illinois (Bloomington: Indiana University Press, 1998).

54. Kate Rousmaniere, "White Silence: A Racial Biography of Margaret Haley," in *Equity and Excellence* 34 (September 2001): 7–15.
55. Julia Wrigley, *Class Politics and Public Schools: Chicago 1900–1950* (New Brunswick, N.J.: Rutgers University Press, 1982); pp. 181–187; George Counts, *School and Society in Chicago* (New York: Harcourt, Brace & Co., 1928), pp. 206–228.
56. Paula Baker, "The Domestication of Politics: Women and American Political Society, 1780–1920," *American Historical Review* 89 (June 1984): 620–647; Maureen A. Flanagan, "Gender and Urban Political Reform: The City Club and the Woman's City Club of Chicago in the Progressive Era," *American Historical Review* 95 (October 1990): 1032–1050.
57. Reid, *Battleground,* p. 113.
58. Gilbert H. Harrison, *A Timeless Affair: The Life of Anita McCormick Blaine* (Chicago: University of Chicago Press, 1979), pp. 84–102 ; Marie Kirchner Stone, "The Francis W. Parker School: Chicago's Progressive Education Legacy," in *"Schools of Tomorrow,"* eds. S. Semel and A. Sadovnik. pp. 23–66.
59. David R. Roediger, *The Wages of Whiteness: Race and the Making of the American Working Class* (New York: Verso, 1991); Noel Ignatiev, *How the Irish Became White* (New York: Routledge, 1995); Ellen Skerrett, "The Development of Catholic Identity among Irish Americans in Chicago, 1880–1920," in *From Paddy to Studs: Irish American Communities in the Turn of the Century Era, 1880–1920,* ed. Timothy J. Meagher (New York: Greenwood Press, 1986), pp. 117–138; and Ellen Skerrett, "The Irish in Chicago: The Catholic Dimension," in *Catholicism, Chicago Style,* eds. Ellen Skerrett, Edward R. Kantowicz, and Steven M. Avella (Chicago: Loyola University Press, 1993), pp. 29–62.
60. Alice Kessler-Harris, "Treating the Male as 'Other': Re-defining the Parameters of Labor History," *Labor History* 34 (1993): 190–204; Elizabeth Faue "Gender and the Reconstruction of Labor History, an Introduction," *Labor History* 34 (1993): 169–177; Dina M. Copelman, *London's Women Teachers: Gender, Class and Feminism, 1870–1930.* London: Routledge, 1996).
61. Ethel Sturgis Dummer to Miss Carrie E. King, 11 May 1916. Ethel Sturgis Dummer Collection, box 21, folder 327, Schlesinger Library, Radcliffe College.
62. Wrigley, *Class Politics and Public Schools,* pp. 181–187.
63. Margaret Haley's autobiographical manuscript, 1934, Chicago, Ill., box 34, p. 220, CTF Archives.
64. Jane Addams to Anita McCormick Blaine, 20 November, 1903, Jane Addams papers, University of Illinois, Chicago.
65. Reid, *Battleground,* p. 102.
66. Jane Addams to Graham Wallace, 1 July 1906, Jane Addams papers, University of Illinois, Chicago.
67. Jane Addams, *Twenty Years at Hull House* (New York: Signet, 1981 [originally published 1910), p. 235.
68. Addams, *Twenty Years at Hull House,* p. 229.
69. John Dewey to Alice Chipman Dewey, 16 April 1907, John Dewey Archives, University of Illinois, Carbondale.
70. Letter to Margaret Haley from James A. Meade, 9 October, 1926, Chicago Teacher Union Archives, box 18, folder 5, Chicago Historical Society.
71. "Catherine Goggin and Margaret Haley" *Journal of Education* 55 (26 April 1902).
72. Margaret Haley's autobiographical manuscript, 1911–1912, Seattle, Wash., box 32, pp. 247–249, CTF Archives.
73. Margaret Haley to Jane Addams, 4 May 1906, in Anita McCormick Blaine papers, University of Wisconsin Historical Society; Reid, *Battleground,* p. 105.
74. Reid, *Battleground,* p. 40.

CHAPTER TEN

ELLA FLAGG YOUNG AND THE CHICAGO SCHOOLS

JACKIE M. BLOUNT

Ella Flagg Young stands as a towering figure in the history of American public schooling. She caught the national eye in 1909 when the contentious Chicago School Board selected her to be superintendent, making her the first women to hold such a position in a large urban school system. She also became the most powerful and best-paid woman executive in the country. A year later, newly enfranchised women teachers elected her president of the National Education Association (NEA), a change that shocked the established power structure of the organization and created permanent meaningful places at the table for women members. Earlier in her career, Young joined the faculty of the newly established University of Chicago, quickly becoming an influential professor who was widely sought by students, many of whom also taught in Chicago schools. At the University of Chicago she worked alongside John Dewey, earning his deepest respect and helping him clarify some of his most enduring ideas about democracy in education. Young provided long-standing leadership to the schools of the entire state by serving on the Illinois State Board of Education for around 20 years before becoming its president in 1910. She maintained an active scholarly agenda in her later years. She edited a series of monographs for the University of Chicago Press and also edited two important educational journals, *The Elementary School Teacher* and *The Educational Bi-Monthly*, each of which stimulated rich discussion among teachers and scholars alike. As principal of the Chicago Normal School, she pioneered a number of rigorous, socially progressive, and intellectually demanding methods for preparing teachers to work in nearby schools. In addition to these strenuous activities, she also assumed substantial leadership in a variety of social, political, and professional organizations throughout her adulthood. Taken together, Young's contributions to public schooling are remarkable, both for their broad extent and their depth.

Beyond these astonishing accomplishments, this well-rounded woman left strong, deeply etched impressions on nearly everyone who encountered her, including John Dewey. His daughter reminisced that "[John Dewey] regards Mrs. Young as the wisest person in school matters with whom he has come in contact in any way. . . . Contact with her supplemented Dewey's educational ideas where his experience was lacking in matters of practical administration, crystallizing his ideas of democracy in the school and, by extension, in life."[1] Social activist Jane Addams concurred, "She had more general intelligence and character than any other woman I knew."[2] And former coworker William Owen, writing for the *Educational Bi-Monthly,* emphasized that Young's skill was so great that "had she been a man . . . [she] would have directed a great corporation, managed a railroad, served as governor of a state, or commanded an army."[3]

Beyond her far-ranging and important accomplishments as a leader, though, Young also left a remarkable legacy of educational ideas. Her published works, although eclectic in their scope, demonstrate a disciplined mind that was quite adept at philosophical inquiry, practical and theoretical curriculum issues, and political analysis. One of her most enduring contributions was her unique ability to integrate theory and practice powerfully in both her leadership and in her writing. Her main book, *Isolation in the Schools* (1900), reveals the carefully crafted philosophy by which she sought to lead the Chicago schools. Unfortunately, her intellectual contributions have yet to be analyzed extensively by contemporary scholars, especially those in *Isolation.* I will briefly examine this work, but first an overview of her life is in order.

YOUNG'S LIFE AND CAREER

Surprisingly, even though Young assumed highly significant and visible leadership positions in education and influenced some of the most powerful social thinkers of the time, the details of Young's family history and friendships remain sketchy. Young and her closest circle of friends carefully guarded her privacy, which offered her some insulation from constant public scrutiny. John McManis, a teacher who worked for her and later became her biographer, explained, "The difficulties encountered in writing the life of another are more than technical difficulties. . . . In the case of Mrs. Young the difficulties are particularly great. Almost no help in such an undertaking can be derived directly from her or from her immediate friends. In the one case, her interest is in her work and not in herself, making it impossible to secure personal touches needed to understand the meaning of her acts; in the other, friends are jealous of relationships and guard them closely."[4] McManis noted that Young was exceedingly reluctant to talk about herself and when pressed to discuss herself personally, she usually shifted the conversation toward public matters, such as the conditions of schools.[5] Young was known to avoid personal interviews by explaining that such triviality robbed the district's teachers and children of her time.[6]

Despite these difficulties, the general outline of Young's life is known. In 1845, just as the common school movement gained national momentum, Ella was born in Buffalo, N.Y. Theodore Flagg, her father, enjoyed the proud reputation as one of the finest, most skillful metal workers in the Great Lakes region. The community knew her mother, Jane Flagg, as a talented caregiver who helped neighbors through all but the most serious illnesses. Ella, the last of three Flagg children, followed the births of a notably intelligent older sister and a healthy, artistically gifted brother.[7] Her mother regarded Ella as a frail, sickly child. Her chief aim, then, was to nurture Ella to full health by getting her outside as much as possible and by watching her closely.[8]

An important consequence of Ella's early poor health was that she did not attend school until she was 11 years old. Instead, her mother tutored her carefully at home.[9] Ella, however, apparently resisted learning some fundamental academic skills in spite of her mother's attention. She did not begin reading until she was eight or nine. Then, according to a vivid story that McManis recounted, she spontaneously taught herself:

> At the breakfast table one morning there was much excitement over an account in the morning paper of the burning of a schoolhouse. Her mother saying, in a horror-stricken tone, 'Think of it, little children of Ella's age threw themselves out of the upper-story windows,' especially impressed Ella! . . . She asked her mother to read it to her. Then, taking the paper into her arms, she went weeping into a room by herself and tried to read it. She remembered the exact beginning, and fitted it with her finger to the words in the newspaper. She soon became aware that she did not know the words after the first few lines, and she went to the kitchen and asked the 'girl' to read it. . . . She became interested immediately in learning to read.[10]

She did not learn to write until a year or two later. In spite of this relatively late start in learning critical academic skills, she quickly made up for lost time and became quite proficient. Once she understood that reading, writing, and other school subjects might prove interesting or useful, she committed herself to achieving full mastery. This remained a central pattern through the remainder of her life.

At age 11, Ella finally enrolled in the neighborhood primary school. Unlike other children of her age, McManis speculated, "The management and the recitations interested her deeply. She did not find the strangers among the children interesting. In short, a child life spent with her mother had resulted in the usual condition—preference for the society of grown people."[11]

Ella had attended school in Buffalo only for two years before her family moved westward to the rapidly expanding Chicago area. She was ineligible for admission to the high school, however, until she first attended a Chicago grammar school for one year. The Brown School, in which she enrolled, left her feeling desperately bored, in part, because the recitations seemed too easy; as a result, she dropped out. She did not stay at home long, though. A friend of hers soon encouraged her to take an exam for teacher certification, which Ella passed. However, because Ella still was too young to teach, she was offered admission to the normal school instead, an invitation she eagerly accepted. Ella quickly proved to be an exceptional normal school student because soon the head teacher had placed her desk by his on the teachers' stage. He had pressed her into service, leading other students through their Lancasterian drills and lessons.[12]

The next few years of Ella's life offered a perplexing mixture of both great opportunity and immense grief. She began teaching at the Foster School at the innocent age of 17. Two weeks later, though, her beloved mother died without warning. Apparently, Ella never spoke publicly about her mother's death. She did throw herself fully into her work, however, staying late after students left her classroom each day, attending weekend teacher training sessions, and studying intensely during the evenings. Within a very short time she became a teacher with legendary skill and knowledge. Her superiors clearly recognized her talents because she moved quickly from teacher to head assistant of the school. The annual reports of the Chicago Board of Education reveal that she also received an unusually rapid series of salary increases for her work. Her exceptional talent even attracted the attention of Josiah Pickard, the district's new superintendent, who offered her the opportunity to serve as principal of the just-created School of Practice. She eagerly accepted this assignment, which essentially entailed the supervision of two elementary classrooms with purportedly difficult students. Ella did not disappoint. Pickard's subsequent annual report praised her poise and great skill with the assignment.[13]

The next few years proved to be quite turbulent for Ella. In 1868, her brother, who had survived military service during the Civil War, died in an unusual train accident. She grieved privately, but continued devoting her energies to her work and, by then, her future plans for marriage. In December of that year, the 23-year-old Ella Flagg married the much older William Young, an old family friend. The following year, tragedy struck again when both her father and sister died of pneumonia. Her husband, too, suffered bouts of poor health, so he traveled to the Southwest to recuperate. Ella remained in Chicago and continued pursuing her work supervising the School of Practice. By Ella's own account, William Young later died in 1873, though there is disagreement among several sources on this date. Ella and William had not lived together for long, however. Young apparently never spoke of her husband afterwards and neither did any of her friends. And as she suffered these successive losses, she quietly devoted herself even more fully to her work.[14]

Her career continued moving forward at a blistering pace. She became principal of the Scammon School in 1876. Then, scarcely three years later, she was promoted to the principalship of one of the three largest schools in all of Chicago, the Skinner School. It was during her years at Skinner that Young met Laura Brayton, a woman who joined the faculty in 1883 to teach English. Young and Brayton soon became dear friends and eventually life partners. As Young's reputation for excellence continued to spread, she eventually was promoted yet again. In 1887, she was appointed to serve as one of five assistant superintendents for the entire district. Young and Brayton maintained their friendship even though they had come to work in different locations. Eventually, they began to share a household, with Brayton formally assisting Young as personal secretary. The two were nearly inseparable from then on. They dined and attended social events together, and generally friends and acquaintances knew them as companions.[15]

Young continued to serve as assistant superintendent of the Chicago schools for a decade. During that time she traveled widely, spoke before enthusiastic audiences of teachers, developed a deeper and broader understanding of education, and generally acquired a national reputation for excellence. In her early fifties, she even found time to take graduate courses at the newly established University of Chicago, where she studied educational theory and philosophy with the young John Dewey. By the end of the century, however, her happiness in the assistant superintendency ended when district reorganization threatened to increase the hierarchical structure of schoolwork with teachers losing powers that she considered essential. She decided to resign in 1899, much to the chagrin of thousands of Chicago teachers and community members who then launched petitions on her behalf. She used her newfound freedom productively, though, writing a dissertation with Dewey's direction. Within a year she completed the work, *Isolation in the Schools,* which the University of Chicago Press subsequently published. Even before she had finished her degree, the University of Chicago asked her to join the faculty, but she refused until she felt that she had fully completed her doctoral degree. In 1900, when she became Dr. Young, though, she relented and quickly became one of the more popular members of the university faculty. Many teachers and other associates who had known her during her career in the Chicago schools eagerly took her courses in the Department of Psychology, Philosophy, and Pedagogy. Her great ability to attract students was not lost on the University of Chicago's ambitious president, William Rainey Harper.[16] By 1904, however, Young resigned her position at the university after having encountered irresolvable differences over the administration of the laboratory school—a conflict that may have put her at odds with Dewey.[17] She spent the next year traveling in Europe and studying the schools of other countries.[18]

Somewhat refreshed and full of ideas about other nations' schools, Young returned to Chicago in 1905. Immediately after her arrival, she was offered the principalship of the Chicago Normal School, a position that she accepted. Young then began implementing changes to improve the quality of teachers who graduated from the program. She designed a more careful screening process for students entering the program. She made sure that the quality of the faculty who taught in the school was high, with each instructor having outstanding teaching experiences, academic skill, discipline, willingness to work hard, and proven talent in training new teachers. During school time she established regular discussions in which faculty could make important decisions about the curriculum and governance of the school. She instituted programs to enhance the manual and industrial arts programs, areas about which she reportedly was quite expert, perhaps from having watched and helped her father so long ago. She hosted many social and professional events at her house for all of the teachers in her school. And among many other remarkable accomplishments, she lobbied diligently for better facilities. Although Young's previous positions in the Chicago schools brought her a large measure of attention and respect, her work at the Normal School made her fine executive qualities abundantly clear, even to her detractors. William Owen, one of her colleagues, explained, "She inspired fear and awe. She created admiration and evoked unquestioning approval. . . . She was a leader. She did not seek leadership, it came to her."[19]

And so, leadership came to her yet again in 1909. Edwin G. Cooley, the superintendent who had been caught in political battles with the school board and the Chicago Teachers' Federation, stepped down after complaining of poor health.[20] The school board, however, was divided on the matter of who should replace him. Factions each preferred different male candidates, and even after much negotiation, the sides came to an impasse. Someone suggested that Ella Flagg Young be nominated because she had provided many years of distinguished service to the district; however, the *Chicago Daily News* reported, "members of the board who do not look with favor on the idea of a woman for superintendent are talking against Mrs. Young under their breath."[21] After considering a large field of candidates and subjecting them to grueling interviews, Young was chosen unanimously on the second ballot. Young had surprised some school board members with her incisive, intelligent, and practical answers to their pointed questions. In particular, she offered great skill in dealing with the Chicago Teachers' Federation, with whom she had worked closely in the past. In the end, the board regarded her skills and experiences as unsurpassed. Perhaps even more surprising, she was offered the same salary as her predecessor—$10,000—making her then the highest paid woman in the country![22]

Young's selection for the superintendency was remarkable in many ways. Clearly, the Chicago School Board included a number of members who either strongly preferred male candidates or who otherwise believed that women should not hold such positions. That Young could prevail nonetheless attests to the extent of her abilities. There were other women superintendents around the country then, too. At the time, hundreds of women in the West and Midwest held county superintendencies, which generally were elected positions. The strengthening suffrage movement easily tapped into the eager support of women who sought to promote women candidates for public positions. The county superintendency was one such position for which women candidates fared well. In fact, suffrage activists often touted the rising numbers of women county superintendents as evidence of women's growing political clout around the nation. City superintendents, however, tended to be selected by school board members, almost all of whom were white, middle- or upper-class, Protestant men. Such board members typically chose men most like themselves.[23] When Young assumed the Chicago superintendency, however; she signaled that the time had come for women to begin leading city school systems—not just county school districts. She contended that her selection marked a turning point in women's power in public schools.[24] Other women teachers around the country agreed and took heart with Young's promotion. Grace Strachan, a prominent New York school superintendent, who also presided over the Interborough Association of Women Teachers, declared in 1910 that "there is absolutely no reason why women should not occupy the executive as well as other positions in the educational system. Chicago proved itself more progressive than New York, when it placed a woman at the head of its immense school system—because she was the fittest one for the place. . . . Fitness should be the only requisite of choice."[25]

For the next three years, Ella Flagg Young enjoyed a period of unprecedented cooperation from the school board, teachers, and the Chicago community. She was able to enact reforms almost immediately. For one, she reorganized the administrative structure of the system, changing many of the positions in the process. She decentralized some of the key system-level functions, such as coordinating substitute teachers. She also insisted that assistant and district superintendents spend more time in the schools. The principalship, which she regarded as highly important, was a position that she endeavored to change from one of rigid accounting and paperwork to one requiring a deep knowledge of educational affairs. Among her many other duties, she juggled district boundaries, worked to upgrade facilities throughout the system, led efforts to coordinate and integrate the curriculum, campaigned for improved and more appropriate teacher preparation, added deans to schools to help counsel students—a remarkable innovation at the time—instituted sex hygiene courses—among the first to be offered in schools anywhere—[26] and worked to enhance the industrial courses available to both girls and boys. Perhaps one of her most vital contributions was the creation of teachers councils, in which teachers collectively made decisions about their working conditions, including curriculum and administrative decisions. She made sure that time was built into the school calendar for teachers to meet regularly in these councils. Interestingly, Young had fleshed out the structure and function of such councils almost a decade earlier in her book, *Isolation in the Schools*. The superintendency offered her the opportunity to bring this remarkably democratic idea to fruition.[27]

Because of the degree to which she brought far-reaching reforms to the Chicago schools while maintaining high morale system-wide, Young's fine reputation spread quickly around the country. She certainly caught the attention of women teachers who eagerly prodded her to seek the presidency of the NEA. The time was right, they reasoned, for an eminent woman to lead the nation's teachers. Though the established nominating committee first forwarded a male candidate, the NEA's overwhelming numbers of suffrage-minded women pressed for Young's nomination. The committee relented and placed her name on the ballot. Then, when an unprecedented high turnout of women teachers materialized at the 1910 national meeting, all demanding that their voting rights be honored, Young won convincingly.

With such powerful support, Young felt emboldened to tackle controversial association business.[28] For example, she discovered that a prominent member of the NEA old guard had invested membership dues in questionable equities. She began to investigate, but he retaliated first by trying to overturn the results of her election and, second, by sending one of his associates to threaten her life. Despite this sobering encounter, eventually Young won the battle over how association dues were invested.[29]

When Young's term as NEA president ended, she continued with her duties as superintendent of the Chicago schools. In time, however, her astonishing record of uniting the community behind sweeping changes would draw to a close. By 1913, the growing conservatism of the school board manifested in increasingly contentious meetings. The cooperative relationship that Young had enjoyed with the school board began disintegrating. The suffrage movement played at least a role in this shift since 1913 was also the year that Illinois women won the right to vote, which in turn triggered a political backlash movement. The powerful political players behind this backlash pressured the mayor to appoint school board members who would resist Young's initiatives and, in turn, those of newly enfranchised Chicago women. One issue that women voters typically held dear was the elimination of corruption in government. Several Chicago school board members allegedly had engaged in several different suspicious, if not illegal, practices and, therefore, were ripe targets for investigation. Some board members regularly accepted kickbacks from publishers seeking exclusive deals to sell textbooks to the Chicago schools. Confronting this practice directly, Young campaigned to select books instead by their quality. A few board members sought to abuse their control of school land by seeking to sell it for personal speculative business ventures. Young intervened once again because she believed that such misappropriated resources not only were selfishly used, but also could have been employed to boost teachers' meager salaries. Her attention to these issues antagonized the board members who otherwise stood to profit. In retaliation, they sought her removal. Young resigned later that year because she believed that she lacked the full support of the board.

Young's avid supporters would not let her go quietly. They mobilized to win back the board's full support of Young's superintendency. Because school board members feared the collective power of women voters, they reconsidered their actions. Soon, they asked her to return to the superintendency—and with strong backing. One of her supporters, Mrs. George Bass, described how some school board members both feared and resented Young's influence in the schools—much as they also feared and resented enfranchised women:

> [T]here was also opposition on the part of the board because of the increasing interest which the women of Chicago took in the schools over which Mrs. Young was placed and they disliked their increasingly definite knowledge of school affairs. It is true that the women of my city have followed every phase of the school situation during the past few years. A proof of their interest was seen when about two years ago, Mrs. Young was put out of her position. The women rose up, held a mass meeting, and packed the great auditorium, calling on the mayor to replace Mrs. Young. There is no need of going through the subsequent events. It is sufficient to say that she was replaced [reinstated]. The bitter feeling of the board increased after this until the opposition to Mrs. Young became open on the part of some of the members who said they would like to get rid of the entire influence of the women.[30]

The women who "rose up" were led in part by Jane Addams, an effective political and, of course, social activist. Their efforts were rewarded when the school board reinstated Young to the superintendency in 1913. Her troubles were now over temporarily, but bitterness remained. One critic explained what had happened: "Four thousand of these teachers, bound to put Feminism and fads on top, met and mutinied and demanded that the School Board should bow down and do their will, and to rescind the election of the man and elect the woman. They squawked and screeched and threatened the Mayor and all his friends with defeat at the polls. As they had an influence on votes, the Mayor obeyed these mutineers."[31] Such hostility found full expression among school board members over the next two years. In 1915, however, Young decided that she had endured enough, and again she stepped down. This time, her supporters could not persuade her to return.[32] Respecting her decision, they turned instead toward staging massive community events to celebrate her accomplishments on behalf of the Chicago schools.[33]

Young decided not to linger. She and Brayton quickly left Chicago and began three years of travel around California. When the United States entered the Great War, Young devoted herself to championing war bonds. She and Brayton traveled widely in these efforts. However, in 1918 as the great influenza pandemic spread across the globe, Young contracted the ailment and died on October 26. She left Brayton all of her personal property, $12,000 in bonds, and two-fifths of the interest from her estate—then deemed a sizeable bequest. The interest from the remainder of her estate was divided among three other trusted women friends. After their deaths, she stipulated that the interest from the whole estate be given to the Chicago Board of Education, the Education Department of the Chicago Woman's Club, the Mary Thompson Hospital, the Chicago Public Library, and the Art Institute of Chicago.[34] Brayton then moved to New York, where she resumed her career as an English teacher before her own death in 1939.

YOUNG'S EDUCATIONAL PHILOSOPHY AND ACCOMPLISHMENTS

An important reason that Young was able to exert such powerful and effective leadership in the Chicago schools is that she had spent many years in disciplined pursuit of professional and personal growth. When she began her teaching career, she devised the meticulous practice of spending three nights a week studying pedagogy, educational philosophy, and other personally important topics. She allowed herself three other nights each week to attend social events and meet people. Sunday, she reserved for church. Young apparently maintained such dogged discipline throughout her life, as McManis explains, "Asked what element of strength lies at the foundation of her success in life, Mrs. Young replies, 'systematic work.' All her life has been molded by continuous application to definite lines of work, not in a haphazard fashion, but in a carefully prepared plan rigidly adhered to from the beginning. Few people have been able to stick to a program more consistently than she has."[35] One of Young's biographers has speculated that Young may have pursued such disciplined study because she might have wanted to deepen her knowledge and understanding to compensate for her own inadequate educational preparation during her earlier years.[36] On the other hand, one admirer simply contended that, "she devotes her evenings to study, that she may be fitted to any position open to her desire."[37]

Her discipline paid off in many ways. During her fifties, Young began to publish her thoughts. In fairly short order, she wrote a number of articles on wide-ranging topics including Horace Mann's pedagogy, elementary school literature, Jane Addams' ideas about peace, the scientific study of education, industrial education, the practice of ethics in schools, secular schools, and John Dewey's *Democracy and Education.*[38] Her books included *Scientific Method in Education* (1903), *Ethics in the School* (1902), *Some Types of*

Educational Theory (1902), *Isolation in the School* (1900), and a coedited work, *Young and Field Literary Readers* (1916).

Perhaps the most significant of these works is Young's published doctoral dissertation, *Isolation in the Schools.* In *Isolation,* Young offers a formidable analysis of the connections between a rigidly compartmentalized educational system and alienated students who fail to find their class work personally meaningful. Young signaled her strong interest in this topic in a brief 1896 piece, "Isolation in School Systems."[39] She developed her ideas considerably during her doctoral studies with Dewey at the University of Chicago. He was so pleased with this and her other work that he arranged for her to publish a series of monographs with the University of Chicago Press, including *Isolation.*[40]

In *Isolation,* Young argued that schools had become mechanized, differentiated institutions that essentially robbed people of their humanity and separated them from their intelligence. This differentiation appeared in many different forms. First, schools had become strictly grade specific, with separate kindergartens, elementary and secondary schools, colleges, and normal schools. Within schools, students attended classes divided by chronological age, with content neatly divided into discrete, strictly defined courses. Students and teachers alike increasingly had been stripped of their capacity to make meaningful decisions about their daily conditions or their assigned tasks. She explained, "The parts have been brought together mechanically, thus making the accepted conception of this great social institution that of an aggregation of independent units, rather than that of an organization whose successful operation depends upon a clearly recognized interrelation, as well as distinction, between its various members and their particular duties."[41]

Young described the process by which this differentiation occurred. She explained that like the supervisors in industry, a new class of administrative personnel in schools was determined to make all decisions for the people who were lower in the increasingly hierarchical structure. This effectively made teachers and students mere operatives in a larger mechanical process with a curriculum that "is mapped out in minute detail, and the time to be devoted to each part, the order in which the steps are to be taken, and even the methods of teaching, are definitely and authoritatively prescribed. As a result the teacher is not free to teach according to his 'conscience and power,' but his high office is degraded to the grinding of prescribed grists, in prescribed quantities, and with prescribed fineness—to the turning of the crank of a revolving mechanism." She decried this growth of the "business of supervision" in schools that demanded uniformity and obedience among teachers and students but not in reciprocal fashion of school supervisors. She attributed this to "the desire of the strong administrative character to guide others rather than to be in the treadmill." Because of the way that administrative layers had been added to the schools, she argued that the highest levels of administrators were isolated from the realities of the schools, understanding less and less of the true problems that arose. Essentially, the more insulated—and isolated—they were, the less effective they could be in managing the school enterprise.[42]

Young maintained that teachers, and in turn students, needed to have much more power in running schools. She believed that it was only when people learned to make decisions together about the conditions in which they lived and worked that they would begin to tap their natural intelligence more fully and regain their humanity and a sense of personal meaning. For teachers to have this power, they would need to have the freedom to create their own ideas and then execute them. First, though, they would need to see more clearly the oppressive conditions in which they toiled.

> In cities where the teaching corps has become aroused to the evils ensuing from a differentiation that means isolation, there are greater possibilities of a healthful readjustment in the organization than in those where the tension is not definitely recognized, for the members are reaching that point of view from which they see that it is not liberty in carrying out, it is freedom and responsibility in origination also, that will make the whole corps a force, a power in itself. To predicate freedom for teachers in the superintendent's position, or for teachers in the principal's or the supervisor's position, is not sufficient to establish freedom as an essential; it must be predicated for all teachers. To prove that some cannot teach unless they possess freedom is not enough; it must be predicated that freedom belongs to that form of activity which characterizes the teacher. The schools will be purged of the uncultured, non-progressive element, the fetters that bind the thoughtful and progressive will be stricken off, when the work is based on an intelligent understanding of the truth that freedom is an essential of that form of activity known as the teacher.[43]

Although the possession of this freedom and responsibility might prove to be a heavy burden for some teachers, Young argued that it would attract the most talented and qualified individuals to teaching while repelling those who might prefer the rote dreariness of an oppressive system. Indeed, she worried that it would not be enough for administrators merely to dribble some limited powers down to teachers, but instead she believed that teachers needed to possess real power from the start. This power required that schools provide time and space for teachers to engage in the intellectual, legislative, and logistical functions of running their schools. This, in turn, would be part of and consistent with a larger social system in which students, parents, administrators, and community members participated similarly in a meaningful democratic process to govern the schools wisely. Short of this, schools would devolve into automated units that slotted students into narrowly defined roles in society, effectively killing their capacity for responsible and free thought and robbing them of their humanity.

Not surprisingly, few other administrators of her time shared her vision of how schools should work. For one thing, they would have needed to come to terms with relinquishing some of their power, which most believed was meager enough. It would have required the mostly male ranks of school administrators to work much more closely with a primarily female cadre of teachers. Because men typically earned and maintained their sense of manhood through their association with other men, and because the masculinity of male workers in a supposedly "woman's profession" was suspect already, this prospect may have been horrifying.[44] Finally, though, Young's vision of how the schools should work would have entailed that all members of the school community engage in the difficult and often messy work of sustained intellectual deliberation.

With *Isolation,* Young mapped out this challenging, but ultimately more liberating vision of schooling as part of a larger democratic society. It was based on her years of significant service to the Chicago schools, the power of her ideals, and her disciplined study of schooling, philosophy, and society. Like much of her other work, it offered philosophical reasoning tempered by careful analysis of her practical experience. Essentially, her work demonstrated a remarkable balance of theory and practice—a balance that also characterized her leadership and her larger life. It would seem that she truly endeavored to minimize the isolating elements in her own practice and instead seek integration through careful deliberation. In fact, years later when she became superintendent, she managed to put many of her ideas in *Isolation* into concrete action. This deep commitment to and success in combining theory and practice alone distinguishes her life and work as among the very greatest of American educators.

YOUNG'S LEADERSHIP AND LEGACY

During the decades following Young's death, the Chicago schools faced persistent corruption, times of poor management, difficult labor relations, and a host of other problems, many of which are well-documented in Mary Herrick's fascinating volume, *The Chicago Schools: A Social and Political History* (1971).[45] Eventually, most of Young's reforms disappeared. The Ella Flagg Young clubs, where teachers and administrators avidly discussed educational ideas, died out as antagonism between teachers and administrators increased. Teacher councils, in which teachers played major roles in governing their schools, were replaced with strictly hierarchical administrative systems, in which principals told teachers what to do. The ranks of women teachers who had supported Young so earnestly found their pay reduced through the years as their working conditions worsened.

Despite these dismal changes, however, much of Young's legacy endures. Many of the teachers, administrators, and students with whom Young worked eventually became educational leaders, scholars, and teachers of teachers. In the course of her association with normal schools in the state, she exerted a profound influence on how teachers would learn their craft. As a direct consequence of her presidency of the NEA, prevailing practice over the decades to follow would mandate that women and men alternate in presidential service, and that teachers and administrators alike would lead the association. Elsewhere, teachers were somewhat emboldened to believe they should share in the power of running the schools, even if they needed to organize and force the matter.

Finally, it is important to note that Ella Flagg Young accomplished much of her life's work in close collaboration with organized groups of women. Suffragists and others involved with the broad women's club movement saw in Young a fine leader whose public successes reflected well on all women. She participated in local suffrage endeavors and actively served in Chicago women's clubs. Diverse women around Chicago, especially teachers, catapulted her into some of her leadership positions. They also supported and comforted her during the dark moments of her career. In turn, Young championed the causes of organized women's groups. She also sought to lead in a manner that was compatible with her beliefs: She involved those around her in making critically important decisions; she structured time and opportunities so that her constituents could discuss and carry out their plans; and she engaged in her work with a spirit of community building. Although this manner of leadership often proved difficult and time-consuming, she sought it as a matter of course. As a result, her relationships with women's communities and the general public were intense, mutual, and ongoing. Without such strong support, she could not possibly have attempted some of the daring programs that she introduced and usually managed to enact. On the other hand, her supporters also probably compelled her to attempt bold, risky innovations or challenges. The mutuality of her relationships with her supporters was so solid that a clear cause and effect probably cannot be determined. Clearly, though, Young was as much a part of the women's movement as she was a symbol of its finest successes.

NOTES

1. Jane M. Dewey, "Biography of John Dewey," *The Philosophy of John Dewey,* ed. Paul A. Schilpp (New York: Tudor Publishing Co., 1951), p. 29.
2. "Necrology," *Journal of Proceedings and Addresses of the National Education Association for the Year 1918,* p. 685. Washington, D.C.: The Association, 1918.

3. William B. Owen, "Editorial," *Educational Bi-Monthly* 10 (December 1915): 185–186.
4. John T. McManis, *Ella Flagg Young and a Half-Century of the Chicago Public Schools* (Chicago: A.C. McClurg & Co., 1916), p. i.
5. McManis, *Ella Flagg Young,* p. 224.
6. Helen Christine Bennett, "Ella Flagg Young," in *American Women in Civic Work* (New York: Dodd, Mead and Company, 1915), p. 270.
7. McManis, *Ella Flagg Young,* pp. 15–18; Joan K. Smith, *Ella Flagg Young: Portrait of a Leader* (Ames, Iowa: Educational Studies Press, 1979), pp. 4–5.
8. McManis, *Ella Flagg Young,* p. 15.
9. Rosemary V. Donatelli, "The Contributions of Ella Flagg Young to the Educational Enterprise" (Ph.D. diss., University of Chicago, 1971), pp. 66–67.
10. McManis, *Ella Flagg Young,* p. 20.
11. Ibid., pp. 22–23.
12. Smith, *Ella Flagg Young Portrait,* pp. 11–14.
13. Ibid., pp. 20–22; Donatelli, "Contributions," pp. 68–69; and McManis, *Ella Flagg Young,* pp. 44–49.
14. Smith, *Ella Flagg Young Portrait,* pp. 23–27; McManis, *Ella Flagg Young,* p. 7; and Donatelli, "Contributions," p. 69.
15. Smith, *Ella Flagg Young Portrait,* pp. 28–36; McManis, *Ella Flagg Young,* pp. 56–72; and Donatelli, "Contributions," pp. 101–02.
16. Kathleen Cruikshank, "In Dewey's Shadow: Julia Bulkley and the University of Chicago Department of Pedagogy, 1895–1900," *History of Education Quarterly* 38, no. 4 (Winter 1998): 373–406.
17. Ellen Condliffe Lagemann, "Experimenting with Education: John Dewey and Ella Flagg Young at the University of Chicago," *American Journal of Education* 104 (May 1996): 180.
18. Donatelli, "Contributions," pp. 136–200; McManis, *Ella Flagg Young,* pp. 101–22; and Smith, *Ella Flagg Young Portrait,* pp. 60–100.
19. William B. Owen, "Editorial," *The Educational Bi-Monthly,* 4 (October 1909): 79.
20. Smith describes how Cooley then accepted a position with the textbook company, D.C. Heath. This move apparently was not a surprise because he was rumored to have cozy connections with textbook publishers for years. See *Ella Flagg Young Portrait,* pp. 140–141.
21. 26 July, 1909, p. 3, cited in Donatelli, "Contributions," p. 281.
22. Donatelli, "Contributions," pp. 279–85; Smith, *Ella Flagg Young Portrait,* pp. 148–53.
23. Jackie M. Blount, *Destined to Rule the Schools: Women and the Superintendency, 1873–1995* (Albany, N.Y.: SUNY Press, 1998).
24. Young's acceptance speech captures the sense of women's political progress and the rightness of their assuming leadership of schools: "Women are destined to rule the schools of every city. I look for a large majority of the big cities to follow the lead of Chicago in choosing a woman for superintendent. In the near future we will have more women than men in executive charge of the vast educational system. It is woman's natural field, and she is no longer satisfied to do the greatest part of the work and yet be denied leadership." From "The Highest Salaried Woman in the World," *Western Journal of Education,* 14 (1909): 15.
25. Cited in David Tyack and Elisabeth Hansot, *Managers of Virtue: Public School Leadership in America, 1820–1980* (New York: Basic Books, 1982), p. 185.
26. Jeffrey Moran, "'Modernism Gone Mad': Sex Education Comes to Chicago, 1913," *Journal of American History* (September 1996): 481–513.
27. Donatelli, "Contributions," pp. 340–429.
28. McManis, *Ella Flagg Young,* pp. 144–155; Smith, *Ella Flagg Young Portrait,* pp. 158–181; and Mary Herrick, *The Chicago Schools: A Social and Political History* (Beverly Hills, Calif.: SAGE Publications, 1971), pp. 114–137. For further discussion of the male cul-

ture of the NEA during Young's life, see David Tyack, *The One Best System: A History of American Urban Education* (Cambridge, Mass.: Harvard University Press, 1974), pp. 264–267; and Marjorie Murphy, *Blackboard Unions: The AFT and the NEA, 1900–1980* (Ithaca, NY: Cornell University Press, 1990), especially pp. 61–79.

29. Smith, *Ella Flagg Young Portrait,* pp. 164–173.
30. Mrs. George Bass, "Mrs. Young and the Chicago Schools," *School and Society* 2 (23 October 1915): 605–606.
31. B. V. Hubbard, *Socialism, Feminism, and Suffragism, the Terrible Triplets Connected by the Same Umbilical Cord and Fed from the Same Nursing Bottle* (Chicago: American Publishing, 1915), pp. 282–284.
32. George Herbert Mead, "A Heckling School Board and an Educational Stateswoman," *Survey* 31 (1914): 443–444.
33. Smith, *Ella Flagg Young Portrait,* pp. 183–231.
34. Ibid., p. 231.
35. McManis, *Ella Flagg Young,* p. 29.
36. Donatelli, "Contributions," p. 82.
37. Josiah L. Pickard, "Introduction of President Young," *Journal of Proceedings and Addresses of the National Education Association for the Year 1911* (Washington, D.C.: The Association, 1911): 74.
38. See, for example, the following works by Ella Flagg Young: "A Review of John Dewey's *Democracy and Education,*" *Journal of Education* 84 (6 July, 1916): 5–6; "The Secular Free Schools," *Journal of Proceedings and Addresses of the NEA,* Fifty-Fourth Meeting (1916): 63–68; "Children Should Have Industrial Education Unless They Go to College," *Journal of Education* 74 (16 November 1911): 514; "The School and the Practice of Ethics," *Journal of Proceedings and Addresses of the NEA,* Forty-Sixth Meeting (1908): 102–08; "Recognition of Scientific Studies," *Journal of Education* 66 (26 September, 1907): 315; "A Review of Jane Adam's Newer Ideals of Peace," *The Educational Bi-Monthly* I (February, 1907): 289–91; "Literature in the Elementary School," *Journal of Proceedings and Addresses of the NEA,* Thirty-Fourth Meeting (1896): 111–17; and "The Pedagogical Side of the Writings of Horace Mann," *University Record* I, 6 (8 May, 1896): 113–15.
39. Ella Flagg Young, "Isolation in School Systems," *University Record* 1, no. 1 (3 April, 1896): 422.
40. Ella Flagg Young, *Isolation in the Schools* (Chicago: University of Chicago Press, 1900).
41. Ibid., p. 7.
42. Ibid., pp. 12–14.
43. Ibid., pp. 16–17.
44. Jackie Blount, "Manliness and the Construction of Men's and Women's Work in Schools, 1865–1941," *International Journal of Educational Leadership* 2, no. 2 (1999): 55–68.
45. Herrick, *Chicago Schools.*

REFERENCES

Bass, Mrs. George. "Mrs. Young and the Chicago Schools." *School and Society* 2 (23 October 1915): 605–606.

Bennett, Helen Christine. "Ella Flagg Young." In *American Women in Civic Work.* New York: Dodd, Mead and Company, 1915, pp. 255–277.

Blount, Jackie M. *Destined to Rule the Schools: Women and the Superintendency, 1873–1995.* Albany: State University of New York Press, 1998.

———. "Manliness and the Construction of Men's and Women's Work in Schools, 1865–1941." *International Journal of Educational Leadership* 2(2) (1999): 55–68.

Candeloro, Dominic. "The Chicago School Board Crisis of 1907." *Journal of the Illinois State Historical Society* 68(5) (1975): 396–406.

Donatelli, Rosemary V. "The Contributions of Ella Flagg Young to the Educational Enterprise" Ph.D. diss.: University of Chicago, 1971.

"Ella Flagg Young, LL. D." *Journal of Education* (16 June 1910): 686–689.

Herrick, Mary J. *The Chicago Schools: A Social and Political History.* Beverly Hills, Calif: SAGE Publications, 1971.

"The Highest Salaried Woman in the World," *The Western Journal of Education* 14 (1909): 515.

Hubbard, B. V. *Socialism, Feminism, and Suffragism, the Terrible Triplets Connected by the Same Umbilical Cord and Fed from the Same Nursing Bottle.* Chicago: American Publishing, 1915.

Lagemann, Ellen Condliffe. "Experimenting with Education: John Dewey and Ella Flagg Young at the University of Chicago." *American Journal of Education* 104(3) (1996): 171–185.

McManis, John T. *Ella Flagg Young and a Half-Century of the Chicago Public Schools.* Chicago: A.C. McClurg & Co, 1916.

Mead, George Herbert. "A Heckling School Board and an Educational Stateswoman." *Survey* 31 (1916): 443–444.

Moran, Jeffrey P. "*Modernism Gone Mad": Sex Education Comes to Chicago, 1913."* Journal of American History 83(2) (1996): 481–513.

Murphy, Marjorie. *Blackboard Unions: The AFT and the NEA, 1900–1980.* Ithaca N.Y.: Cornell University Press, 1990.

"Necrology." *Journal of Proceedings and Addresses of the National Education Association for the Year 1918,* p. 685. Washington, D.C.: The Association, 1918.

Owen, William B. "Editorial." *Educational Bi-Monthly* 10 (December 1915): 185–186.

Pickard, Josiah L. "Introduction of President Young." *Journal of Proceedings and Addresses of the National Education Association for the Year 1911,* p. 74. Washington, D.C.: The Association, 1911.

Smith, Joan K. *Ella Flagg Young: Portrait of a Leader.* Ames, Iowa: Educational Studies Press, 1979.

Tyack, David. *The One Best System: A History of American Urban Education.* Cambridge, Mass.: Harvard University Press, 1974.

Tyack, David and Hansot, Elisabeth. *Managers of Virtue: Public School Leadership in America, 1820–1980.* New York: Basic Books, 1982.

Young, Ella Flagg. "Isolation in the Schools" Ph.D. diss.: University of Chicago, 1900.

———. *Ethics in the School.* Chicago: University of Chicago Press, 1902.

———. *Some Types of Modern Educational Theory.* Chicago: University of Chicago Press, 1902.

———. *Scientific Method in Education.* Chicago: University of Chicago Press, 1903.

———. "A Reply." *Addresses and Proceedings of the National Educational Association,* 1916, pp. 356–359.

Young, Ella Flagg and Field, Henry, eds. *Young and Field Literary Readers.* Boston: Ginn and Co, 1916.

CHAPTER ELEVEN

LAURA BRAGG AND THE CHARLESTON MUSEUM*

LOUISE ANDERSON ALLEN

At the end of *Gone with the Wind,* Rhett told Scarlett that he was returning to Charleston, where there was a little bit of charm and grace left in the world. Forty years later, novelist Henry James saw the city as "a place of ruins and widows—of gardens and absolutely of no men—or of so few that, save for the general sweetness, the War might still have been raging and all the manhood at the front." Should either man have sailed into Charleston Harbor with Laura Bragg in September of 1909, he would have found the city little changed, for although it possessed some grace and charm, there was little else of appeal.[1]

Bragg recalled being on a Clyde steamer, the Arapahoe, when she saw this southern town in the autumn light. Landing just after lunch on September 1, she found the city bathed in sunlight and the crape myrtle in bloom. And the city's air was probably a "melange of the odors of heavy salt, pluff mud" and the above-water-level sewer drains. These smells "befouled the air" inasmuch as the primary means of human waste disposal in the city were some 12,000 privies whose contents leeched into the soil. Charleston's narrow streets were a patchwork of mostly dirt roadways, and only about one-third were paved in granite, brick, or asphalt. Some of the streets in the oldest section of the city were still made of gravel, oyster shells, cinders, or cobblestones. Certainly, it was a place in which time had stood still—a city almost untouched by modernization. A travel writer visiting the city at the time declared, "Charleston is perhaps the only city in America that has slammed its front door in Progress' face and resisted the modern with fiery determination."[2]

In recalling the large mansions built along the seawall of Charleston's High Battery, Bragg remarked in an interview, "There wasn't a gallon of white paint in the city. There was a shabbiness about it that was simply charming." Charlestonians were said to be "too poor to paint and too proud to whitewash." Fortunately, the homes were built of heart pine and cypress and had survived the Civil War, the hurricanes, an earthquake, and the Great Fire of the century just past. The city that Bragg saw had also endured, but as a "static industrial backwater since the Civil War." She also saw an adventure and an opportunity for new experiences. "The obvious ignorance was divine. They needed her. Like young Alexander the Great before her, [she] saw worlds to conquer," a local reporter wrote after an interview with Bragg when she was 74.[3]

Bragg had been hired by the Charleston Museum's director, Dr. Paul Rea, to be the librarian in the oldest museum in the Western Hemisphere. Initially founded by the sons of the planter aristocracy in 1773, this scientific and natural history museum, located in the main building of the College of Charleston, was used only by the educated upper classes of the city. When Bragg arrived, one of her first duties was to move the collections of plants, bones, shells, and books four blocks away to the Thompson Auditorium, which had 35,000 square feet of floor space. Here, she was to create new exhibits for these collections that had been stored away and unused for a number of years because there was little display area at the college.[4]

Fundamentally understood to be educational institutions, early museums in other regions of the country had been open to those who did not have the benefit of an extended education. Through objects and pictures, museums could teach those who were not skilled in teaching themselves. Then, around 1860, specialized societies restricted public access to museums, and education was abandoned as a museum function. Rather, their emphasis became entertainment. It would not be until the beginning of the Progressive Era before museums would once again become "necessary institutions in social education." During this time, the primary motivation for moving the museum became a need to instill literacy in the growing numbers of immigrants crowding American cities. As social institutions, museums assumed the task of uniting society as a common place where the social classes could meet on a level ground. As museums in the Northeast embraced this role, they experienced a reawakening—one that would eventually extend to the Charleston Museum.[5]

A native of Massachusetts, Bragg traveled back to her family's New England home each summer. During these trips, she visited other museums, such as the Boston Museum of Natural History and the Metropolitan Museum of Art in New York City. There she learned how to make her museum live. Instead of being mausoleums for the collections of inanimate objects, Bragg came to understand from her visits north that museums must seek to educate the public through their well-thought-out exhibits. Bragg's new installations at the Charleston Museum, with original and illustrative labels, transformed what had been a scientific, natural history museum for the academic elite into a public institution.[6]

Although the museum was changing with a new employee, building, and exhibits, the city did not embrace progress easily. It was one of the last places in the South where business still began after ten o'clock in the morning in the financial and legal district along Broad Street. Families with names like Middleton, Pinckney, Rutledge, Manigault, Ravenel, Gibbes, Waring, and Heyward continued to dominate the economic, social, and political elite, just as they had in previous generations. The Civil War, Reconstruction, and financial disasters had not diminished their control over local affairs. Although the advent of progressivism had shaken the system, this group of white men formed political alliances with the few newcomers who also "recognized family name, gentlemanly behavior, and a South-of-Broad address as minimum requirements for playing the game."[7]

With much of the planter aristocracy still in power and overtly thwarting economic growth, the city had gained a reputation as an anti-industrial area. The gracious and charming southern gentlemen literally scared off new people with new money and new ideas, thus saving the city and its citizens from anything that might have challenged the prevailing social and racial mores of the ruling class. Whereas other cities of the New South were infused with new money, new ideas, and new people, Charleston proudly embraced its "genteel poverty," and society became more inflexible, inward looking, and provincial. There were few new jobs; most women still stood mute on the pedestal; and blacks, kept functionally illiterate by white supremacists, provided cheap and plentiful labor.[8]

The great challenge for the white patriarchy for many years after the war and well into the twentieth century was to maintain its superiority over silenced women and blacks who were subordinated to second-class citizenship. Here, family and not money was the chief determinant of southern identity. Without family connections when she first arrived, Bragg made her place in and among the city's elite initially through her choice of church membership and then through the relationships she nurtured with the younger members of the Junior Natural History Society. Once Bragg moved into the home of the aristocratic Belle Heyward, she had the Heyward family connections to support her efforts at the Charleston Museum. And once she was named director in 1920, she forged alliances with powerful and influential men whom she recognized as beneficial for her goals at the museum.

Bragg's intent as a museum educator and administrator was to create an intellectual revolution for all—black and white, rich and poor, men and women. First as the curator of books and public instruction, and then as the director, Bragg shook the power structure, socially and, in the long view, politically, by providing the same educational services to both black and white city schools beginning in 1913. These services were then extended to all of the county's schools, and the services continued throughout her tenure at the museum until 1931. Bragg flaunted societal conditions in a bolder way when she opened the museum to black patrons in 1921. This was at a time when the white power structure, both in Charleston, and throughout the South, was systematically denying blacks educational and political opportunities. As an outsider and a northern "new woman," Bragg did not know her place, nor did she understand, initially at least, that blacks also had one.

THE SOCIAL REFORMER

Born in Northbridge, Massachusetts, on October 9, 1881, Laura Bragg was the eldest of three children. At age six, just after her younger sister's birth, Bragg contracted scarlet fever, which left her with progressively increasing deafness. Her father, Reverend Lyman Bragg, refused to consider this a handicap and consulted with Alexander Graham Bell's School of Elocution in Boston, seeking a cure or a treatment. Although there was no remedy, Bragg did learn to read lips quite successfully. Her father refused to allow her deafness to impede her intellectual growth, so the rest of the family followed his lead, treating her as if she could hear. Smart, quickly perceptive, insatiably curious, and with a remarkable memory, she responded to her father's teaching and found wonder in his world of books.[9]

Lyman Bragg was educated at a local academy and then a seminary in Vermont, graduating from Middlebury College in 1875. He intended to become a physician; however, he attended Boston University School of Theology, finishing there in 1878 with a master's degree in theology. He was fully ordained as a Methodist minister in 1880. Bragg's mother was the former Sarah Julia Klotz, who attended a music conservatory, and Bragg recalled that her mother was an excellent clergyman's wife.[10]

It is not surprising that her family members were reform-minded, given their education and northern middle-class heritage. Bragg's mother was a member of the Woman's Christian Temperance Union. Her paternal grandmother, for whom she was named, Laura Church Bragg, read voraciously and, at the age of 61 graduated from the Chautauqua Literary and Scientific Circle. And her father joined the American Association of Colored Youth while the family lived in Holly Springs, Mississippi.[11]

Soon after his daughter lost her hearing, Reverend Bragg's health failed in 1890. He then left the pulpit and joined another form of Methodist ministry when he became a mathematics professor at Rust University in Mississippi. Founded by the Freedman's Aid Society of the Methodist Episcopal Church in 1866, it was the first attempt to provide higher education for blacks in the state, though it offered primarily elementary and secondary instruction. Bragg's Methodist group, along with the other societies, believed that it was their responsibility to lend a helping hand to blacks through classical education, which was seen as a means to liberate the black man, thus achieving racial equality.[12] The family remained in Mississippi for two years and then returned home to New England. Although Bragg's siblings did visit her in Charleston, her parents never came back south.

Not only was Bragg influenced by these experiences as a missionary's daughter in the racially divided South, she was also representative of the northern "new woman," a concept that grew out of the Progressive Era. Coming of age at the turn of the century, she was distinguished by her independent spirit. Wanting a life beyond her minister father's home, she eagerly sought a career, first as a librarian and then as a professional museum administrator. Relatively young when she came to the city (she was 28) and unmarried, Bragg dismissed whatever societal conventions were imposed on her. She had more formal education than her mother and southern peers, having earned a degree in library science from Simmons College in its first graduating class of 1906. When she came to Charleston in 1909, she was 1 of fewer than 50,000 women in the country with a college diploma.[13]

Even before she arrived in Charleston, Bragg had asserted her independence in different ways: living hours away from home while she attended college; visiting the museums and art galleries of Boston; and working as a librarian on Orr's Island, Maine, and at the New York City Public Library. Her freedom of movement had been assisted first by the invention and widespread use of the bicycle and then the automobile, which offered her both "masculine adventure and feminine glamour." Bragg was an irreverent figure, who thought her opinions worth listening to, and she disliked household duties. Without the benefit of a chaperone, she enjoyed friendships with both women and men. Smoking cigarettes from a long holder and comfortably having drinks with men, Bragg met all on equal footing. She conversed with them easily and freely on every topic.[14]

Bragg's counterpart in the South, however, was vastly different. The South's new woman was still "surrounded by years of tradition regarding appropriate gender roles and expectations," with the pedestal remaining a stumbling block for many women in the area. Expected to "fill their natural mission in life," women were not meant to enter college, become professionals, or speak in public. In Charleston, most women of a certain class lived a remarkably sheltered existence in 1909; if they worked outside of the home, social mores dictated that a man would accompany them home.[15]

It would take involvement in the woman's suffrage movement before the political (and public) life of southern women changed. However, most southern new women found their power in the public sphere through women's clubs. Through these organizations, they established libraries, expanded schools, investigated labor conditions, secured the passage of child labor laws, organized settlement houses, and even worked for interracial cooperation. Though a generation behind their northern sisters, southern club women were a recognizable political force by 1910. Even the regional press acknowledged their accomplishments by "publicly rubbing their eyes in astonishment."[16]

In her pathfinding work about the Southern lady ideal, Anne Firor Scott found that even during the time of women's political progress, southern women understood that it was still necessary to maintain the demure demeanor that southern men expected if they wanted to secure a hearing before the men in power. Well into the 1920s, the image of the "Southern Lady" (beautiful, gracious, with charming and winning ways, deferential and submissive to men, and churchgoing) survived intact and remained relatively unchanged. One Charleston author, Herbert Ravenel Sass, was still writing as late as 1932 about the "romantic ideal of Woman as being enshrined and set on high." Although maintaining this image was still considered to be good politics in the South, there were a few of the more radical new women, such as Mary Munford in Virginia, Jessie Daniel Ames in Texas, and Sallie Cotten in North Carolina, who fought what one called "chivalric nonsense."[17]

One aristocratic Charlestonian and a contemporary of Bragg's, Susan Pringle Frost, was active in the early club movement of the city. With family ties to the political and cultural power structure in the city, she scandalized them all in 1919 by setting up a real estate office on Broad Street. Most office buildings at that time in Charleston did not even have female toilet facilities. In contrast, by this time, Bragg had developed the Charleston Museum's educational programs for city and county schools, traveled extensively throughout the state and Northeast on museum business, and gained a reputation as "The General." Many Charlestonians and others across the state recognized that it was Bragg, the woman, who directed the museum, while a man had the title.[18] If Sue Frost, a daughter of the Old South, was an embarrassment, how did the social and power elite view Bragg?

Bragg was an anomaly in the South. She was aggressive and showed initiative and determination in her role as a museum administrator. Not knowing that her place was in the home, she acted as if she were the equal of any man. Unmarried, Bragg devoted all of her time and energy to promoting her work with the museum, nationally through her membership in the American Association of Museums; on the state level in her work with the state teachers' association; and locally, through her community involvement with the Charleston Chamber of Commerce and the Civic Club.[19] It was through her association with community leaders in these organizations, along with her ties to the social elite via her partnership with the aristocratic Belle Heyward, that she garnered support for her work at the Charleston Museum.

SOCIETAL CONDITIONS IN CHARLESTON

Emancipation, Appomattox, Reconstruction, and the turn of the century had also not altered to any great degree black participation in Charleston's social, political, and economic structures. Their roles were virtually unchanged from antebellum times, based on the myth of a mutually beneficial relationship with reciprocal responsibilities. Whites still held a sense of noblesse oblige "for the dependent helpless race," whereas "blacks were expected to remain content with their lowly lot in life."[20] There was no pretense of equality, and their shared sense of reciprocity was dependent on both knowing and accepting their prescribed roles in dealing with one another.

Until 1920, blacks formed the majority population of the city. Many of them lived in squalid and insanitary conditions in the dependencies, servants' quarters, and carriage houses that were located in the alleyways or backyards of the mansions of the aristocracy. This housing pattern was still in existence well into 1940 when sociologists Karl and Alma Taeuber conducted a study that found Charleston to be the least racially segregated of 109 American cities. With such close proximity between the races, personal and familiar contact was nurtured. Many white homes still employed black women as live-in domestics, thus continuing a similar paternalistic relationship that had existed before the Civil War. By design, the social elite of Charleston had no opportunity to mix either socially or in the business world with members of the black population and did not seek any.[21]

Most blacks understood that they would be laborers, domestics, or custodians—all jobs not far removed from the antebellum plantation days of slavery. Fearing an educated slave, whites had enacted laws and supported customs that denied blacks an education. Once slaves were freed, and for nearly a half-century after Reconstruction, whites had a greater fear—an educated black. In actuality, schooling became the key to the maintenance of the white race's dominance over blacks. And, as with all other institutions of power, blacks were not permitted active participation in the educational system once the state's 1895 constitution was ratified. Although education offered the poor white masses deliverance from apathy and ignorance, the major problem for the state's politicians was how to make sure that blacks did not share equally in the educational system. South Carolina did not neglect black schooling entirely, though educational inequality was institutionalized long before World War I.[22]

The state government had been under pressure since 1877 to reduce educational expenses, especially in black schools. White citizens distrusted education for several reasons: "It was begun by the Radical Reconstruction government; it implied racial equality; and it might 'spoil' good Negro field hands and improve [their] political potential." By 1895, the state government's systematic financial starvation of the black schools was apparent in the education superintendent's annual reports. For example, the state had raised expenditures per white student to $3.11, at the same time as it reduced expenditures for blacks to $1.05. By 1915, "the ratio of expenditures for the schooling of white and black children was about twelve dollars to one." And by 1920, although well over half of the state's students were black, only 11 percent of the state's education money was spent on them—with white children receiving $26.08 and black children receiving only $3.04.[23]

The result of this blatant discrimination and financially unsound policy was an astounding literacy rate for black South Carolinians. From 1880 through 1930, the illiteracy rate for the country declined from 17 percent to about 4 percent. At the same time, the state's illiteracy rate also declined from 55 percent to about 15 percent. But that rate in 1880 in South Carolina had included more than 78 percent of the black population, which also declined to almost 27 percent by 1930. Although state funding to black schools slowly increased (though not at the same rate as it did for white schools), agencies other than the public school system contributed to this decline of the illiteracy rate. These agencies included Northern philanthropic organizations, such as the Rosenwald Fund and the General Education Board, along with private black schools like Charleston's Avery Normal Institute. Still, it is truly a miracle that the literacy rate for blacks improved any, given their mistreatment by whites.[24]

Without money, education, or training, blacks could not expect to participate in the city's social and political institutions. Although the state's General Assembly enacted a series of laws segregating the races between 1896 and 1906, it was not until 1912 that Charleston's City Council passed its first Jim Crow laws, segregating blacks and whites on Charleston's trolley cars. Other city ordinances followed this one, prescribing absolute segregation in all public places, including the museum. During the Jim Crow era, the two races coexisted, generally peacefully, under the formula arrived at by Charleston's Progressive Era leaders: "Humanity but not equality, economic but not political opportunity."[25]

THE MUSEUM'S EDUCATIONAL PROGRAM AS PROGRESSIVE SOCIAL REFORM

At this time, southern schools were, in the words of Southern historian Vann Woodward, "for the most part miserably supported, poorly attended, wretchedly taught, and wholly inadequate for the education of the people."[26] Woodward's words could have been written about South Carolina, as another historian, William Link, identified the state as one of three Southern states (the others were Alabama and Mississippi) whose educational officials were the most reluctant in modernizing the system in the early 1900s. Even South Carolina's education superintendent understood this, writing in an annual report, "It is a misnomer to say that we have a system of public schools. In the actual working of the great majority of the schools in the state, there is no system, no orderly organization. . . . Each District has as poor schools as its people will tolerate—and in some Districts anything will be tolerated." County and state supervision had just begun by the 1910s for white schools, whereas the custom of state supervisors for the black schools was to shut their eyes to the miserable conditions.[27]

With a poor school system, poorly trained teachers, and a lack of professionalization across the South, middle- and upper-class women, such as Bragg, were provided with a theater for their activities. Thinking that the societal conditions of a rural population were obstacles to social reform, they believed that the best technique in dealing with these issues was through mass education. These women also shared common goals: to continue economic development and material progress while at the same time working toward a "more orderly and cohesive society." But it is important to understand that southern women, or even northern women such as Bragg, who wanted to create a better society, had to confront the pedestal that continued to define their activities while also treading cautiously around racial issues.[28] During her first decade in Charleston, neither issue seemed to present a problem for Bragg.

As a result of these conditions, reforming education across the South took a different route than in the urban North. In the South, education was distinctive because the basic unit of administration remained the local school district. It controlled all facets of education with little adherence to research, standards, or systematic management. For example, in Kentucky there were eight thousand school districts all located in rural towns. And, in Charleston, most of the 22 school districts, with 104 elementary schools, were located in the rural part of the county. Thus, progressives saw towns and their services as modernizing influences; the goal, then, for southern reformers was to reduce local control over schools by increasing state and county supervision.[29]

This was best accomplished by the consolidated school. Bragg understood the importance of this organizational structure in such a rural state and county. Writing to the state superintendent of education, she claimed, "These [traveling] exhibits are particularly important for the smaller isolated schools where the teacher has few helps. In

Charleston County they have proved an important factor in creating the desire for consolidated schools and better equipment."[30]

An educational historian, Lawrence Cremin, has written in his *Transformation of the School: Progressivism in American Education, 1876–1957,* that progressive reformers, such as John Dewey, sought to "democratize intellectual culture to the point where it could be made available to all." Bragg's goals for the museum's educational program meshed with those of Dewey. Like Dewey, Bragg saw the schools and education as agents of socialization. It would be through the school and its teachers that democracy could be promoted and preserved. Here, students were prepared for life by providing instruction in responsible political behavior and training in vocational skills.[31]

In South Carolina, Bragg could do social missionary work by practicing reform through the museum's educational program and the traveling school exhibits. Her educational program was the first of its kind in the South, offering both black and white students and teachers a window on the world, beyond the city, town, or crossroads in which they lived. Her role at the museum had quickly expanded from librarian in 1909 to curator of books and public instruction by 1911. In that two-year time span, she had developed the first educational program in a southern museum. Her report in the December 1912 *Bulletin of The Charleston Museum* outlined the 12 types of assistance given to both public and private schools of the county. These included a lecturer and lecture room, a guide to the general collections of the museum, a reading room, field trips, copies of the Charleston Nature Study Curriculum (developed and written by Bragg), and 25 traveling school exhibits (also called Bragg boxes), each packed in an attractive portable wooden case; the subjects included animal and picture exhibits, and an iron and steel exhibit.[32]

The museum's work with the public schools had so impressed the County Board of School Commissioners that they passed a special resolution in 1914, asking for further extension "of the educational advantages of the Charleston Museum to the city schools" and they sought a formal affiliation between the board with the museum. By 1916, every primary teacher was required to use Bragg's Nature Study Curriculum. The traveling school exhibits were matched to Bragg's curriculum, and the teachers had to bring their students to the museum for classes. Bragg taught summer school for teachers at the museum on home geography, nature study, and local history as well, with credit given as if the course were provided by a normal school or a university summer school. Another indicator of Bragg's success with the education program was how the daily newspaper reported her work at the Charleston Museum, calling it a university and her its head officer. The museum and Bragg had found their way not only into the school system but also into the public consciousness.[33]

Bragg learned, through her development of Bragg boxes, that if the public would not or could not come to the museum, then she could make patrons aware of what the museum had to offer them. The exhibits represented a method of social reform, educating those who might not otherwise be exposed to the information contained within them. Bragg boxes delivered a service (education) and, in turn, alerted potential patrons in the community and in the rural areas to the value of the museum and the other services it offered. It was with this educational extension work that the museum first touched the lives of black and white students alike in Charleston, encouraging them to visit the museum.[34]

Sent out on the back of a horse-drawn wagon in Charleston and then later in the back of a truck, the exhibits were packed in green wooden suitcases. With handles and hinged doors, the traveling Bragg boxes served as windows on the world. Built like small stages, they displayed, when opened, scenes that helped children to understand the animals around them, or the people, customs, and background of other cultures and countries. The exhibits also included pictures (usually from a government bulletin or an issue of *National Geographic*), items that exemplified the topic of the exhibit (such as stamps or post cards), and a teacher's story written by Bragg or another museum staff person.

Unlike most of the other educational services in Charleston, South Carolina, and throughout the rest of the South, Bragg boxes went to all of the schools in the city—both black and white—beginning in 1913. Grace Dobbins was a student at Simonton in 1916, an overcrowded black school. She clearly recalled the exhibits and the teacher's story: "The exhibits came on a weekly basis. The opened suitcase would sit on the teacher's desk while she read us the story. . . . If there was an animal inside [of the exhibit], we would touch it."[35]

Soon after the first 1912 Jim Crow law was passed by the city, Paul Rea, the museum's director was successful in gaining separate status for the museum from the College of Charleston, by incorporating it with a Board of Trustees in 1915. The museum then became a municipal institution, and with this shift, its emphasis focused on an educational function, bringing with it social implications. Now, The Charleston Museum reached into the community, finding new patrons first in the different social classes and ultimately among the black citizens—though not at first. In 1917, the museum's trustees affirmed an executive committee decision to allow admittance of classes of black students to the museum when accompanied by a teacher. Within that same policy, however, the trustees decreed that black adults were to be denied admission, even black maids accompanying white children under five.[36]

It is possible that this policy of admitting black students was due to the efforts of Benjamin F. Cox, the principal of Avery Normal Institute. He wrote to Paul Rea, within days of the trustees' new policy, acknowledging the "privileges to be received at the Museum." Located within three blocks of the museum, this private school for blacks was founded by the American Missionary Association in 1865. Once the museum moved from the college into the old Thompson Auditorium on Rutledge Avenue, many of Avery's students walked past the museum on their way to and from school each day. Some recalled being fascinated with the turnstile, the whale skeleton suspended from the ceiling, and the other wonders that the vast building contained.[37]

Others, like Dobbins, who had transferred out of the city's public school system, remembered walking with her class from Avery to the museum. She recalled that there was always a guide who was "cordial . . . [the] Museum was pleased to have us." Dobbins also recalled that when the black students were at the museum, there were no white classes in attendance. But, then, the policy was not to have two white classes at the same time either. "We never care to have classes from two white schools at the same time, as the attention of the staff would then have to be divided," Bragg wrote to another progressive museum administrator in Mississippi.[38]

Avery's curriculum was in marked contrast to that of the black public schools. Here, the students were expected to follow a course of study, " . . . along New England classical lines, which included French, Latin, English, the sciences, and mathematics." The school was a cultural center for the black community as well, with a strong music department and a library open to the public. (There was no free library in Charleston until 1931.) The public schools, however, believed that it "was important to give the colored pupils training along industrial lines . . . [as this is] the education which the Negro needed most. . . ." Adopted by the white Board of School Commissioners in 1910, Booker T. Washington's plan of industrial education had already been denounced four years earlier by Charleston's growing black middle class, many of whom had attended Avery.[39]

It was this school and the efforts of Benjamin Cox and his wife Jeanette that helped to create the growing black middle class in Charleston. Along with skilled artisans, many of the city's black doctors, merchants, ministers, and teachers were first educated at Avery. As the College of Charleston did not admit black students, they had no other alternative but to leave the city for higher education. Even with the existing political and societal conditions, many chose to return. Barred from most occupations, this growing middle class existed and worked among their own people.[40]

With the election of John Grace as mayor, first in 1911 and then again in 1919, the black community found a friend and a supporter. A second-generation Charlestonian, Irish Catholic, and a populist, Grace believed in "justice regardless of race, color, or previous condition of servitude. . . ." His first administration was characterized by a balanced budget with an application of business principles in the conduct of the city's business. But it was in his second administration that Grace attempted to bridge the color line in his role as mayor, seeing himself as serving all of the people and not just the white populace.[41]

There is much evidence to endorse Grace as a friend of the black community. Opposed to the state's Jim Crow laws, he argued against legal segregation on the city's streetcars. He appointed the first black citizens as commissioners and signed a federal antilynching petition. Blacks also gained Grace's support of a petition to have black teachers in the city's black schools in 1920. Grace backed their cause, pointing out that "blacks paid their taxes and contributed to the wealth of the community."[42]

From the time of the Civil War, all public school teachers within the city schools had been white. Black educators were prevented from teaching black students and had to work in the rural schools of the county. Many of the teachers prepared by Avery were forced out into these primitive one-room schools. It was due to the conditions faced by the teachers in the rural schools, as well as the treatment of black students by white teachers in the city schools, that the city's chapter of the National Association for the Advancement of Colored People (NAACP) sought a change within the system through a petition drive. This was an important political accomplishment and marked a milestone of solidarity for the black community.

The local chapter of the NAACP had more than 800 members by 1919, and it was this group—made up of people like Edwin Harleston, an artist, and Septima Clark, a teacher and later civil rights activist (both Avery graduates)—who collected almost 5,000 signatures on the petition. The delegation led by Harleston submitted the petition to the state's General Assembly, which passed legislation outlawing white teachers in the city's black schools, beginning with the 1920 school term. This change in the teaching force in the city schools did not affect the continuation of Bragg's educational programs in the black schools. " . . . [C]olored teachers replaced white. Naturally we continued to send the exhibits as we had done before."[43]

THE ROARING TWENTIES AND THE SEARCH FOR FUNDING

With blacks asserting their political power more than ever and southern women like Sue Frost becoming businesswomen, Charleston was beginning to stir itself out of a self-imposed slumber. Though comparatively still a conservative and provincial backwater,

Charleston's brand of the 1920s probably roared a little differently than it did in other cities. "Doin' the Charleston" meant more than a dance; it meant having a good time. "Blind tigers," or Charleston's bars, generally ignored the state and federal prohibition laws. And Charlestonians and the city's officials passionately opposed "their enforcement and both the citizens and the city itself were noted as being wringing, sopping, dripping wet." One could always find a drink or a good time in Charleston. Bragg thought that the 1920s in Charleston were the greatest and most exciting years of her life.[44]

During the four years of Grace's second term, Charleston went from an economic boom to a bust. The wartime economy extended into 1921, and federal money flowed into the city. There was a high demand for farm goods, even though the boll weevil had destroyed 90 percent of the Sea Island cotton crop in 1919. This eventually led to a depressed cotton industry and many bank closings. Wartime prosperity ended as quickly as it had begun, with a postwar depression lasting several years. Some 4,000 Charlestonians fled the city's pessimistic economy. By the end of his term in 1923, Grace had increased taxes to provide additional city services. These had added more than $3,000,000 to the city's municipal debt—a move that put the city at risk of bankruptcy.[45]

Industry, which was not welcome in the city, found a home in the "Neck," or north area of the county. Two new industries, Standard Oil Company and an asbestos plant, employed 1,000 mostly white workers. Because few blacks were hired by either company, segregated mill villages developed in the surrounding areas of the Neck, which was almost six miles from the city. The city itself, though, remained racially mixed. Many blacks still resided in the servants' quarters of many of the fine South-of-Broad homes. Regardless of location, however, housing for blacks, unlike that for whites, was usually "devoid of the minimum of sanitary conveniences, and their lanes and alleys were open sewers." Even so, Charleston was fast becoming a city, with a population of almost 100,000, who owned nearly 3,500 cars. With the automobile came traffic problems, for the city's streets remained unpaved and were too narrow to accommodate opposing lanes of traffic. Bragg had one of those cars for which the museum had paid half the cost as a part of her salary.[46]

In August 1920, Bragg was named the director of the Charleston Museum, earning a place in history as the first woman in the country to attain such a position at a large scientific museum supported by public funds. She notified the Simmons College newsletter of her promotion, writing, "I am not directly in library work, but I will find my library training of service." It proved to be so, as one of the major goals for her first full year in office was the further development of the museum's reading room into a full-service library. Prior to 1900, libraries and museums operated as common centers. Providing library services in cities without free libraries continued to be common practice for small museums, such as the Charleston Museum. Soon after being named director, Bragg broached the subject of a full-service library to the museum's trustees in 1921 and would continue to serve as the impetus behind the drive for a free library until one was created in 1931.[47]

Because Bragg believed education to be the dominant ideal of every museum, equalizing "opportunities between the rich and the poor," she continued to seek additional involvement with the city's and county's educational departments. She dealt directly with the school superintendents, meeting with them at the beginning of the school term to discuss the use of the traveling exhibits in conjunction with her Nature Study Curriculum for each grade. The schedule for the traveling exhibits was prepared collaboratively with teachers in the lower grades, and instruction was provided for the upper grades at the museum.[48]

Other components of her program included weekly lectures on various topics that ranged from seashells, birds, and trees to theories of evolution, China's history, customs, and legends. Lectures were for the adults, but Bragg was most interested in the children—so there were story hours for them. They heard about days of knighthood, Merlin's gift of the Roundtable, the Norse sagas, prehistoric animals of South Carolina, shells, clouds, Greece, Egypt, Japan, Uncle Remus stories, and other legends. The museum also provided classes in drama, modeling clay, art, and etching, conducted by local artists Ned Jennings and Elizabeth O'Neill Verner. Most popular of all were the field trips, which were a part of the Natural History Society activities. These were taken by boat to plantations and coastal islands, with Bragg using each field trip to teach Charlestonians about their city: its history, nature, and geography.

As educationally sound as the exhibits were, funding for them (and for the museum itself) was a problem for Bragg that was never to be solved. Initially, even Mayor Grace had his doubts about Bragg's ability ("I was very skeptical when Miss Bragg took up the work when he [Rea] left off") to run the museum and deal with the funding issue. And he admitted that he was "by no means originally a friend of the Charleston Museum. While I have been always a friend of education, I did not at first regard the museum . . . as such a necessary part of our educational program as to be worthwhile." In her first appearance before the city's Ways and Means Committee, seeking the museum's yearly appropriation, Bragg had been unable to answer Grace's questions about the division of funds from the museum's three funding sources: city, county, and general. She had been afraid to appear before the committee at the outset and then wished that she had not. "I feel so helpless because I cannot talk and I have no opportunity to show you . . . ,"she wrote to Grace, "I can show people what I am doing at the Museum but I do not know how to talk for it."[49]

In this same letter to Grace, she also sketched out her plans for the museum's library, which was a key element in her intellectual revolution. Hoping to make it the children's center in Charleston, Bragg recalled, "I was brought up in the north where there are public libraries everywhere, and though my father had a modern library of several thousand volumes, I know that a big part of my education is due to the public library." She pointed out to Grace, "Perhaps you have always had plenty of books. . . . When . . . [the] story hours are over, the children clamor for books about the stories. . . . Do you know most of the children who come to the Museum for the story hours know nothing of the stories that are the heritage of our race, the old Greek myths, the stories of King Arthur, German folklore, Irish legends, and the Norse sagas."[50]

The success of the traveling school exhibits (Bragg boxes) and the museum's other educational opportunities led Bragg to a great discovery. She came to the realization that the museum was "an engine of social change, an opportunity to do good for ordinary people on a massive scale." Indeed, Bragg called museum work "the work of the people." Although this understanding was relatively new even to the museums in the Northeast, it was almost unheard of in the South, because of the poverty, isolation, and racism that so predominated in the region. What few museums there were in the South were cemeteries for relics and open only to whites. But her museum, the Charleston Museum, was not a relic and she was intent on opening its doors and broadening horizons for the citizens. The times and people were almost ready for Bragg's next change at the museum—that of crossing the color line and bypassing the Jim Crow laws.[51]

In many respects, Grace was far more liberal than his fellow Charlestonians. He had once been described as "an urban politician in a rural state, a Roman Catholic in the most Protestant area of the nation, a liberal Democrat in an aristocratically conservative city, a nationalist in the most rabidly sectionalist area of the country." Time and place brought Bragg—the social reformer—and Grace together. She surely could not have had a more enlightened mayor in such a southern town to support her goal of using the museum as an engine for social change. Together, Grace and Bragg set the world of black-white relations atilt when she approached him about opening the museum to black patrons in 1921.[52]

"At that time (1921)," Bragg wrote in a later letter, " . . . we found that an occasional negro [*sic*] fair enough to pass for white came into the Museum with the white people. We then decided to appropriate Saturday afternoon . . . for negro [*sic*] attendance." By December, Bragg was corresponding with Avery principal Cox, informing him of the museum's Saturday hours for black patrons. And her "Report of the Director of the Museum for the Year 1921," was published in the *City Year Book,* where she included her first notation of black attendance at the Museum: "Total colored attendance . . . 2,057."[53]

Believing in equality, Bragg had first provided the same educational services to black students, and by opening the museum to black patrons, she was offering them a greater measure of parity. But crossing the color line in such a blatant way was "beyond the pale" and was certainly not celebrated or noted in the local press. However, notice did appear in a national paper. The *Christian Science Monitor* noted in a long article, "Museum Here is Pioneer," that "Encouragement is given irrespective of color or creed, and Saturday afternoons are set apart at the museum for Negro citizens."[54]

Her conduct, however, provoked many white Charlestonians who understood and revered the color line. One of Bragg's later associates recalled how upset people were when she opened the museum to black patrons. Albeit that there were those who disapproved of her behavior; others in both the black and white communities took a different viewpoint. Searching for information about the museum's patrons in 1922, Bragg employed a social worker to complete a survey among residents in the downtown area. In Mary Preston's report to Bragg, she wrote, " . . . but when one ran across the colored, one was struck by their satisfaction when they heard they were entitled to visit the museum. With encouragement many more than do now would attend. . . . Here in passing I should say that I heard no adverse criticism (from whites) of the fact that negroes [*sic*] were admitted."[55] But these comments were never published in Bragg's director's report for 1922, thus sparing the white power structure from the impact of new thoughts and ideas.

Mayor Grace remained a friend of the museum and of the black community throughout his second term. He began the summer of 1923 in high hopes of an easy reelection, unaware that Thomas Stoney, a Broad Street lawyer with ancestry dating back to the seventeenth century, would be a surprise entry in the primary. Bragg thought that Grace would be reelected, but, she wrote to a museum friend, "he will have to go through a bitter mud slinging because of his religion." She was correct. That primary election was bitter, with the National Guard called out on Election Day to maintain order. And there were charges on the part of each candidate's supporters during the campaign about Grace's Catholicism and Stoney's alleged association with the Ku Klux Klan. Stoney's supporters solicited the support of the Klan's membership, and he benefited from the association. Bragg, however, was wrong about Grace's reelection; "much to everyone's surprise," he lost to Stoney in the primaries.[56]

Stoney, the youngest mayor in the city's history, faced tremendous problems. The region's postwar prosperity was ending for Charleston; the price of cotton had dropped sharply, leading to layoffs in the cotton mills, and civilian workers were also being laid off at the Navy Yard. Living costs and taxes rose while job opportunities declined. Times were lean, and money for the museum and its programs became more difficult to find in the city budget. As the Charleston economy deteriorated, Mayor Stoney worked to develop the tourist trade. Although he was a member of the planter aristocracy, himself, and understood how members of his kind felt, and although Yankee visitors from "off" may well have been the last people most Charlestonians wanted roaming their streets and peering into their gardens, tourism became the city's biggest industry by the late 1920s. One very aristocratic Charlestonian remarked somewhat prophetically, "Nothing is more dreadful than tourists, whether grasshoppers, boll weevils, or money-bagged bipeds. They will make Charleston rich and ruin her."[57]

Stoney's election and the area's economic depression did not immediately affect Bragg's intellectual revolution. Her work with the public schools had expanded far beyond those first traveling exhibits sent out only to city schools in 1912. By 1925, the county schools received the Bragg boxes, as well as the traveling libraries. The service to the city schools had become integrated with the teacher's curriculum. Individual students from the city's 12 schools came to the museum for classes and lectures, and they also used the museum's library as a public reference library.[58]

A year later, Bragg's Nature Study Curriculum continued to be taught in the city schools, and through her efforts, every school in the city (black and white) was a part of the educational program at the museum in some way or to some degree. Her educational focus was not limited to the students but extended to county teachers also. She continued to teach a two-week summer course on nature work, home geography, and local history and arranged for college credit to be earned.[59]

Even though the museum had demonstrated its usefulness as an educational institution with a 300 percent increase in attendance during the previous decade, the city had not paid its museum appropriation in 1926, and the county millage had also been reduced by four mills. In fact, each year after 1924, the museum "had had a constant shrinkage" in its allocation. Consequently, Bragg was forced to conduct the museum's first drive for paying members in its history. There are several possible explanations for this decrease: the county's and city's deteriorating financial conditions, the political reaction from Stoney's Ku Klux Klan supporters to the Saturday openings, Bragg's gender, which framed the all-male council's reaction to her funding requests, or the inability of the politicians to understand and accept the museum as an educational institution.[60]

With local funding insecure, it seemed to Bragg that the logical move was to appeal to the state superintendent of education for help in funding the traveling school exhibits, inasmuch as they were now being used by teachers across the state. The cost of building the portable display cases, as well as the additional costs of assembling the display materials, writing the teacher's story, and paying the salary of the curator of public instruction had been borne by the museum through its city appropriation.[61]

Because the local appropriation had become unreliable and the amount received from it was less each year, Bragg also went outside the state to organizations such as the Carnegie Foundation and the General Education Board. Dealing with the issue of funding for the school exhibits and other museum activities was a continuous worry for Bragg. For the next four years she was unsuccessful, as she continued to petition her museum friends for political support, which they willingly gave, and to bargain with the state superintendent of education, who played politics with her request. After just one business lunch at the Country Club of Charleston, Bragg and Superintendent James Hope engaged in a war of wills. She wrote to him, telling him what she would do, and what she would not do as the museum's director. And she also informed him what the Charleston County Delegation would do at her request. Rarely did Hope respond to her letters, and he frequently used intermediaries to communicate with her. Quite possibly she was unlike any woman he had ever known, southern or not.[62]

Whether dealing with the state superintendent of education, the mayor, the school board, or the museum's trustees, Bragg was never one to give up or be daunted by adversity. Even though there was little money for the Charleston Museum and its educational programs, she continued to seek the expansion of the traveling exhibits. She wrote to the director of the Educational Museum in St. Louis in June of 1927 that she was "planning for a series of traveling exhibits to go to all the states of the South and to serve as an initial step in providing a system of traveling school exhibits for each state."[63]

In 1926, there were 147 traveling school exhibits (Bragg boxes) going out to 30 city and county public schools, as well as 9 private schools, along with 100 traveling school libraries. By the next year, traveling school exhibits went out to all of the schools in the city and the county, and for the first time, they were systematically circulated to the black county schools. Charleston was still a rural county in what remained an extremely isolated state. The days of slavery and Reconstruction were not that far in the distant past, as a photographer from Maine noted when she visited the South Carolina Low Country in the late 1920s as she took pictures of many young black children. Bragg also understood the effects of isolation and poverty, and expressed sorrow over the condition of southern people who were, in her estimation, "in need of a better comprehension of modern science, a greater understanding of foreign peoples living today, and some approach to an appreciation of the fine arts as applied to life. In spite of the educational efforts of the last few years, people outside the centers are more backward than those of the north, and there is at present, along with an awakening of spirit and desire for knowledge, a reactionary tendency which is inclined to direct this spirit conservatively."[64]

Within two years, the museum had increased the number of Bragg boxes circulating regularly in the city schools (white, black, and private), as well as in the county schools. They were also used in more than 22 other towns across the state, as well as being utilized to prepare teachers in three of the state's normal schools. The Wake County Teachers' Association in North Carolina sent representatives to the museum, where they were trained to construct their own exhibits. Because of Bragg's work with other southern museums, Bragg boxes became a part of the educational programs at the Birmingham Museum in Alabama, the High Museum in Atlanta, and the Witte Museum in San Antonio. However, in a letter to a prospective member, Bragg wrote what were prophetic words: "The demands which people are making on us are now so much greater than I can meet, that if I did not have a good deal of faith in the future, I would want to throw everything up and go north. Beside the need of new members, the city's not being able to pay anything for two months makes our condition rather more serious than it would be otherwise and you will be helping very definitely, if you are able to give us your generous subscription immediately."[65]

THE FREE LIBRARY

The funding of the exhibits and, ultimately, the free library, was due to Bragg's professional association with Clark Foreman, whom she had assisted in 1927 in the development of a small biological museum in Highlands, North Carolina, where his Atlanta family summered. Two years later, he wrote to Bragg that he was working with the philanthropic Rosenwald Fund of Chicago, which is "particularly interested in education and health." Given that the fund provided financial support to southern educational and social service institutions, Foreman believed there was a match with the fund's goals and Bragg's work. Because he and other fund officials would be in Charleston, Foreman wanted them to hear from her about the museum's library service and the educational program. He also asked that she bring together some of the "more outstanding supporters in the white group and some of the more educated colored leaders" to meet E. R. Embree, the president of the fund.[66]

On this visit, Foreman and Embree, saw the museum and also visited Dart Hall, which contained the personal library of Reverend J. L. Dart. They met Susie Dart Butler, who managed her father's collection, making it available to the black community. Embree was impressed with both efforts at the museum and Dart Library. Not only did he pledge support of a free library for the county (with certain conditions as to matching funds from the legislators), but he also recognized the "great educational value of the [traveling school] exhibits and pledged one thousand dollars of the [library] funds in any one year or each year used for extending the Museum exhibit service." As wonderful as this news was to Bragg, she was discouraged in her efforts to receive some direct help for the museum. She knew that the library would only add to the financial burden and responsibilities of the already struggling museum. So she asked for a "few thousand dollars to extend what we are already doing."[67]

Tying the library and museum money together was not the only issue with which Bragg was unhappy. Even though Foreman had written in the last sentence of his February letter, almost as an afterthought, " . . . the Rosenwald Fund does not limit its help to either race," Bragg clearly did not want to deal with the "colored problem" any more than she already had. She wrote to Foreman, "As it is, I very greatly appreciate your good will and what may come from this, I nevertheless feel that the burden of handling the colored problem is much a serious one that the harm to the Museum in the struggle will more than offset any benefit."[68]

This struggle for political, social, and cultural parity during the Jim Crow era encouraged black patrons to seek even greater equality. Several years after the Saturday openings began, "certain tax paying negroes [*sic*] wrote to our Mayor and remonstrated because they were limited to Saturday afternoon. . . . We gradually let leaders among the negroes [*sic*] know that admission could be had by appointment," Bragg wrote to another museum administrator. As a result of this request, blacks were admitted by appointment on days other than Saturday. By 1929, Bragg had ceased reporting "colored attendance" in her yearly report to the mayor and city council. The last reference had come three years earlier in her 1926 director's report in which she noted that it had decreased. With black Charlestonians confronting the color line more openly, local as well as national conditions had deteriorated for them. Handling the "colored problem" had become more than she could deal with, given the funding problems confronting the museum. It was possible that Bragg had come to understand that blacks were expected to remain subservient to white men.[69]

Bragg ended her May letter to Foreman with a plea: "Why is it that the big funds make it so hard for us to get help? The additional tax for the library will probably reduce the Museum's appropriation because the city literally has not the money. Therefore I am exceedingly discouraged but I would not be selfish enough to stop working for the library and I just trust that the Museum is taken care of. Mr. Foreman, you people from the foundations have no idea of the courage you could give people and the actual help you could give if you would be willing to spend a few thousand dollars here and there without conditions." Though she was later to call their help unsolicited, Foreman and Embree heard her appeal, giving Bragg $5,000 for the traveling school exhibits. Once the museum received the funds, Bragg reported to Embree how the method of servicing the black county schools had improved.[70]

Mayor Stoney supported Bragg's efforts for the free library, and both believed that the 1929 library proposal from the Rosenwald Fund was too good for either the city or the county to turn down. Unfortunately, the Charleston legislators disagreed, thinking that the city did not need a library, especially one that gave service to both blacks and whites. By December, Bragg wrote to Helen McCormack, her protégé, "Everything possible has gone wrong" with the library proposition for the county.[71]

This battle for the free library had been a long one. It began when Bragg first arrived in the city in 1909 and was a major focus of her first year as the museum's director. She was able to persuade others to support the cause over the years, and on two occasions, the women's organizations had offered her their highest praise by calling her a Charlestonian for her efforts for the free library and the museum's programs. They rallied around the issue again once the delegation refused to support the proposal. They, along with Bragg and other community leaders, mounted a campaign to obtain several thousand signatures in support of the library, proving to the delegation that the city both needed and wanted one. Prominent Charlestonians and educators supported the funding of the library, and local leaders not only conducted the petition drive but also successfully "concentrated their efforts on the County Delegation." After a year-long campaign, the delegation eventually approved the library appropriation in 1930. Once that happened, the way was cleared for the establishment of the library within the museum. Bragg's own philosophy was a match with the Rosenwald Fund's requirement, "that the library shall give service to both white and colored people with equal opportunities," and she worked with Susie Butler in bringing Dart Library into the system. When the library opened within the museum on January 1, 1931, Bragg was one of six trustees, as well as an incorporator and its first librarian, a role that allowed her to use the library skills learned so long ago at Simmons.[72]

The year 1931 was a difficult one for both the museum and the community. There were huge financial burdens due to the depression that had hit Charleston with a vengeance. With a large municipal debt by 1931, the city teetered on bankruptcy. Bragg wrote former Director Paul Rea in late January that even though City Council had given the museum the same appropriation as last year, there was a "great agitation on the part of the citizens for a revision of the city budget and a reduction all along the line. The tax payers are determined on state retrenchment and so naturally, the legislators are starting with the educational institutions and have brought in a bill to cut the teachers' salaries 20 percent. The Museum has not yet been mentioned but even members of our own board of trustees, Mr. Hagood, for example, say that they are willing to have the Museum's quarter mill cut if it will lower taxes."[73] Because Rea had also dealt with City Council over the lack of funding when he was the museum's director, he well knew what Bragg was facing. If the museum had any less money in this budget, there would probably have to be not only a reduction in staff, but also a reduction in services.

Not long after writing this letter, in March of 1931, Bragg was offered an incredible opportunity to return to her native Massachusetts. Z. Marshall Crane wanted her to come to Pittsfield, where she would have carte blanche to turn his father's "old man's attic of a museum" into an educational and cultural institution. In undertaking this work, Bragg saw the inducement as the chance "to do the reorganization work which I have so much enjoyed in Charleston and in doing over the Valentine Museum and in helping other museums in the south and particularly the prospect of building up a large extension work with traveling school exhibits up and down the Housatonic Valley."[74]

Although she accepted Crane's offer to return to Massachusetts in July 1931, Bragg intended to return to her adopted home. Though from "off" and very much the bold new woman, Bragg was well-respected as the director who had brought national recognition to the Charleston Museum. Consequently, the trustees granted her request for a five-year leave of absence, guaranteeing that she would assume her position upon her return. Bragg's intellectual revolution, her educational programs, and other activities continued at the museum. However, it would be a number of years before blacks entered the Charleston Museum on any day at any time, without an appointment or just on Saturday.

CONCLUSION

Hired by a man, it was Laura Bragg, the woman, who significantly shaped the educational direction of the Charleston Museum for more than 20 years, even though she held the title of director for only 10 of those years. When Rea hired Bragg, the Charleston Museum was like many other museums across the country. It was beginning to stir itself out of years of disuse and lack of involvement with the general public. Around 1860, education had been abandoned as a museum function, and it was not until the turn of the century, at the beginning of the Progressive Era, that museums saw their primary motivation as a need to instill literacy in the American public.[75] The Progressive Era in the South presented special problems because of the rural population as well as the extreme poverty of the region. Bragg's traveling school exhibits and libraries were relatively inexpensive to build and easily transportable; thus, they were an ideal way to deal with both of those issues. They provided education to the masses, who, in the minds of progressives such as Bragg, were apathetic and ignorant.

With Rea filling faculty positions at both the Medical College and at the College of Charleston, Bragg was presented with the perfect opportunity to transform a scientific, academic museum into a public institution that was dedicated to education. In the long view, there are some who could say that she used the museum as an instrument for social change in Charleston. She was either doing good for ordinary people on a massive scale or she attempted an intellectual revolution through the educational programs offered by the museum.

Revolutions "have always been limited by the social settings in which they take place." In Charleston, Bragg's revolution was framed by class, race, and gender. Race had not been dealt with at the museum until 1917, when the trustees allowed classes of black students into the building for instruction. However, Bragg began providing the same educational services to black and white students when her traveling school exhibits first reached the city's black schools in 1913. These black schools were overcrowded, substandard to the white schools, and offered fewer textbooks and other educational materials for students and teachers. It was here that they were of the most significance educationally—where they were truly windows on the world.[76]

As in most cities, socioeconomic class determined who belonged to the museum and participated in its activities. Because the social class of the ordinary citizen differed from the museum's trustees and Bragg's associates, the museum's educational programs under her leadership were termed by her as missionary in nature.[77] In effect, she was attempting to offer the promise of American life, as Bragg understood it from her New England perspective, to the ordinary citizen through the Charleston Museum. Her educational programs as extension activities for the school system were created to provide the masses with educational opportunities. Opening the museum to black patrons during the Jim Crow era came during the most repressive period in the country's history of race relations. It was a bold step for an outsider and even bolder for a woman.

A product of northern progressivism, Bragg first saw herself as a social missionary, who recognized the ignorance and apathy of the population. The librarian/curator became an accidental educator with her Bragg boxes. She came to see them as a means of achieving education of the masses through the museum's extension service. She was practicing social progressivism with its own brand of education, and in time, Bragg became a cultural missionary, reshaping the cultural and social landscape of Charleston through the museum.

Through her Bragg boxes, windows on the world, she was able to expose children to countries, cultures, and nature that they would never have experienced. As a means to an end, the boxes accomplished her primary goal, which was to increase the attendance at the museum. But viewed from the perspective of more than 80 years later, Bragg's actions are examples of progressive social reform. She was attempting to end the apathy and ignorance that she believed predominated in the region. She hoped the that students—black and white—would become educated adults who would visit the museum and bring their children.[78]

Her traveling school exhibits, and thus her intellectual revolution, could have reached more children if the male power structure of the General Assembly and State Superintendent of Education James Hope had not chosen to stonewall her every effort to seek state funding. Bragg's gender and the role expectations for southern women bounded whatever she attempted at the museum. Her determination, her forthrightness, and her vision of educating all children as a progressive new woman distinguished her from many other women in the state. She was aggressive in setting forth her agenda and making the necessary connections in both the museum and political worlds to secure its success. By not fulfilling the prescribed gender role of knowing her place, Bragg represented a threat to the power structure. Ultimately, her inability to become the southern lady affected her ability to raise funds within the state for the museum.

Historically, South Carolina's lawmakers were slow to reform anything, and with the exhibits also being sent into black schools, racism could very well have been a part of the politician's game.[79] In passing her request between the State Department of Education and the county delegation, these men (except for two Charleston legislators) did not want to deal with Bragg and the traveling school exhibits. In the four years between 1926 and 1930, State Superintendent Hope frequently used other people to communicate with Bragg, rarely corresponding with her directly. Even though she finally received a concurrent resolution in support of the exhibits, the state never provided the museum with financial support.

With the Jim Crow era and the "colored problem" at the forefront in Charleston, Bragg's actions must have shaken the community's social structure as determined by white men. Her director's reports included "colored attendance" for five years, with the figure reported at about 2,000 in each of those years. Her first report of it was set apart from the rest of the text. But in later years, the manner of reporting had changed. "Colored attendance" was mentioned at the end of a paragraph, and finally, in her last notice of it, she wrote that it had decreased, with no explanation offered. It is only in her

1929 letter to E. R. Embree that we are offered a glimpse of Bragg's understanding of the "colored problem" and what impact the library proposal, coupled with the Rosenwald Fund's requirement of serving both blacks and whites, would mean to the museum. At that time, Bragg had lived in Charleston for two decades, and although she, the woman, never learned that she had a place, politics taught her that blacks in Charleston did.

Self-identified as a missionary, Bragg was also a pragmatist. The museum needed money, and she was determined that Charleston would have a free library. Accepting the Rosenwald Fund grant for both the exhibits and the library ensured the museum's participation in granting an even greater degree of equality to black citizens during the Jim Crow era. As it was her intent that the museum and its educational programs would awaken "spirits and a desire for knowledge," Bragg again willingly bridged the color line in accepting the money, thus ensuring that her intellectual revolution would continue.[80]

NOTES

* Parts of this chapter were previously published in chapters 1–4 of Louise Anderson Allen, *A Bluestocking in Charleston: The Life and Career of Laura Bragg* (Columbia: University of South Carolina Press, 2001). Reprinted with permission of the University of South Carolina Press.

1. Henry James, *The American Scene* (New York: Harper Brothers, 1907), p. 414.
2. Laura Bragg, interview by Miriam Herbert, 2 June 1972, tape recording, Laura Bragg Papers, South Carolina Historical Society (hereafter referred to as SCHS), Charleston; Robert Molloy, *Charleston, a Gracious Heritage* (New York: D. Appleton-Century Company, 1947), p. 2; Walter J. Fraser, *Charleston! Charleston! The History of a Southern City* (Columbia: University of South Carolina Press, 1989), p. 344; and Mildred Crum, *Old Seaport Towns of the South* (New York: Dodd, Mead & Company, 1917), p. 125.
3. Bragg-Herbert interview, SCHS; Frank Gilbreth, *Loblolly* (New York: Crowell Press, 1959), p. 63; Molloy, *Charleston, a Gracious Heritage,* pp. 6–7; Eleanor P. Hart, "Weighing Her Merits," *Preservation Progress,* January 1965, Hinon Clippings, Charleston Library Society, Charleston (hereafter referred to as CLS).
4. "City Year Book," 1907, pp. 35–36, South Carolina Room, Charleston County Public Library (hereafter cited as CCPL); *Information for Guides of Historic Charleston* (Charleston: Tourism Commission, 1985) pp. 390–391.
5. Jean Weber, "Changing Roles and Attitudes," *Gender Perspectives: Essays on Women in Museums,* ed. Jane Glaser and Artemis Zenetou (Washington, D.C.: Smithsonian Institution Press, 1994), p. 34; Albertine Burget, "A Study of the Administrative Role of Directors of Education Departments in Non-School Cultural Organizations," (Ph.D. diss., Loyola University, 1986), p. 19; and Steven Conn, *Museums and American Intellectual Life, 1876–1926* (Chicago: University of Chicago Press, 1998), pp. 4–6.
6. Paul Marshall Rea, *The Museum and The Community* (Lancaster, Pa.: The Science Press, 1932) p. 18.
7. Stephen O'Neill, "From the Shadow of Slavery: The Civil Rights Years in Charleston," (Ph.D. diss., University of Virginia, 1994), p. 75.
8. Ibid., p. 57; Idus A Newby, *Black Carolinians: A History of Blacks in South Carolina From 1895 to 1968* (Columbia: University of South Carolina Press, 1973), p. 82.
9. Louise Anderson Allen, *A Bluestocking in Charleston: The Life and Career of Laura Bragg,* (Columbia: University of South Carolina Press, 2001), p. 6.
10. Ibid., p. 7.
11. Memoirs, Official Record of the 99th Session of the New Hampshire Annual Conference of the Methodist Episcopal Church (Lancaster, 11–15 April 1928), vol. XII, part 1, pp.

78–81, General Commission on Archives and History The United Methodist Church, Drew University (hereafter cited as DU).

12. Memoirs, pp. 78–81, DU; Neil R. McMillen, *Dark Journey: Black Mississippians in the Age of Jim Crow* (Urbana: University of Illinois Press, 1989), pp. 98–99; James D. Anderson, *The Education of Blacks in the South, 1860–1935* (Chapel Hill: University of North Carolina Press, 1988) pp. 240–242.
13. George Thomas Kurian, *Datepedia of the United States, 1790–2000: America Year by Year* (Lanham, Md.: Bernam Press, 1994), p. 145; Sheila Rowbotham, *A Century of Women: The History of Women in Britain and the United States* (New York: Viking, 1997), p. 95.
14. Laura Bragg, to Lyman Bragg, before Thanksgiving 1907, Laura Bragg Papers, 11/80/1, SCHS; Gregg Privette, interview by author, 9 July 1997, tape recording, Sears Papers (hereafter referred to as SP), Duke University; Gene Waddell, interview by author, 9 August 1996, tape recording, SP. See also Mary Ryan, *Womanhood in America: from Colonial Times to the Present* (New York: New Viewpoints, 1975), p. 256; and Rosalind Rosenberg, *Beyond Separate Spheres: Intellectual Roots of Modern Feminism* (New Haven: Yale University Press, 1982), p. 54.
15. Rosenberg, *Beyond Separate Spheres,* p. 176; J.J. Duffy, "Charleston Politics in the Progressive Era" (Ph.D. diss., University of South Carolina, 1963), pp. 352–353; Nancy Smith, interview by author, 9 June 1997, tape recording, SP; Edmund Drago, *Initiative, Paternalism, and Race Relations: Charleston's Avery Normal Institute* (Athens: University of Georgia Press, 1990), p. 276; and College of Charleston Faculty and Administrative Manual, 1994, p. 2.
16. Anne Scott, *Making the Invisible Woman Visible* (Urbana: University of Illinois Press, 1984); Mary Martha Thomas, *The New Woman in Alabama* (Tuscaloosa: University of Alabama Press, 1992); and Sidney Bland, *Preserving Charleston's Pats, Shaping Its Future: the Life and Times of Susan Pringle Frost* (Columbia: University of South Carolina Press, 1999), p. 35.
17. Anne Scott, Foreword, *The Southern Lady,* 25th anniversary ed. (Charlottesville: University of Virginia Press, 1995), p. 225; Estelle Freedman, "The New Woman: Changing Views of Women in the 1920s," *Journal of American History* 61 (1974): 374–395; Herbert Ravenel Sass, "The Lowcountry," in *The Carolina Low-Country* ed. Augustus Smythe, (New York: MacMillan, 1932), pp. 3–4; and Scott, *Invisible Woman,* p. 224.
18. Bland, *Preserving Charleston's Pats, Shaping Its Future,* p. 90; Louise Allen, "Laura Bragg: A New Woman Practicing Progressive Social Reform as a Museum Administrator and Educator," (Ed.D. diss., University of South Carolina, 1997), p. 165.
19. Scott, *Invisible Woman;* Allen, "Laura Bragg," 20–21.
20. O'Neill, "From the Shadow of Slavery," p. 20.
21. Karl E. Taeuber and Alma F. Taeuber, *Negroes in Cities: Residential Segregation and Neighborhood Change* (Chicago: Adline Publishing, 1965), pp. 45–47; Allen, "Laura Bragg," p. 234; O'Neill, "From the Shadow of Slavery," p. 42.
22. Leon F. Litwack, *Trouble in Mind: Black Southerners in the Age of Jim Crow* (New York: A.A. Knopf, 1998), pp. 52–113; Fraser, *Charleston! Charleston!,* p. 329; and Jack Temple Kirby, *Darkness at the Dawning: Race and Reform in the Progressive South* (Philadelphia: J.B. Lippincott, 1972), pp. 100, 103.
23. Ernest McPherson Lander, *Perspectives in South Carolina History, The First 300 Years* (Columbia: University of South Carolina Press, 1973), pp. 127–128; Kirby, *Darkness at the Dawning,* p. 103; Newby, *Black Carolinians,* p. 86.
24. Lander, *Perspectives in South Carolina History,* p. 123; Asa Gordon, *Sketches of Negro Life and History in South Carolina* (Columbia: University of South Carolina Press, 1971, 2d ed.), pp. 106–108.
25. O'Neill, "From the Shadow of Slavery," pp. 54–55, 63; Litwack, *Trouble in Mind,* pp. 233–234; Duffy, "Charleston Politics in the Progressive Era," p. 22.

26. C. Vann Woodward, *Origins of the New South, 1877–1913, A History of the South,* vol. 9, eds., Wendell Holmes Stevenson and E. Merton Coulter (Baton Rouge: LSU Press and the Littlefield Fund for Southern History and the University of Texas, Austin, 1951) p. 398.
27. William A. Link, "Privies, Progressivism, and Public Schools: Health Reform and Education in the Rural South, 1909–1920," *Journal of Southern History* 54 (November 1988): 641; South Carolina State Department of Education, *Thirty-Second Annual Report, 1900* (Columbia: South Carolina State Library), pp.12–13; and *History and Development of Negro Education in South Carolina, Division of Instruction* (Columbia: South Carolina Department of Education, 1949), p. 7.
28. Link, "Privies," pp. 637, 641; and Joseph F. Kett, "Women and the Progressive Impulse in Southern Education," *The Web of Southern Social Relations,* ed. Walter J. Fraser et al. (Athens: University of Georgia Press, 1985), pp. 168, 173–174.
29. Kett, "Women and the Progressive Impulse," pp. 171–173; H.O. Strohecker, *Present Day Public Education in the County and City of Charleston 1929* (Charleston, S.C.: Charleston County Board of Education), p. 21.
30. Laura Bragg, letter to J.H. Hope, 8 October 1928, State Department of Education File, Charleston Museum Library Archive (hereafter referred to as CMLA).
31. Martin S. Dworkin, ed. *Dewey on Education* (New York: Teachers College Press, 1959) p. 43; William A. Link, "Privies, Progressivism, and Public Schools: Health Reform and Education in the Rural South, 1909–1920," *Journal of Southern History* 54, no. 4 (November 1988): 637; Herbert M. Kliebard, *The Struggle for the American Curriculum, 1893–1958,* 2d ed. (New York: Routledge, 1995) p. 233.
32. *Bulletin of the Charleston Museum,* 8, no. 8 (December, 1912): 70–71, CMLA.
33. Ibid. (January 1915): 8; Ibid. 11, no. 3 (March 1915): 21–26, CMLA; Edward Alexander, *Museums in Motion: An Introduction to the History and Function of Museums* (Nashville: American Association for State and Local History, 1979), p. 13; Board of Trustees Minutes of the Charleston Museum, May 1917, CMLA; "4,000 Students in University Here," 10 May 1915, *Post,* Museum Clipping Book, CMLA; "Nature Study at Museum," 3 July 1921, *American,* Museum Clipping Book, CMLA.
34. Laura Bragg, interview by Gene Waddell, 30 October 1971, tape recording, SP; Kenneth Yellis, "Museum Education," in *The Museum: A Reference Guide,* ed. Michael Shapiro (Westport, Conn.: Greenwood Press, 1990), p. 169.
35. Laura Bragg to Wallace Rogers, 18 June 1930, Attendance File, CMLA, Charleston; Drago, *Initiative, Paternalism, and Race Relations,* pp. 175–176; Grace Dobbins, interview by author, 10 February 1998, transcript.
36. *Bulletin of the Charleston Museum* 11, no. 3 (March 1915): 21–26, CMLA; Edward Alexander, *Museums in Motion: An Introduction to the History and Function of Museums* (Nashville: American Association for State and Local History, 1979), p. 13; Board of Trustees Minutes of the Charleston Museum, May 1917, CMLA.
37. Benjamin F. Cox to Paul M. Rea, 17 May 1917, Avery Institute File, CMLA; Eugene Graves, March 1998: personal communication.
38. Dobbins interview; Bragg to Rogers, 18 June 1930.
39. Fred L. Brownlee, *New Day Ascending* (Boston: Pilgrim Press, 1946), pp. 135–136; Burchill Richardson Moore, "A History of the Negro Public Schools of Charleston, South Carolina 1867–1942," (Masters thesis, University of South Carolina, 1942), p. 40; Drago, *Initiative, Paternalism, and Race Relations,* p. 119.
40. Duffy, "Charleston Politics," p. 18.
41. Doyle Willard Boggs, "John Patrick Grace and the Politics of Reform, 1900–1931," (Ph.D. diss., University of South Carolina, 1963), pp. 151–152.
42. Duffy, "Charleston Politics," pp.18–20.
43. Drago, *Initiative, Paternalism, and Race Relations,* pp. 175–176; and Duffy, "Charleston Politics," p. 18.

44. Fraser, *Charleston! Charleston!,* p. 361; *Charleston News and Courier,* 28 January, 1940.
45. Fraser, *Charleston! Charleston!,* p. 368; Boggs, "John Patrick Grace," p. 245.
46. Fraser, *Charleston! Charleston!,* p. 348; Duffy, "Charleston Politics," p. 28; Boggs, "John Patrick Grace," p. 165; Board of Trustees Minutes of the Charleston Museum, 7 December 1921, CMLA.
47. *The Simmons College Review,* 3, no. 2 (December 1920): 88, The College Archives, Simmons College, Boston; Paul M. Rea, *The Museum and the Community* (Lancaster, Pa.: Science Press, 1932), p. 183; *Memorandum on the Report of the Advisory Group on Museum Education* (New York: Carnegie Corporation, 1932), p. 7; *Bulletin of the Charleston Museum,* 18 (Jan.–Feb. 1921): 1, CMLA.
48. Helen von Kolnitz to Richard F. Bach, 15 April 1921, Metropolitan Museum File, CMLA.
49. John P. Grace to F. A. Whiting, 12 May 1923, Cleveland Museum of Art File, CMLA; Laura Bragg to John P. Grace, 19 February 1921, John P. Grace File, CMLA.
50. Bragg to Grace, 19 February 1921; Grace to Whiting, 12 May 1923; Laura Bragg to Alice Kendall, 4 May 1921, Indian Work File, CMLA.
51. Yellis, "Museum Education," p. 170; Laura Bragg to A. J. Buist, n.d., Undated Correspondence, box 1, Laura Bragg Papers, CMLA.
52. Duffy, "Charleston Politics," p. 206; Waddell-Bragg interview, SP.
53. Bragg to Rogers 18 June 1930; Laura Bragg to B. F. Cox, 20 December 1921, Avery Institute File, CMLA; "City Year Book," 1921, "Report of the Director of the Museum for the Year 1921," p. 371, CMLA.
54. Waddell-Bragg interview, SP; Drago, *Initiative, Paternalism, and Race Relations,* p. 197; 1923 Museum Catechism, CMLA; *Charleston Evening Post,* 30 December 1926.
55. Barbara Belknap, interview by author, 29 May 1997, tape recording, SP; Undated document, Bragg-Preston Correspondence, box 1, Laura Bragg Papers, CMLA. This information was not published in her 1922 Director's Report in the "City Year Book."
56. Laura Bragg to F. A. Whiting, 19 April 1923, Cleveland Museum of Art File, CMLA; Duffy, "Charleston Politics," p. 385; Boggs, "John Patrick Grace," p. 203; Fraser, *Charleston! Charleston!,* pp. 369–370, Laura Bragg to F. A. Whiting, 2 November 1923, Cleveland Museum of Art File, CMLA.
57. Fraser, *Charleston! Charleston!,* pp.370, 374; Boggs, "John Patrick Grace," p. 166.
58. Laura Bragg to John Cotton Dana, 9 November 1925, Newark Museum Association File, CMLA.
59. "City Year Book," 1926, pp. 217–219, CCPL; von Kolnitz to Bach, 15 April 1921, CMLA.
60. *Bulletin* (Jan.–Feb. 1921): 1, CMLA; Rea, *The Museum and the Community,* pp. 74, 274; Laura Bragg to Harold Madison, 11 April 1921, American Association of Museums (hereafter referred to as AAM) 1921, CMLA; Laura Bragg to Harold Madison, 14 April 1921, AAM 1921, CMLA.
61. Laura Bragg to E. R. Hardy, 21 April 1926, AAM 1926 File, CMLA; Laura Bragg to J. W. Ott 16 March 1927, J. W. Ott File, CMLA; Allen, "Laura Bragg," p. 230.
62. Allen, "Laura Bragg," pp. 227–242.
63. Laura Bragg to C.G. Rathmann, 10 June 1927, AAM 1927 File, CMLA.
64. See 1926 and 1927 "Report of the Director of the Charleston Museum," "City Year Books," 1926 and 1927 CCPL; Marius B. Péladeau, *Chansonetta: The Life and Photographs of Chansonetta Stanley Emmons, 1858–1937* (Waldoboro, Me.: Maine Antique Digest, 1977) no page ; Laura Bragg to General Education Board, 5 May 1927, General Education Board File, CMLA; *Fifteenth Census of the United States Taken in the Year 1930,* State Compendium for South Carolina (Washington, D.C.: Government Printing Office, 1931), p. 3.
65. Louise Barrington to Ralph Smith, 13 December 1928, AAM 1928 File, CMLA; Laura Bragg to Newark Museum Association, 9 October 1928, Newark Museum Association

File, CMLA; Laura Bragg to John Maybank, 16 November 1925, John Maybank File, CMLA; "City Year Book" 1928, p. 248, CCPL.

66. Clark Foreman to Laura Bragg, 8 February 1929, Rosenwald Fund File, CMLA.; Allen, "Laura Bragg," p. 240; Lois Simms, *Profiles of African-American Females in the Low Country of South Carolina* (Charleston: Avery Research Center for African American History and Culture, 1992), p. 15.
67. Letter from E. R. Embree (copy), 28 February 1929, Rosenwald Fund File, CMLA; Laura Bragg to Clark Foreman, 14 May 1929, Rosenwald Fund File, CMLA.
68. Foreman to Bragg, 8 February 1929, CMLA; Bragg to Foreman, 14 May 1929, CMLA.
69. Bragg to Rogers, 18 June 1930; Drago, *Initiative, Paternalism, and Race Relations,* p. 193; Report of the Director of the Museum for the Year 1926, "City Year Book," 1926, p. 246, CMLA.
70. Bragg to Foreman, 14 May 1929; Laura Bragg to E. R. Embree, 25 September 1929, and 17 September 1930, Rosenwald Fund File, CMLA; E. R. Embree to Laura Bragg, 28 September 1929, Rosenwald Fund File, CMLA; Laura Bragg to E. R. Embree, 28 February 1930, Directors' Correspondence, CMLA.
71. Laura Bragg to H. E. Wheeler, 5 March 1929, Library File, CMLA; Clark Foreman to Laura Bragg, 30 March 1929, Rosenwald Fund File, CMLA; Laura Bragg (unsigned) to Helen McCormack, 26 December 1929, AAM 1929 File, CMLA.
72. "Miss Bragg Highly Praised," 3 December 1921, Newspaper Clipping Book, CMLA; Minutes of the 1923 AAM Annual Meeting, pp. 32, 142, AAM 1923 File, CMLA; Sidney Rittenberg to Clark Foreman, 14 May 1930, Bragg, Laura 1930, box 2, Laura Bragg Papers, CMLA; *The History of the Charleston County Public Library* (Charleston: Charleston County Public Library, 1981), p. 4; Laura Bragg to Paul Rea, 31 October 1930, Paul Rea–Carnegie File, CMLA; Rittenberg to Foreman, 14 May 1930, CMLA.
73. Fraser, *Charleston! Charleston!,* p. 378; Laura Bragg to Paul Rea, 29 January 1931, Paul Rea-Carnegie File, CMLA.
74. Laura Bragg to L. V. Coleman, 29 May 1931, AAM 1931 File, Directors' Correspondence, CMLA.
75. Yellis, "Museum Education," p. 169.
76. Rosenberg, *Beyond Separate Spheres,* p. 245; Drago, *Initiative, Paternalism, and Race Relations,* pp. 125–126.
77. von Kolnitz to Bach, 15 April 1921, CMLA.
78. *Memorandum on the Report of the Advisory Group on Museum Education* (New York: Carnegie Corporation, 1932), p. 9.
79. Link, "Privies," p. 641; Mary Crow Anderson, *Two Scholarly Friends* (Columbia: University of South Carolina Press, 1993), p. 60; Dewey W. Grantham, *Southern Progressivism: The Reconciliation of Progress and Tradition* (Knoxville: University of Tennessee Press, 1983), p. 59.
80. Bragg to General Education Board, 5 May 1927, CMLA.

CHAPTER TWELVE

CHARL WILLIAMS AND THE NATIONAL EDUCATION ASSOCIATION

WAYNE J. URBAN

Charl Ormond Williams began her career in education by working her way through the ranks in the rural schools of West Tennessee, from teacher to principal to normal school instructor to county superintendent. Moving, then, from the local to the national stage, Williams served as a notable leader of the cause of women teachers in the National Education Association (NEA) for over a quarter of a century. Williams was neither the founder of a particularly notable progressive school nor was she identified with pedagogical innovation that commonly has been or is referred to as progressive. Yet, her work as a successful, change-oriented rural school teacher and administrator prior to coming to the NEA, her long career as a prominent staff member of the NEA, her links to various women's groups in pursuit of her NEA activities, her defense of public education as a cornerstone of American democracy, her political savvy and experience as both a lifelong Democrat from the South and a nonpartisan advocate of NEA causes, and her constant devotion to the cause of women as teachers and administrators in the public schools all mark her as a worthy candidate for inclusion in a volume on "founding mothers."[1]

WILLIAM'S EARLY LIFE AND CAREER

Williams was born in Arlington (Shelby County), Tennessee, in 1885. She graduated from high school in Arlington in 1903. Later in that same year, she began her teaching career in rural schools. The following year, she became principal of the secondary school in the rural area of Bartlett and then taught at the high school in Germantown—a larger community with a larger high school—until she became principal of that institution in 1912. She next served for two years teaching mathematics at the West Tennessee State Normal School (which became Memphis State University and now is the University of Memphis), until she was appointed superintendent of the Shelby County schools—the relatively affluent, but rural, district surrounding the city of Memphis.

Williams replaced her sister, who was getting married, as Shelby County superintendent, and she served in this position successfully, bringing reputability to the schools and the community, until she began to work for the NEA in 1922. As superintendent, she pursued reforms such as consolidation and improving school facilities, as well as relating the school curriculum to rural life. She was instrumental in reducing the number of one-teacher white schools in the county from 16 to 3. She was especially interested in African American schools, advocating support for them by funds from the county, which in turn was used to secure money for new schoolhouses from the Julius Rosenwald Fund. She appointed several Jeanes supervisors for African American schools and received support for teacher training and industrial education in these schools from the General Education Board.[2]

In addition to her school leadership, Williams also was active in women's political work in the Volunteer State, leading its campaign for ratification of the Nineteenth (woman suffrage) Amendment to the U.S. Constitution. Her rapid rise in the schools of Shelby County was due, in part, to her family's political connections. The famous "Crump" political machine was in firm control of politics and political patronage in Memphis and in surrounding Shelby County, and Williams's family enjoyed good relations with E. H. Crump and his political operatives. Further indication of both Williams's political astuteness and her ties to the Crump machine were her being selected as a delegate to the Democratic National Convention in 1920 and her being chosen as the first woman to become national vice chairman of the Democratic Party.[3]

Because of both her educational accomplishments and her political connections, as well as her work on various committees, Williams was elected to the presidency of the NEA in 1921. She was the youngest woman, the first rural woman, and the first southern woman to be chosen for the NEA's top elected office. She threw herself into her presidential duties during her year in office, traveling throughout the country on behalf of the NEA in its effort to win the allegiance of the women teachers of the nation.

Williams's election occurred in the midst of a most important period in the association's history. Founded in 1857, the NEA had functioned for its first six decades largely as a debating society that provided a platform for the ideas and platitudes of the leaders of American education, who were almost all male. In fact, women were prevented from speaking at NEA conventions until the end of the nineteenth century. In that century's last decade, however, women began to agitate for a greater role for themselves within the NEA and for the association to pay more attention to the conditions of the teaching force, the majority of which was becoming increasingly female.[4]

The election of Williams to the presidency occurred at the first convention in which the association was governed by a Representative Assembly. This form of governance was substituted for the town meeting format under which the NEA had formerly operated. The change in governance was adopted in the midst of a larger make-over of the NEA into an association that claimed to speak for all of those who worked in the schools. Women teachers had exerted some substantial influence on the NEA under the old town meeting format. They had managed, by the middle of the second decade of the twentieth century, to install a regular gendered rotation in the presidency. This pattern, with a man and a woman elected in alternate years, continued until the early 1970s. They also had institutionalized a concern for teachers' salaries and other issues such as tenure and pensions in the NEA, through the establishment of a committee that researched those topics and published its results. Because of these accomplishments, women teacher activists feared that the Representative Assembly would dilute their influence, which depended in large part on their ability to turn out a great number of voters at the annual convention. With the vote tied to representation from local or state bodies rather than simply to attendance, the women balked at approving the new governance system. At the same time that the NEA leadership overrode the opposition and engineered the adoption of the new arrangements, they understood that they could not afford to alienate women teachers. They clearly needed this constituency if they were to be successful in creating a powerful new occupational association.[5] Electing a noted woman educator such as Charl Williams as president helped the new NEA to earn and retain a positive image within the ranks of women teachers.

Williams's year as president, from July 1921 to July 1922, furthered the NEA's goal of effective outreach to women teachers. Her picture was featured in the middle of an article outlining the new NEA arrangement by which teachers would join and pay dues to the local, state, and national association. According to the article, this plan enabled the creation of "a complete and effective organization of the teaching profession, and the closest co-operation between local, State, and National organizations." After declaring the success of the new Representative Assembly form of governance in making the NEA "the instrument of the teachers of the nation," the article concluded with an announcement by President Williams of a prize for the state that enrolled the largest number of members in the forthcoming year.[6]

Williams, herself, issued a published call to service to "educators of all classes and ranks" in an article in the NEA's new national magazine. In this call, which was placed within a larger article on better salaries for better teachers, Williams indicated several challenges facing the NEA: enormous improvement in elementary education, particularly in rural areas; the universalization of secondary education "for every boy and girl;" and appropriate growth in higher education to meet the challenge of providing leadership in a democracy. She concluded that professional unity of all teachers within the NEA was required to meet these and other challenges.[7]

Prior to becoming president of the NEA, Williams had served as chair of the association's Committee on Tenure. In an article published during her presidential year highlighting the work of the Committee, Williams laid out the NEA's argument for increased tenure protection for teachers. She invoked principles of both civil service employment and professional stability as the grounding for that protection. She then noted the normal probationary period that preceded the granting of tenure, as well as the appropriate appeal procedure that should be attached to the process of granting tenure to teachers. She identified tenure as the foundation of professionalism in teachers, salary stability, and the occupational improvement of the teaching force. In all of these particulars, Williams linked the interests of teachers directly to the program of the NEA, concluding with a plea to support the efforts of NEA affiliates in every state to achieve tenure legislation.[8]

In several addresses to the NEA convention that culminated her presidential year, Williams sounded themes that reinforced the association's relationship with the teacher. Her presidential address, titled "The Democratic Awakening and Professional Organization," touted the recently enacted organizational changes as the cause of the increase in the number of NEA members from 50,000 to 100,000 in the past year. Moving to larger themes, she tied the professional improvement of the nation's teaching force to the task of democratic awakening throughout American society. After brief mention of the special problems of rural schools, she concluded with an invocation of public education as a vehicle for national greatness: "We are here to dedicate ourselves anew to . . . further improvement . . . to the end that education of the people, by the people, and for the people shall make good the glorious promise of democracy."[9]

In another speech to the NEA convention at the end of her presidential year, Williams sounded themes that reassured the women teachers who had joined the NEA that their investment in the association was well spent. After noting that the "classroom teacher is the very heart of the public-school system of America," she spent considerable time on the improvements that were necessary in teachers' salaries for the heart of the system to be maintained in a healthy state. Particularly important were pension legislation, minimum salary provisions, and a single salary scale to the educational improvement of the nation. The single salary scale was especially significant for women teachers, as it offered "the same salary to all men and women of equal training and equally successful experience, regardless of the grade or department in which the work is done." The single salary scale would not be realized fully until the time of Williams's retirement from the NEA at mid-century. She made support of this equity reform a constant theme of her work for the association.[10]

In two other speeches to the NEA convention in 1922, Williams decried the low average salary of the American teacher of $700 per year, bemoaned the fact that many teachers received even less than this paltry sum, and outlined the provisions of a model teacher tenure law.[11] Williams almost always combined her program for improved teacher welfare with a call for increased professional training, establishment of codes of ethics or other devices to ensure professional practice, and larger enrollment of teachers in local, state, and national educational associations. Like other NEA leaders, she realized that the success of the NEA rested on its ability to attract teacher members through appeals to both their material and professional interests.

EARLY WORK FOR THE NEA

After Williams's presidency of the NEA ended at the 1922 convention in midsummer, she returned to the Shelby County school system but only for a brief period. By November of that year, she was chosen to be a member of the NEA's permanent staff at its Washington headquarters. Williams's official title was that of Field Secretary, and her duties were to publicize and develop the NEA program with lay audiences.[12] Within two years, Williams also took on the role of chief NEA liaison to groups within the profession. She retained her position as head of the NEA's field office for the next quarter of a century and was the NEA's most visible female presence in these years.

Although Williams had a wide-ranging assignment as the NEA's field secretary, the bulk of the work that she did in her first decade on its staff was as a lobbyist for a federal Department of Education. Within the NEA's larger crusade to become the leading professional organization in the nation, the campaign for a federal department had an important role. A U.S. Department of Education would increase the visibility of the nation's educational effort, which was obscured by the place of the federal Bureau of Education within the Department of the Interior. More prominence and significance for education in Washington would, in turn, bring more respect for the NEA from the nation's political leaders. And this, in turn, would enable the NEA to pursue its program of professional and occupational improvement more effectively.

Williams had been involved in the NEA's effort to establish a federal department of education during her presidential year. As the elected leader of the association, hers was the first signature on "A Petition for a Department of Education," which was also signed by the leaders of 13 other national organizations. Prominent among these organizations were six women's groups (several with which Williams either had served or would serve as an officer); the American Federation of Labor (AFL); the American Council on Education; and national groups in the library and music fields.[13]

Williams worked assiduously on behalf of the federal department during her first ten years at the NEA. In that effort, she interacted effectively with the political leaders of women's groups and labor, as well as with NEA officials and other educational leaders and politicians who favored the cause. In analyzing the sources of her own commitment to the NEA program, the parallel between her earlier work for women's suffrage in the state of Tennessee and the NEA's political campaign deserves special notice. In the context of a report on her efforts for the federal department, she, herself, remarked to the 1927 NEA convention, "What we really need is some of the old spirit of the suffrage campaign." She added that many of the women at the NEA convention were familiar with that earlier struggle and that the decades-long campaign for suffrage that eventually bore fruit was a perfect model for the NEA's effort.[14]

As part of her work with women's organizations, Williams held office in a number of those groups, including the National Federation of Professional Women's Clubs and the National Congress of Parents and Teachers (NCPT). She served as a vice president of the parent-teacher group, and as director of its Department of Education. In that capacity, she wrote several articles in the NCPT magazine, *Child Welfare,* which supported the establishment of a federal Department of Education as a major step in accomplishing the improvement of education desired by the organization. She performed similarly in her role as an officer in the National Federation of Professional Women's Clubs, serving as spokesperson for the cause of the federal department within the organization, and as a writer of articles advocating the cause and other aspects of the NEA program in its publications.[15]

Williams's interactions with the AFL effectively harnessed that group to the NEA's campaign for a federal department at the same time that it helped maintain a wedge between the federation and its own constituent body representing teachers—the American Federation of Teachers (AFT). Support from Williams and the NEA for a child labor constitutional amendment, which was a cornerstone of the AFL's program, as well as Williams's successful interactions with AFL leaders at a personal level, cemented the alliance between the educators and the craft union leaders of the AFL. Marjorie Murphy has commented on the effectiveness of Williams in hampering the AFT. Furthermore, she contrasted Williams's "free rein" within the NEA to pursue affiliation with other women's groups and to appeal directly to women teachers with the plight of women activists in the AFT. The union women had to contend with the male dominance of their own organization, as well as with the conservatism of the larger AFL toward women's rights and other social and political reforms.[16]

Murphy also highlighted Williams's effective cooperation with Republican women, a critical alliance in the 1920s when three different Republicans held the presidency. Some of these women believed that Williams, herself, might become the first secretary of the new Department of Education when it was established. In addition to women Republicans, Williams interacted on behalf of a federal education department with several other more conservative groups that might have been expected to oppose it. Williams's reports and correspondence during this period reveal political alliances in support of the federal department with a group of southern governors, the Women's Christian Temperance Union, the Daughters of the American Revolution, the Scottish Rite Masonic Order, and several different national Protestant religious groups. Many Protestant denominations were conspicuous in their support of the federal department, whereas

Catholics were prominent opponents of the legislation.[17] Catholic fears of the federal department were linked to their campaign for support for, and their fear of government control of, their own private schools.

Williams led a substantial lobby for a federal Department of Education that acted throughout the 1920s into the early 1930s. Although the effort to establish the federal department was unsuccessful, Williams's political accomplishment in establishing and maintaining the political coalition that supported it was unprecedented in educational circles.

The depression and the election of 1932 boded ill for the NEA, as its own leadership, with the notable exception of Williams, was Republican. Although a Democrat, Williams herself, was able to work with the various Republican groups that supported a federal department, many for reasons allied to their fear of Catholics and the working classes. The candidacy of Franklin Roosevelt in November of 1932 was a cause for fear by the NEA and other advocates of a federal department, as FDR and the Democrats were linked to Catholic and other working-class interests that had little sympathy for the department or the NEA. Roosevelt's election in November of that year meant a substantial interruption in the NEA's effort on behalf of a Department of Education.[18]

Roosevelt's victory in 1932 also meant a decline in NEA influence in Washington for the next decade, in great part because of the overwhelmingly middle western, Republican, and Protestant background of the NEA's male leadership.[19] Unlike the male NEA leadership, Williams's southern roots linked her to the Democratic Party. Her work at the NEA lasted through all four of Roosevelt's administrations and into that of his successor, Harry Truman. It did not feature national lobbying for a federal Department of Education, however. While not abandoning national politics completely during the 1930s, Williams turned to other interests that had roots in her own past. It is those activities to which we now turn.

DEMOCRATIC EDUCATION AND PROFESSIONAL DEVELOPMENT

The onset of the depression and its deepening in the 1930s meant a time of rethinking for the NEA and a change in direction in the work of its staff. The NEA headquarters turned toward addressing the intensifying financial crisis that was increasingly affecting American schools. The association response was to sponsor a number of studies in the area of taxation, using them as a forum to push for more equity in state school finance. The traditional nature of the topic of school finance ensured that the association bodies that considered it would be overwhelmingly male. Charl Williams accepted this reality in her own work life and chose to stress her activities in women's groups and the parent-teacher organization, rather than try to break into the closed male circle of NEA finance reformers.[20] As already noted, Williams had multiple affiliations and responsibilities outside of the NEA, all of which served to complement her professional NEA endeavors.

In the middle- and the late- 1930s, Williams edited two volumes for the NCPT. Both books dealt with challenges presented to the public schools. The initial volume considered the role of democratic public schooling in the depression, and the second looked at the challenge presented to democratic education by the ideologies of fascism and communism.[21]

Although there was much overlap between the two volumes, there also was some difference. In the first, the economic crisis of the depression held center stage, and various problems created for the schools by the depression, such as poor financial support due to decreased tax revenues, were considered. The main theme of the volume was that the public schools should be supported and improved as a democratic bulwark against the dismal social, political, and economic climate caused by the depression. In the second volume, with the depression having receded somewhat by the latter part of the 1930s, the political moved to the fore. The first article in the book was by the noted Social Reconstructionist educator George S. Counts.[22] In it, he sketched out the major articles of the democratic beliefs that undergirded the American public school in a time of crisis. Equality and freedom of inquiry were the cornerstones for Counts that needed to be protected if the public school, and American democracy, were to survive.[23]

Williams's exact position on Counts's ideas is unclear. It is doubtful that she embraced the more radical aspects of the Columbia University Teachers College professor's educational ideas. However, she never went on record as opposing Counts and the other educational radicals. Her inclusion of a chapter by Counts in the volume for the NCPT is an indication both of the extent to which that lay body would consider educational radicalism as a solution to the crises of the 1930s and the willingness of Counts to present his ideas in ways that might make them amenable to an audience of lay supporters of public education.

Another chapter in the second volume edited by Williams in the 1930s related much more directly to an interest that she was just beginning to develop. Herman Donovan, a Kentucky educator, produced a chapter on "Making Teaching a Profession." In it, he considered the wide variety of problems standing in the way of a truly professionalized teaching force. In presenting various cures for the ills besetting the profession, he stressed the variety of in-service educational opportunities that were currently available to teachers. Most notable among these opportunities was the existence of the wide variety of courses and programs made available to teachers at summer schools, and their taking advantage of these opportunities in increasingly large numbers.[24]

This invocation of in-service and summer sessions as an occasion for professional development dovetailed nicely with a new program that Williams sponsored, beginning in the late 1930s. In the fall of 1938, she helped organize the first of a series of Institutes on Professional Relations, this one at George Peabody Teachers College in Nashville, Tennessee, the capital of her home state. In an article describing the conference, Williams remarked that she enjoyed the chance to hear speakers from several established professions recount their own problems, which were remarkably similar to those faced by teachers and the NEA. The differences, however, were as important as the similarities. Teachers had a social heritage that, in Williams's words, "has ridden our shoulders like an incubus." Part of this was due to teachers' status as public employees, which meant that they were more closely dependent on public support than traditional professionals. A major goal of this and future institutes was to develop a public relations emphasis for teachers' associations that would communicate the importance and significance of their work to the community at large.[25]

Williams went on to note that the rapidly expanding body of pedagogical knowledge was a boon to the movement to professionalize teaching. Similarly, the development of codes of ethics for teachers was a positive sign. The cause of professionalization was put at risk, however, because, as Williams noted, "teaching is not considered a life career by thousands of men and women who enter its ranks." She went on to bemoan the high rate of annual turnover among teachers. She then noted that "teaching can never truly attain professional status until certain discriminations are removed. Prominent among these are the regulations in many places concerning the married woman teacher." Herein Williams referred to the customary practice in many school districts of removing a woman from the teaching force if and when she married. The NEA was then beginning to highlight this discrimination against married women in an effort both to eliminate the policy and to win substantial membership support from women teachers. Williams thus effectively harnessed her own emphasis on increasing professional development to the NEA's larger thrust of increasing teacher membership.[26] She spent much of her last decade at the NEA developing, promoting, and attending NEA-sponsored Institutes on Professional Relations. Although they did not always consciously espouse the cause of women teachers—married or unmarried—Williams rarely passed up an opportunity to tie these institutes to those larger commitments.

The Institutes on Professional Relations were almost always held in the summer, on college or university campuses. The clientele was the teachers who were students in the summer schools at the various higher education entities. The official goals, as stated by Williams herself, were threefold: to unify the profession by encouraging cooperation among various organizations that trained and worked for teachers, to democratize the teaching profession by encouraging teacher participation in school administration, and to facilitate problem solving in education by encouraging "frank discussion of vital problems."[27]

During the years after the institutes were first developed, Williams was tireless in her advocacy. For example, at a national seminar on building strong educational organizations, held at the 1940 NEA meeting, Williams reported on the three-day-long-program. In addition to acquainting all attending with the concept and practice of the Institute on Professional Relations, she stressed the emphasis on public relations, which characterized many of the institute agendas. She added, however, that "placing the emphasis . . . on public service [in institutes and in the seminar] . . . did not suggest that loyal efforts toward better salaries, tenure, retirement, and other teacher benefits be diminished." Instead, she stated that such efforts should be "increased, believing that a profession cannot be evolved in poverty" and noted that the institutes served the public welfare "by giving the teacher a more secure place in society."[28] Williams then went on to describe a situation that hampered the NEA's membership efforts—the weakness of the tie between the local, state, and national association. Although professional relations institutes did little to directly address this relationship, they could at the least alert the teachers who attended the institutes that the NEA was interested in better organizational development in all of its existing organizations.

Two years later, another national seminar was held that was specifically devoted to building stronger local and state associations. In reporting on that seminar, Williams noted the twofold purpose of local associations as "the promotion and advancement of education and the promotion and advancement of teacher welfare." The work of the local associations was so important that the 1942 seminar resolved to concentrate on that topic alone in the seminar session at the next year's meeting. In her local association work, as well as in her professional institute work, Williams was effectively linking a variety of professional policies and commitments to the material interests of teachers.[29] Probably because of her popularity with women teachers, Williams was the NEA staff member assigned to work with all of these seminar groups. Her commitment to the strengthening of NEA local associations, linked with all of her work on behalf of women teachers, remained prominent throughout her remaining years at the NEA.

In addition to her work with the professional institutes and on behalf of strong local associations, Williams continued her earlier work with various women's groups. The wartime conditions of the early 1940s added some new dimensions to that work. Internationally, women's organizations were becoming increasingly interested in peace as the war heated up. Williams attended one international meeting in 1940 at which the topic of world peace was addressed in some detail.[30]

Federal relations also returned to a front burner for the NEA, as well as for Williams, during the war. In 1941, Williams reported on working for an education plank in the Democratic Party platform. One year later, she added that she was working assiduously on behalf of the NEA's push for federal aid to public education—a goal that was becoming increasingly important as the nation's school systems struggled under war-imposed financial burdens.[31]

Throughout the war years, however, Williams continued to work on behalf of professional institutes and local associations. In conjunction with the former, she wrote an article on teacher ethics for the Illinois NEA affiliate. In the article, she summarized the NEA's work on a code of ethics, dating back to the 1920s, and concluded with an endorsement of professional relations institutes as the proper arena in which to address ethics and other professional problems.[32] She gave a feminist twist to the institutes—one that came close to seeing women teachers and male administrators as having divergent interests—in an article published in the NEA's magazine in 1943. In that article, titled "Yes Mr. Rawlings," Williams began with the male administrator's admonition to Miss Giles, an elementary school teacher, that her first and most important task was "to do the best possible job in her classroom." Although Miss Giles agreed with her superior, she also understood that there were other priorities for teachers besides effective student contact. Williams then spoke of the professional relations institutes as the appropriate vehicle through which teachers could address their out-of-class concerns. The theme of conflict between woman teacher and male administrator was, at best, implicit in the article. No NEA publication would have dared to describe direct antagonism between teachers and administrators or between men and women. Nevertheless, the choice of Mr. Rawlings as the administrator in the article and Miss Giles as the teacher, as well as the endorsement of Miss Giles's belief that out-of-class concerns were a proper focus for her own energies, indicated that Williams was pushing the NEA to recognize that the occupational subordination of the woman teacher was a problem to be acknowledged in educational work and an issue to be addressed in the development of the NEA program.[33]

One year later, Williams highlighted the theme of teachers' legitimate concern with out-of-class issues in an article published in the magazine of the NEA's Department of Elementary School Principals. Here, the focus was on "Miss Bonny," whom legend said spent so much time on the nuts and bolts of her work of washing pots that she ignored the uses to which the clean pots were to be put. She scrubbed the pots so diligently and so repetitively that she even ignored her "Prince Charming" when he came to "cure her from her work." The result of this misplaced single-mindedness was that "her own fair face came to look—horror of horrors—like one of her pans." Williams went on to note the resemblance between Miss Bonny and those teachers who prepared their students diligently for the inspection of the supervisor or principal, who "beams on the automatons lined up for his scrutiny and devoutly hopes that Teacher will continue the good work, keeping herself well within the bounds of the schoolroom, where she belongs." Williams countered this scenario with the admonition that the teacher had "duties and obligations beyond the walls of the schoolroom" and that the "wise principal will know this situation" and take steps to facilitate appropriate teacher activity directed to those ends. Williams concluded that Institutes on Professional Relations were a most effective vehicle for teachers to develop their external relations, and she concluded that the Department of Elementary School Principals had made a great contribution in its support of the movement for such institutes.[34] When writing for the women elementary school principals, Williams used feminine imagery that was more traditional than she had used in the article in the NEA's magazine, perhaps because she was addressing an audience of older women who had served long enough to gain a principalship. Whether using more or less traditional language, Williams was consciously speaking to women in the educational profession and alerting them to the gains that women teachers could make through participation in larger educational affairs.

Williams made her boldest statement on the potential of the professional institutes as a vehicle for realizing the occupational goals of women teachers in a 1946 article in the NEA magazine. Perhaps reflecting the militance that would soon cause teachers to strike in several of the nation's cities, Williams began the article with the description of a teaching force with an 80 percent woman majority. The morale of those women, according to Williams, determined "in great measure the success of efforts to make teaching truly a profession." She went on to enumerate the obstacles to high morale for women teachers. First, the marriage bar meant that unless it were altered as the NEA desired, "marriage will automatically end the career" of those who married. Women teachers, according to Williams, "should be able to lead normal women's lives and still continue their career—indeed they should be encouraged to marry and have children and yet continue teaching." Additionally, women "should have a chance to take administrative positions in school systems and in their professional organizations which they largely support." These problems, and others, were peculiar to women teachers who were unable to "speak their minds freely without fear of reprisal." All of these matters were "appropriate to the alive Institute on Professional and Public Relations, . . . the place that any teacher could sit down on an equal basis with the administrator, the supervisor, . . . or the college president, to discuss problems in professional and public relations common to all members of the profession. . . ." Williams then described the progress that had been achieved in many states in establishing professional relations institutes and other innovations such as professional relations workshops and leadership schools. In these schools, as in the other NEA-sponsored forums, "each worker 'counted one' regardless of sex or teaching position." She commended the professional relations institutes "to every teacher who believes in such general welfare objectives as equality of opportunity in education, cooperation, and organization as a means for curing existing ills" and concluded that institutes were for "all who believe in these goals, or even have thoughts on them."[35]

PERSONAL AND POLITICAL PROMINENCE

Williams's increased boldness in raising women's issues in an NEA forum was characteristic of her activism on several women's issues in the last years of her service at the association. Late in 1944, as the end of World War II was drawing near, readers of the NEA magazine were apprised of the activities of women in the peace movement. At a White House Conference on Women called earlier in the year, which was chaired by Charl Williams, plans were made for women to be involved in "all councils dealing with post-war policy making." These plans and the other activities engaged in at the conference were termed "a milestone in the historic march of women" by first lady and intellectual leader of independent women, Eleanor Roosevelt. A speaker at the conference noted that there had been "an encouraging change of approach in the matter of women's participation in public affairs" since the early suffrage days. This change meant that government had the right "to have the special services that women have to give." Those services included, of course, education in which the great majority of the teaching force was made up of women. The conference speaker warned that "it would be a costly oversight to omit from leadership in this field those who know best the mind of childhood and how it may be led forth." After calling for a roster of women qualified in many fields to aid in postwar planning, the article ended with a return to the special role of women teachers: "To the development of an enlightened public opinion in their own communities, teachers can make substantial contributions and can lead in a mobilization of women to their obligations as citizens with a stake in the future."[36] The prominence of Charl Williams as convener of this White House conference, as well as the highest ranking woman on the NEA staff, linked her firmly in the minds of the women teachers who read this and other articles in the NEA magazine to the cause of women teachers as well as to the advancement of American women in general.

Later in that same year, Williams was instrumental in the establishment of another White House conference, this time devoted to another topic that was of long-standing personal concern to her as well as to the NEA—rural education. In the publication arising from the conference, Williams wrote an introductory essay that profiled the history of earlier White House conferences and detailed the steps engaged in to prepare for the current one. Williams paid homage to President Franklin Roosevelt for his support of the conference. She used more space, however, in acknowledging the effort of Mrs. Roosevelt, who had responded immediately when asked for support. Williams recounted the idea for the conference as being hatched as she sat in "the mezzanine section of the Republican National Convention" earlier in that year. Positive response to the idea from both Republican and Democratic politicians emboldened her to proceed with the planning and to vet her plans to Mrs. Roosevelt "during an overnight visit to Hyde Park." Eleanor Roosevelt's positive response meant that the conference became a reality, fueled by the participation from NEA leaders at national and state levels. The conference and its ten study groups produced reports that would receive nationwide distribution. Although she demurred from predicting what the resulting recommendations would be, Williams ended her essay with a final homage to the president and his wife: "I cannot close my remarks without expressing my deep and abiding appreciation to Eleanor Roosevelt, without whose cooperation and vision this Conference could not have been held, and to our great President—my friend of long standing—Franklin Roosevelt, whose talk to us this afternoon will give significance to the cause to which this group of men and women have devoted their lives."[37]

Williams's political connections with the Roosevelts clearly had been instrumental in establishing this conference and her high-profile role in the earlier conference on women. Her interactions with women's clubs and with the NCPT, as well as her work as an advocate of the NEA's various federal involvement schemes, culminated in her personal triumph in the two White House conferences. Her long-time activism in the Democratic Party, as well as the president's affinity for the South and southern Democratic votes, probably also contributed to her political influence. Given the male, midwestern, and western NEA staff leadership's affinity for conservative causes such as Prohibition, as well as its enmity for Franklin Roosevelt, Williams was one of the association's most effective communicators with the White House during Roosevelt's terms in office.[38] Her personal relationship with Eleanor Roosevelt, forged to a great extent through Williams's work with the various women's clubs, reinforced her image as a formidable political leader of the NEA.

CONCLUSION

In spite of her influence and the considerable energy with which Williams pursued her NEA duties, she was unsuccessful in achieving the NEA's programmatic goals of a federal Department of Education in the 1920s. She had substantial success in other ways, however. Personally she became a powerful figure in the women's movement, particularly in Washington, D.C. Her long-time service in various women's clubs no doubt set the stage for that influence. It peaked in her last years on the NEA staff when she was closely identified with Eleanor Roosevelt and her husband. The memorial article published at the time of Williams's retirement in 1949 highlighted the Roosevelt connection, pointing out the two White House conferences discussed above, as well as Williams's personal presentation in 1947 to Eleanor Roosevelt of several mementos of their interactions. Additionally, the article noted Williams's friendship with former Secretary of State and Mrs. Cordell Hull, with then President Calvin Coolidge in the 1920s, and with numerous other Washington luminaries.[39]

Williams's affiliation with the Democratic Party, as already noted, contrasted with that of most of the male NEA leadership. This difference raises the issue of other possible cleavages between the NEA's leading woman staff member and the rest of the association hierarchy. On the surface, all was well, as indicated in the formal tribute paid to Williams by NEA Secretary Willard Givens, at the time of her retirement.

> You have given the best years or your life to a great cause—that of helping to improve and strengthen our system of free, public, tax-supported schools. . . .
>
> The officers and members of the Association are deeply indebted to you for the fine, constructive work which you have done. The members of the headquarters staff value your friendship.
>
> You leave the active service of this Association with our best wishes for health, happiness, and a deep sense of satisfaction which comes from knowledge of important work well done.[40]

In spite of these official sentiments, there is some evidence of gender-related conflict between Williams and other NEA leaders. A female long-time NEA staff member, who recorded a lengthy interview as part of an effort to document the history of the association, spoke at length about Charl Williams. According to this source, Williams was looked up to by most of the women on the NEA staff and by many women teachers. Furthermore, Williams was conscious of her role as women's leader and conducted herself both personally and professionally to maintain it. Williams took part enthusiastically in the campaign to develop local teachers' associations in the 1940s, knowing that women had a great opportunity to build careers in the NEA and in the school systems by winning office in those associations. Williams's enthusiasm contrasted with the reticence of NEA Secretary Givens to commit enthusiastically to the local organizing efforts. Additionally, tension between Williams and Givens existed over her desire for travel arrangements that met her high personal standards and conflicted with his inclination to control costs.[41]

This conflict was never allowed by either party to boil up to the surface of NEA affairs. Yet it was known to some of the women on the NEA staff, and, in turn, given the connections between staff and membership, to many of the female members of the NEA.[42] It does not seem too great of a stretch to conclude that Williams was the visible leader of an informal, but formidable, women's caucus within the NEA staff and membership. Although this caucus functioned informally and chose seldom to reveal itself or to advocate its own priorities apart from those of the larger NEA, it functioned to keep the male leaders of the NEA in tune with the association's expressed commitments to women teachers.

Like male NEA leaders, Williams identified strongly with school administration. Also, like other female NEA staffers, Williams was far removed from the classroom. For these two reasons, she was unlikely to favor any women's rights–oriented teacher moves for independence from school administrators. Yet the situation in the teachers' union—the AFT—was worse for women than that in the NEA, according to Marjorie Murphy.[43]

The informal women's caucus within the NEA, then, with Charl Williams as its unelected but visible spokesperson, can be seen as the major vehicle available to women teachers for the pursuit of their occupational and political objectives, particularly in the late 1930s and 1940s. Rather than her work on behalf of the NEA's desire for stronger federal involvement in the schools, her commitment to professional institutes, or even her substantial political reputation, this unofficial but very real and often effective women's caucus represents the most significant contribution made by Charl Williams during her NEA career. Combined with her constant devotion to the causes of equal pay for women teachers through the single salary scale and her defense of the married woman teacher, Williams can be seen as a true pioneer in the women teachers' movement of the twentieth century.

NOTES

1. Material related to Williams's career is found in several boxes in the NEA archives, reached through the association headquarters in Washington, D.C. Biographical details are in an undated (1949?) special issue of the *NEA Journal*, in box 462 of the NEA papers.
2. Mary Hoffschwelle, *Rebuilding the Rural Southern Community: Reformers, Schools, and Homes in Tennessee, 1900–1930* (Knoxville: University of Tennessee Press, 1998), pp. 43, 65, 81.
3. Charl Ormond Williams and Carroll Van West, eds., *The Tennessee Encyclopedia of History and Culture* (Nashville: Rutledge Hill Press, Tennessee Historical Society, 1998), pp. 1063–64. Crump had helped Williams in supporting African American education. In turn, he and his political machine received substantial electoral support from African Americans in elections. See Hoffschwelle, *Rebuilding the Rural Southern Community*, p. 81.
4. The best history of the NEA in its early years, though by no means an adequate one, is Edgar B. Wesley, *NEA: The First Hundred Years* (New York: McGraw Hill, 1957).
5. Wayne J. Urban, *Why Teachers Organized* (Detroit: Wayne State University Press, 1982), chap. 5; Erwin Stevenson Selle, *The Organization and Activities of the National Education Association: A Case Study in Educational Sociology* (New York: Bureau of Publications, Teachers College, Columbia University, 1932), p. 59.
6. "Effective Professional Organization," *NEA Journal* 10 (September 1921): 119–120.
7. Charl O. Williams, "The Call to Service," *NEA Journal* 10 (October, 1921): 135.
8. Charl Williams, "Tenure—An Important Problem," *NEA Journal* 10 (November, 1921): 151–152.
9. Charl Williams, "The Democratic Awakening and Professional Organization," *Journal of Addresses and Proceedings of the National Education Association* 60 (1992): 208–210 (hereafter cited as *NEA Proceedings*).

10. Charl Williams, "Actual Results of the Year," *NEA Proceedings* 60 (1922): 482–484.
11. Charl Williams, "The Hope and The Result of American Education," *NEA Proceedings* 60 (1922): 378–381; idem, "The Improvement of the Teaching Profession Through Tenure Legislation," *NEA Proceedings* 60 (1922): 685–688.
12. See insert, with photographs and description of C. Williams in *NEA Journal* 11 (November 1922): 371.
13. NEA Press Release (October 1921), including a copy of "A Petition for a Department of Education," addressed "To the President of the United States," box 323, NEA Papers.
14. Charl O. Williams, "The Policy of the National Education Association Towards Federal Legislation," *NEA Proceedings* 65 (1927): 152–156 (quotation, p. 156).
15. For example, see Charl Williams, "The Challenge," *Child Welfare* (July–August 1931) pp. 662–663; idem, "A Wise Economy in Education," *Child Welfare* (May 1932): 531–532; idem, "Are You Posted on Committees? Department of Education," *Child Welfare* (October 1932): 88–89; and idem, "A Message for American Education Week," *Independent Woman* 17 (November 1934): 338.
16. Marjorie Murphy, *Blackboard Unions: The AFT & the NEA, 1900–1980* (Ithaca, N.Y.: Cornell University Press, 1990), p. 115.
17. Box 594 of the NEA papers contains substantial correspondence to and from Charl Williams in regard to the creation of a federal department. The signers of these letters constitute a veritable who's who of American educational leadership in this period. The contents of the letters give fascinating insight into the politics of a federal lobbying effort by Williams and the NEA. For a listing of the national organizations supporting a federal department, see Williams, "Report of Legislative Division. National Education Association, *NEA Proceedings* 64 (1926): 1139–1140.
18. Murphy, *Blackboard Unions,* p. 114. Charl Williams acknowledged her own affiliation with the Democrats, and the unfavorable consequences that a Democratic victory foretold for the federal department, in an exchange of letters with an official of an educational film organization in 1932; see Thos. E. Finegan to Williams (20 September 1932) and Williams to Finegan (21 September 1932), box 594, NEA Archives.
19. See David Tyack, Robert Lowe, and Elisabeth Hansot, *Public Schools in Hard Times* (Cambridge: Harvard University Press, 1984), and Edward A. Krug, *The Shaping of the American High School, 1921–1940* (Madison: University of Wisconsin Press, 1972).
20. Marjorie Murphy has noted the depression years as a time of particular difficulty for women teachers in both the NEA and the AFT. Her account of the NEA reads as if Williams had left the association and no comparable advocate remained. This was not true. Williams remained with the association, but her profile was reduced substantially. See Murphy, *Blackboard Unions,* p. 172.
21. Charl Ormond Williams, ed., *Our Public Schools* (Washington, D.C.: National Congress of Parents and Teachers, 1934); idem, compiler, *Schools for Democracy* (Chicago: National Congress of Parents and Teachers, 1939).
22. Counts, a professor at Teachers College, was a leader in the movement to reform the public schools to enable them to deal with the economic and political crises of the 1930s. Earlier in the decade, he had electrified the Progressive Education Association, with his call to replace child-centered progressivism with a more socially oriented version. See George S. Counts, *Dare the School Build a New Social Order?* (New York: John Day Company, 1932).
23. George S. Counts, "Our Articles of Faith," in *Schools for Democracy,* Williams, compiler, pp. 13–25.
24. Herman L. Donovan, "Making Teaching a Profession," in *Schools for Democracy,* Williams, compiler, pp. 71–86 (especially pp. 81–82).
25. Charl Ormond Williams, "How Professional Are Teachers?," *Peabody Journal of Education* 16 (September 1938): 118.

26. Ibid., p. 119.
27. Charl Ormond Williams, "Field Service," *NEA Proceedings* 78 (1940): 921.
28. Willie A. Lawton and Charl Ormond Williams, "National Seminar on Building Stronger Professional Organizations," *NEA Proceedings* 78 (1940): 104–105.
29. Joe A. Chandler and Charl Ormond Williams, "National Seminar on Making the Teaching Profession More Effective Thru [*sic*] Local, State, and National Associations," *NEA Proceedings* 80 (1942): 70.
30. Charl Ormond Williams, "Field Service," *NEA Proceedings* 79 (1941): 876.
31. Ibid.; Charl Ormond Williams, "Field Service," *NEA Proceedings* 80 (1942): 479.
32. Charl Ormond Williams, "Teacher Ethics and Professionalization," *Illinois Education* (March 1942): 198–199.
33. Charl Ormond Williams, "Yes, Mr. Rawlings," *NEA Journal* 32 (March 1943): 82.
34. Charl Ormond Williams, "The Legend of Miss Bonny," *The National Elementary Principal* 23 (June 1944): 41–42.
35. Charl Ormond Williams, "Professional Institutes," *NEA Journal* 35 (January 1946): 29.
36. "If the Women of America," *NEA Journal* 33 (September 1944): 149.
37. Charl Ormond Williams, "Background of the Conference," *Proceedings of the White House Conference on Rural Education* (Washington, D.C.: National Education Association of the United States of American, 3, 4, and 5 October 1944): 27, 28.
38. On the enmity between Franklin Roosevelt and the NEA leaders, see Tyack, Lowe, and Hansot, *Public Schools in Hard Times,* chap. 3 (especially pp. 106–110).
39. Special Number, *NEA Journal* (n.d., 1949).
40. Ibid.
41. Hazel Davis Interview (17 June 1988), box 3117, NEA Papers. This interview was undertaken by a consultant hired to conduct a number of interviews with former NEA staff members. It consists of three tapes that have not been transcribed. For more information on Davis, see Wayne J. Urban, *Gender, Race and the National Education Association: Professionalism and Its Limitations* (New York: Routledge Falmer, 2000), pp. 70–73, 171–175, 194–196.
42. NEA staff members were routinely assigned to liaison duties with constituent groups and committees of the association. The Department of Classroom Teachers, the major institutional link to women teachers, was the main responsibility of its director, NEA staff member, Agnes Winn. Winn was an associate and admirer of Williams.
43. Murphy, *Blackboard Unions,* p. 115.

CHAPTER THIRTEEN

AND GLADLY WOULD SHE LEARN: MARGARET WILLIS AND THE OHIO STATE UNIVERSITY SCHOOL

CRAIG KRIDEL

And gladly wolde he lerne, and gladly teche.

—Chaucer, Prologue

Can educators live successful, gratifying careers while remaining in the classroom? Can teachers emerge as school leaders without ever taking administrative reins? The career of Margaret Willis (1899–1987) causes one to examine these questions carefully, as well as to reconsider our conceptions of feminist leadership and progressive education.[1] With recent efforts to establish career ladders that permit teachers to "move up the rungs" yet not to climb out of the classroom, the professional career of Margaret Willis becomes all the more relevant. Willis represents "the" classroom teacher—one who was satisfied to work with secondary school youth throughout her 47-year career. Thirty-six years were devoted to one specific institution—the Ohio State University School. She arrived as a member of the first faculty in 1932 and retired in 1967, with the close of the University High School (after three years of protest against the decision to discontinue the school), just one month before the final high school commencement exercises.

This volume features the careers of founders of child-centered, progressive schools, yet Margaret Willis and the Ohio State University School counter these terms somewhat. First, Willis was not the founder of this school. Due to its unique conception, however, the Ohio State laboratory school was never really "founded" in a conventional way. College of Education Dean George Arps and Professor Boyd H. Bode began discussing the importance of a laboratory school as early as 1922. Arps's periodic lobbying of the Ohio General Assembly led to funds for construction and equipment in the spring of 1929, and the school, established in 1930, finally opened in 1932. However, a "founder," in the traditional sense, would have too strongly determined predefined ends, that is, absolutes, which are anathema to Bode's vision of education.[2] This school experiment was conceived in a Deweyan sense of democracy, in which structure emerged. No "founder," who would have been assigned a role of power by an outside entity, would be permitted to exert that degree of influence during the initial stages of the venture.

Willis helped to "form" the school throughout the years. Her leadership emerged as the Ohio State University School permitted individuals to define themselves and their place in the social order. With an emphasis on the cooperative expression of ideals and the role of collective intelligence in the process of social change, Willis embodied the spirit of democratic leadership and community that came to represent the Ohio State University School. Few metaphors accurately describe her presence—she was not "the brains of the operation" nor was she the "heart" of the school. Describing Willis in his recent history of the Ohio State University School, school historian Robert Butche, quite accurately states, "she would prove to be a learned and complex person not so easily labeled."[3] Perhaps "guide" or "usher" would be a better designation; yet, none really suffices. Many other educators represented the school, and many spokespersons described its countless innovative projects. Yet, when the topic of the University School arose in conversation, Margaret Willis's name typically materialized at the outset.

Secondly, the Ohio State University School belies somewhat the commonly accepted perspective of progressive education as a child-centered endeavor. This school sought to explore "the manner and the order of presenting subject matter and the ways in which the different subject matter is interrelated and integrated into a progressive educational program."[4] Yet, the educational program did not have its roots in the child-centered, private day school movement of the 1910s nor in the Project Method of the 1920s. The Ohio State University School, a public school for the State of Ohio, opened in time to participate in the Progressive Education Association's (PEA) Eight Year Study[5]—a curriculum development, staff development, and evaluation project that took place between 1930 and 1942 and included hundreds of teachers and thousands of students. Participating schools were extended the freedom to develop the secondary school curriculum. The purpose of the University School, however, was to explore the implications of pragmatism and the conception of "democracy as a way of life," as defined by Boyd H. Bode.[6] The school's focus became the development of an integrated educational program and the establishment of a school community that did not "reflect" society but, instead, served to embrace the best attributes of society. From the carefully selected inscriptions above the building doors ("Prize the Doubt," "The Old Order Changeth," and "New Occasions Teach New Duties") to the relations among teachers and students, the school sought to articulate the practice of "a democracy which is not a matter of forms and election ritual, but rather of finding effective ways for the administration, faculty and student body to cooperate in the search for answers and solutions."[7]

Although the Ohio State University School emphasized aspects of education other than a child-centered curriculum, one overlooked dimension of progressive education emerges quite clearly from the school and Margaret Willis's career—namely, the dynamic, active intellectual life of the classroom teacher. This theme, teacher as learner, while certainly not universal within the ranks of all progressive educators, appears with regularity in the progressive literature throughout the 1930s and constitutes the theme of this chapter.[8] Too often, progressive education is defined by the curricular activities of students. We have not, however, fully considered how progressive educators, themselves, were prepared or examined how they continued to be professionally nourished in their educational settings. In fact, only recently have we begun to realize the importance of examining the lives of teachers. "Teachers and teacher education students have lives—personal histories that demand attention because they influence how they interpret the world and determine what is learning in teacher education."[9] A teaching career centering on learning as well as teaching, intellectual adventure, experimentation and community—such are the insights of examining this individual life. Margaret Willis's story is not representative nor generalizable. Her style of leadership is not easily reduced to a series of traits nor does it ascribe to an essentialist image of a feminist selfless nurturer. Yet, as one becomes aware of Willis's work—a career distinguished yet not uncommonly different from many unrecognized classroom teachers throughout the decades—one becomes enraptured by the power of biography "to inspire comparison. Have I lived that way? Do I want to live that way? Could I make myself live that way if I wanted to."[10]

A LIFE OF CURIOSITY AND ADVENTURE

Margaret Miriam Willis's educational background was not uncommon for women who emerged as academics in the early part of the twentieth century.[11] Willis stated that she "did not plan to teach; I just wanted to know what was going on in the world and why."[12] Following her graduation from Wellesley College in 1919 with a B.A. degree in history and economics, she spent two years in Japan as governess for an American Embassy staff member. Her appointment was arranged through the Wellesley College placement bureau, and although Willis did not view herself as an educator, teaching offered her the opportunity to travel. She returned to her home in Mount Vernon, Washington, with no plans; however, she stated, "my interest in the world, its past, present, and future, had grown."[13] Willis taught high school history at the local secondary school for the next three years, from 1921 to 1924. As Willis later reflected on this experience, "I learned something and the children something, but none of us very much. When it got to be too monotonous I resigned."[14] The use of the word "monotonous" is noteworthy and helps to define a leitmotif for Willis's quest for the adventure of ideas. Her ultimate embrace of progressivism, her social agency to community, and her never-ending thirst for travel all embody this sense of adventure.

There were many reasons for Willis to remain in Mount Vernon. Her father, Herbert L. Willis, arrived in Skagit County in 1905 and quickly rose to distinction as county farmer and community leader. The daughter of a teacher, Margaret could easily have made a career in the community as educator or journalist-historian. In fact, she developed this role upon her return to Skagit County after years of service at the University School. Curiosity reigned supreme, however, she stated, "When any job has become so familiar that understood routine bulks large, I have always gone on to something else, taking what chance threw at my feet, my only demand of the experience being that it should be new."[15] New York City became the next chance thrown at her feet.

In 1925, Willis enrolled at Columbia University for an M.A. degree in political science (history—her thesis topic addressed imperialism in Korea) and then moved on to Maryland State Normal School (Towson State). She describes her Maryland State interview: "I spent all the time explaining how little I knew about education and educational theory, and when the director persisted in offering me the job, I accepted out of sheer amusement."[16] After two years, Willis left for Istanbul, where she eventually taught at the Constantinople Women's College for Girls for a four-year period, returning in 1931 to teach at the Bennett School in Millbrook, New York. No doubt Willis could have continued moving from one institution to another, demanding only that the "experience should be new," and living a satisfying, fruitful educational career. She was offered, however, a teaching post at the Ohio State University School, a new experimental demonstration school. This laboratory school furnished Willis with new experiences for the next 36 years.

Willis's quest for adventure, although satisfied domestically throughout her tenure at Ohio State, was still sought out in international settings. Along with periodic summer travels throughout Africa and the Middle East, she served as principal of the Cairo School for American Children from 1946 to 1948. Willis wrote, "I have great admiration for the teachers who have devoted their lives to educational work in some other nation, and for the things which their schools have accomplished. For myself, however, the variety of briefer contacts in widely separated areas of the Far East, the Near East, and Africa has helped me, as an American, to understand Americans and our relations with our world neighbors."[17] And, as Willis defined herself as a social studies teacher at the Ohio State University laboratory school, travel became part of her professional development. "I hope some of that understanding [of other nations] and some of my continuing curiosity about the fascination with the infinite variety of human attitudes and institutions carries over into my teaching."[18] Indeed it did, as Willis established herself as a beloved teacher during her career. What becomes important, however, is the school venue in which her teaching was housed. By 1932, her career comprised 13 years at seven different institutions; she had not spent more than 4 years at any one school. Margaret Willis proceeded to spend the next 36 years at one institution—the Ohio State University School. Not all schools would have provided sustenance for Willis and, quite frankly, not all schools would have welcomed her as part of the teaching faculty. This school, however, was quite special. It offered a richness of experience that never proved monotonous. Moreover, it provided an openness for the free exchange of ideas, different points of view, the clash of ideals, and a public space in which individuals could search for a greater good. At the end of Willis's teaching career she was in the midst of a battle with, among others, the associate dean of the College of Education, David Clark. She wrote to him saying, "I seem to have so many rejoinders lately. . . . I try to hold my tongue for two days, and then if the thing still has to be said for my peace of mind, I say it."[19] "Say it" she did throughout those 36 years. The dynamic aspects of this democratic community kept Margaret Willis at this school for the remainder of her professional career.[20]

Willis describes her abilities as teacher prior to her employment at the Ohio State University School in quite self-effacing terms. Of her two years in Japan, she actually admits that her teaching "was pretty mediocre and that it is an understatement to surmise that I learned more than the children did."[21] Perhaps she was merely teaching as a way to provide freedom and mobility. This all changed with her arrival in Columbus, Ohio. In 1940, she informed Wilford Aikin, director of the Eight Year Study, "My own career in teaching had been largely accidental and pretty haphazard before I came to the University School. I had managed to accumulate a great many experiences, but had done little toward thinking them through. I have found [the Eight Year Study] immensely stimulating and gained enormously from the experience."[22] Willis became swept up in the seriousness of the progressives' endeavor, the life of teacher and learner, the adventurousness of the classroom, and the camaraderie of colleagues and students. Her professional role found a venue in which it could flourish. "When [Eight Year Study] teachers conceive of teaching as a creative art, based upon scientific method and knowledge, they are transformed from routine workers to creative adventurers."[23] And creative adventurer she became as a member of the secondary school social studies program for this quite extraordinary progressive school.

PROGRESSIVE EDUCATION: EXPERIMENTATION AND THE PLAY OF IDEAS

To define the Ohio State University School as a progressive school is rather simple since the faculty so closely worked with the PEA. To define the term "progressive education" becomes much more complex. In fact, the PEA seemed to have its own difficulties. At the annual 1938 PEA meeting, a committee reported on its efforts to define "progressive education." Although a statement was produced, nearly the entire committee disagreed, noting that progressive education is not a definition but "a spirit." Too often, progressive education has been defined by slogans, such as "learning by doing" and "freedom to develop naturally," and viewed primarily, if not solely, as a child-centered curriculum, focusing on the "needs of students." Such impressions overlook the complexity of the classroom and the controversy within the PEA over student needs. As we recall the disagreements among progressives towards "dare the schools change the social order,"[24] the debates over "determining needs" proved as volatile. Boyd Bode and V. T. Thayer waged war over these distinctions and their implications for progressive practices. Bode did not support a curriculum based on student needs and, as a result, resigned from Thayer's PEA Commission on Secondary School Curriculum with the objection that "needs is a weasel word." In fact, many Eight Year Study participants saw the child not as the focus of a school program; building democratic communities and establishing the interests of the group proved to be the teacher's quest. I raise this point solely to suggest the complexity of those practices that fell under the description "progressive education." Many individuals were defining progressive education in many different ways, and even the PEA was viewed to have competing coalitions among its constituency.[25]

Standing among the factions was Eugene Randolph Smith, past president of the PEA and headmaster of a private Boston day school. Smith recognized the importance for school reform while placing substantial faith in the scientific method and achievement tests; he defined progressive education in the following way: "It seems to me that truly "progressive" education must continually be tested by two questions: Does it keep itself fitted to present day requirements, changing as necessary with changing living conditions and changing needs? Does it keep apace with investigation and discovery in the educational field, particularly with advances in teaching methods, and with the most recent discoveries in child psychology?"[26] Although I venture not to determine "the spirit" of progressive education, Smith's statement features one neglected aspect—investigation and discovery were primary missions of progressive educators. Lest we forget, Dewey's laboratory school was conceived as a place to experiment with educational practice. Were there such places in the 1920s and 1930s? Certainly many schools identified themselves as laboratory schools; however, often these demonstration schools served merely to provide a convenient locale for the supervision of student teachers. Smith's definition brings forth a profoundly distinctive theme of progressive education—namely, the importance of experimentation. This proves to be the most distinguishing aspect of the Ohio State University School and one reason for Willis's long tenure there. Child centeredness and "learning by doing" have obscured the rather remarkable pursuit of adventure and experimentation.

Dewey saw education as "the laboratory in which philosophic distinctions become concrete and are tested." And, as a laboratory, the Ohio State University School was designed for school experimentation, in Smith's views, to attend to a "changing society and to the investigations and discoveries in the education field."[27]

How would a spirit of experimentation appear? We often overlook the play of ideas, the curiosity, the delight of discourse, conversation, and exploration of ideas. Such teachers as Willis abhorred the monotonous and were engaging, exploring, and wondering about new ways of instruction, new forms of curricular materials, and new ways to build community. Teachers were involved in their own pilot demonstrations—creative classroom adventurers—engaged in experimentation and wondering if others would find their ways useful. For example, H. Gordon Hullfish, Willis's colleague, describes listening to a conversation between John Dewey and Boyd Bode in the late 1920s as they were engaged in an "informal exploration of ideas. In the presence of these giants, a young man would have little to say, being grateful, as I was, for the opportunity to listen. Eventually, however, I did make a comment. I was amazed when Dewey asked me to repeat it and so overcome with excitement when he said, 'I never thought of that before, let's explore it,' . . . My comment related to a procedure I thought a teacher could use fruitfully in helping students both express and examine their beliefs. Our unhurried exploration, under Dewey's guidance, deepened my understanding of an idea I had thought I controlled and suggested ways in which I could further test it. The discussion then turned to other questions and, as the day progressed, Dewey and Bode went on to [other] matters. . . . Within ten days, however, a letter from Dewey said that . . . he wanted to report what had happened when he tried out my idea in class."[28] Discussion, exploration, reflection, the playfulness of ideas—for Dewey, Hullfish, Bode, and Willis—constituted a spirit of progressive education. Perhaps trite, perhaps prophetic, this anecdote is certainly indicative of educators learning and thinking about teaching.

OHIO STATE UNIVERSITY SCHOOL: A PLACE FOR THE DEMONSTRATION AND EXPLORATION OF IDEAS

The University School served as a demonstration center for the Ohio State College of Education and the state of Ohio. In this respect, the school's charter was somewhat different from other progressive schools. Funded by the General Assembly, the University School was conceived as a public school and an experimental and observation center for the state of Ohio. Upon opening in 1932 as an integral part of the College of Education, the school faculty and students permitted its organization to emerge. "In line with the policy of setting up no organization in advance of need,"[29] guidelines, rules and regulations, and mission were drafted at the end of its first year.

The school was not a conventional neighborhood school; approximately 400 elementary, middle, and high school students from throughout the city were enrolled. Although the majority of pupils came from the middle and upper-middle classes, a socioeconomic, racial, and intellectual cross-section was sought. "Every effort has been made to maintain a 'typical' school situation and to prevent the student body from being a 'selected' group." The student population included 10–20 percent African American pupils—and IQs, while never used to screen for admission—averaged approximately 115, with some students in the 80s or below.[30]

As a demonstration center the school hosted annually tens of thousands of visitors. Through the 1930s and 1940s, the school averaged 15,000 visits per year. As late as the 1950s, a time period not known for interest in progressive education practices, 40 visitations per day were common (during the 1953–1954 year, over 8,700 individual observations were recorded).[31] The experimental dimensions of the school involved a rather remarkable exploration of core curriculum and the role of teacher/pupil planning as "equal" members of an educational community. These themes were developed from countless experiments and demonstrations through the years.[32] The University School helped to define further curricular and instructional practices in mathematics (the "nature of proof"), reading and language arts ("free reading" programs), unified science programs, and other experimental projects in guidance, music, dance, and fine/industrial arts. The school was perhaps best known, however, for its integrated core programs and its efforts to relate the various separate subject areas. Social studies and language arts were closely linked, but interestingly, Willis objected to efforts to "merge" those two fields. Willis recalled, "I remember with special vividness a June meeting of the Thirty Experimental Schools about 1934 or 35 at which a spinsterish lady of uncertain age spoke glowingly of the happiness which had resulted in her school from the marriage of English and social studies. They have never been married in the University School. . . . English and social studies work together where their aims are best realized by such cooperation, and having them work separately where their objectives are distinct."[33] Classroom experimentation was conducted but not at the expense of knowledge and student learning.

Although the curricular experimentation was radical, the University School administration was commonsensical. Willis commented that the school "tried to avoid arbitrary rules and meaningless routines while insisting on the stability and order which are necessary for effective work. The small size of the school plus the general atmosphere of freedom with responsibility facilitated this. When children arrived at school in the morning they could go to their rooms whether or not the teacher had arrived. During the day they went to rest rooms, or lockers, or library, or on some errand without a system of hall passes since the level of self-direction was high."[34] Similarly, administrative traditions were continually reconceived as the school broke down established grade levels and artificially constructed subject areas. The school followed portfolio grading, and while tests and examinations were administered periodically, students did not receive conventional letter grades. Teachers stayed with classes for more than the conventional one-year period (now called looping), and four "houses" with approximately equal membership from every class were organized for extracurricular activities. The lunch period was seen to serve educational ends "not only from the dietetic but from the social point of view." Students were seated in natural social groupings with designated classmates serving as hosts.[35] Even administrative staffing proved unusual as the teaching staff conducted searches for their director and submitted a short list, in order of preference, to the dean of the College of Education.

WILLIS AS LEARNER, TEACHER, AND LEADER

Within the context of the Ohio State University School, Margaret Willis *began* a career as teacher. This is also where she emerged as leader. She did not seek power; it seemed to evolve as she lived an active, independent, and curious life. "I am satisfied only when performing an active deed, when I am living for the present in a daring and adventurous deed. I dislike cautious foresight and relaxed ease. . . . I think that life tends to become sluggish, too comfortable, unchanging because of too much thought and no action. . . . I am eager for constant activity."[36] Such daring and independence were contagious. Robert Butche attributes this sole trait as the defining characteristic of Willis's leadership; "her independence caused others to act independently. They wanted to be around her because they felt this independence. And her force of personality comforted those who were not yet ready to act independently."[37] Willis's "democracy as a way of life" became a balance of independence with cooperation. Clearly, democracy did not mean mere group decision making. "The practice of democracy can be made unutterably wearisome and wasteful," Willis wrote, "if the whole group undertakes to make every decision and to perform every function."[38] She described the fall of 1932 when the University School faculty met almost daily for a month before the school opened: "We held endless faculty meetings, evaluating, criticizing, planning, making new schedules, and deciding every question, major and minor, only after long group discussion. How the faculty lived through the first few months is still a mystery, but somehow the sense of working together at something important carried us through."[39] This was not a democracy of "forms and election ritual" but a democracy for independent thinking, cooperation, and thoughtful discourse.

This demonstration school became the space in which the complexity and oscillation of Willis's feminist-pedagogical traits took form. She believed in discourse, cooperation, and the democratic process, yet she was not long-winded and at times seemed quite gruff. Impatience caused her to change subjects with her common expression, "We can't go any further with this." Stories abound of Willis "no-nonsense, no-further" demeanor.[40] A colleague of Willis described a special meeting in the early 1960s called, at the behest of certain faculty, to establish a student dress code: "Margaret Willis arrived at the faculty meeting with her usual stash of manila folders clutched firmly in hand. The look on her face, however, foretold her contempt for the business at hand. 'I've been here since the School opened,' she announced in her sometimes faltering voice. 'Fads come and fads go, and so far we've survived them all.' With that, Margaret Willis gathered up her papers and left the room. Recognizing there was nothing more to be said, everyone else soon followed."[41] On another occasion in the 1950s, the parents of a graduating high school senior objected to their daughter being paired with an African American graduating senior for the commencement processional into the amphitheater. Since the class included only two African Americans, male and female, the parents requested that those students proceed together and that another male student be assigned to their daughter. Willis was called to the meeting with the parents and, when she asked, "what is this all about," the parents described their concerns. Willis looked at them and said, "I don't see that as a problem." She then stood up and walked out. The meeting concluded with no comments, and the processional pairings were not changed.[42]

With this no-nonsense behavior, her compassion—the role of selfless nurturer—also appeared as she returned to the basics of community and democracy. Yet, the question becomes, What was she nurturing? True to her progressive spirit, she was not content to nurture merely the "needs" of students. Margaret Willis, who "suffered no fools," nurtured the intellect. As was the case with her colleague, Lou LaBrant, both women "combined a sincere concern for the student while rigidly holding each to their idealized standards of academic rigor; LaBrant and Willis bent neither to soft emotionalism nor to a static curriculum."[43] The embodiment of feminist pedagogy and leadership, conceived in this volume as more humane, more democratic, more concerned with caring, is placed in juxtaposition with this display of fluidity. The curiosity of an authoritarian

Margaret Willis demanded all to join her in the "academic chase," the "intellectual adventure," and the delight of new experiences.

Students loved this; and, yes, some students mocked her, but all colleagues and students respected her.[44] Although she was never described as charming, her kindness and humanness embodied a leadership style that proves as complex as any description of progressive education. Willis was tough and gentle; her interests, however, remained focused on learning and teaching and providing the best experiences for her students and colleagues. Problems became occasions to instill learning opportunities for others. Butche remembers one such occasion:

> I only saw Margaret Willis fearful of losing control once and it was in my Senior year in 1954. Our class was going to take the traditional senior trip to the big apple and Washington. A few days before departure, *the Goose,* as she was often called by her students, called me into her office with bad news. She had assigned me to room with the major troublemakers and disciplinary problems in our class. I complained loudly that I wanted to room with my friend and class president, or if that wasn't possible with a girl with whom I was, let us say, very infatuated. That was not to be, Willis told me, without so much as acknowledging the humor in my alternative suggestion. I was stuck with those I least wanted to be around. I threatened to not go on the trip.
>
> The next day, Willis invites me to her office again, but it was a far different person than the authoritarian, "you will do as you are told" teacher of the previous day. This was a Willis I had never seen before and saw only once again in the 1970's. Willis had not changed her mind about the rooming arrangements, but, in an effort to get my willing compliance, she choose to explain why. She told me that I was mature enough to apply my influence where needed. And Willis truly needed me to influence and control these classmates for she believed she might lose control of them on the trip. I immediately saw Margaret Willis in a new light. She did not walk on water. She did not always get her way. She was not cast in stone. She was fearful of the unknown just like all of us. I agreed to baby-sit the troublemakers and I was successful. I learned something important about Willis—beyond her vulnerability—and it was delegation of authority to achieve goals. I willingly served in *loco parentis* for her, kept the troublemakers in check, and in the doing learned how to take responsibility, seize control, and drive the ship of life.
>
> I revisited this episode with Willis later in life, and I was quite surprised that she barely remembered it. For little had I imagined that it wasn't the first time Willis trumped the hand of an emerging student and, in so doing, propelled them to independence.[45]

Selfless nurturer, thoughtful pedagogue, firm taskmaster. All of these characteristics were oriented toward helping students. She avoided entangling herself in their lives, preferring to help students become independent thinkers, work together in a spirit of cooperation, and realize that they were ready to participate in the adult world. Perhaps the most notable example occurred when Willis accompanied the graduating class on their senior trip to New York City. On a specified afternoon, students were shuffled into a hotel conference room, wondering and awaiting the appearance of a special guest. Willis's long-time friend, Eleanor Roosevelt, arrived and discussed national and world issues with the students. Some thought that this was Willis's way of showing the thrill of New York or to corroborate her friendship with Roosevelt. That was far from her intent. Willis wished to place her students in situations where they would converse, as equals, with adults. There was no "silly talk," and Roosevelt did not speak down to the youth. While some students would be overtaken with fame, others would engage in the discussion of important issues. The experience was "strengthening as students were placed in new environments and asked to take on new roles as young adults."[46] This was a form of her leadership, as Willis sought out new opportunities for the educational community of which she was a member. Willis was leader and creative adventurer, but students were to speak for themselves.

THEY WERE GUINEA PIGS

One of University School's distinctions was developing this balance of independence and cooperation among teachers and students. And what ultimately proved most unique was a gesture made by its students on behalf of themselves, their teachers, and their school. The graduating class of 1938 published a book about their six years at the Ohio State University School, entitled *Were We Guinea Pigs?* This was the first group to have completed their entire middle and secondary school years and, as they stated, "We feel that it is important for us to examine and evaluate our six years' experience. We are becoming more conscious of the principles behind all that we have been doing."[47] Two faculty members were more directly involved with the project, yet Margaret Willis became best known as "their teacher."

To have a group of 55 secondary students publish a book with a well-known New York publisher, Henry Holt, might be viewed as a gimmick if done today. In 1938, this clearly was no stunt. The students stated in their opening paragraph that all work was completed by them. "No teacher was even present at the final revision of the manuscript." Yet, what proves more interesting is their one sole reason for writing the book: "Our school is often misunderstood. We have been criticized and opposed because we are 'progressive.' So many attacks are made."[48] Throughout the 300-page book, the students describe in remarkable detail, made possible by their portfolio system, their many interesting curriculum experiences and the truly unique learning community of which they were members. Their emphasis, however, was on the quality of teaching throughout their school years. "We feel that our faculty is one of our greatest assets. Contrary to the situation found in many public schools, our teachers are free to express opinions of their own without fear of losing their jobs. We are fortunate in having teachers with a wide variety of backgrounds and opinions. . . . We enjoy talking with the teachers because our interests are similar. Our discussions are usually frank and open-minded."[49] Ironically, in 1936, Willis had summarized her two professional hopes and ambitions to include clarifying "my own thinking by writing up some of the phases of our experience here. . . ." and "maintaining enough contact with them (then the 10th grade, the graduating class of 1938) to watch or help work out their activity program."[50] These hopes were more than fulfilled and, in so doing, Willis perhaps most directly established herself nationally as "the beloved teacher" of these rather remarkable students.

Any educator who has become closely involved with a group of pupils inevitably asks, What happened to these students? Margaret Willis found out. Willis, who had stayed in touch with many of her students, including those members of the class of 1938, researched and published a follow-up study of this "guinea pigs" group. Her 340-page work, titled *The Guinea Pigs After Twenty Years,* reviewed in notable journals including The *Saturday Review,* explored the "impact" of secondary education. "The thesis of this study is simple. If basic high-school curriculum reorganization is worth the effort, it should have results which are apparent in the adult living of the students who experience it. This follow-up attempts to look at all the "guinea pigs" to find out how successful they are in their living nearly twenty years after high-school graduation, and to see what connection, if any, can be established between the nature of the high-school experience and the kind of adult living discovered."[51] Willis acknowledges the difficulties of defining "success in adult living" and recognizes the many variables that cause one to hesitate from making direct causal relations between school and adult outcomes. The project, however, reveals the continuation of Willis's curiosity and the school's quest for experimentation. Moreover, Willis turns the merit of the project back to her students, as she explains, "The idea for this book goes back to 1938 and is not original with me. When the seniors of the University School were writing *Were We Guinea Pigs?*, they hoped and expected that it would be followed up at some later date. The faculty shared the hope, but no definite or specific plans were made or responsibilities assigned."[52] I interviewed the "guinea pigs" in 1988 at their fiftieth school reunion; they assured me that if a follow-up study were to occur, they knew Margaret Willis would be the researcher.[53] Willis conducted interviews with all of the students, traveled to their homes throughout the United States and asked them to spend hours filling out questionnaires. "Only for Margaret Willis" was the comment many of them said to me. And Willis approached the task as scientific researcher, teacher, and learner.

I wish not to patronize the class of 1938 by summarizing their book nor do I plan to make unwarranted generalizations from Willis's claims. Yet, here is where Willis distinguished herself and, clearly, brought national recognition to the students and the schools. My point is not what Willis discovered nor what special experiences the students underwent between 1932 and 1938. This class worked alongside their teachers, and one particular teacher took the time to learn what became of her class. Willis "found the work on this study a most rewarding experience. To see what kinds of men and women yesterday's boys and girls have become is a dream of every teacher."[54] Yet, how many teachers actually do this. Willis's daring, adventurous deed was as commonsensical as any administrative change that occurred at the University School. Any teacher can track down former students after 20 years, and all teachers dream of what happened to their students. Willis initiated this discovery in her very fluid roles of nurturer, creative adventurer, teacher, learner, academic, and taskmaster. Reading today *Were We Guinea Pigs?* and *The Guinea Pigs After Twenty Years,* one senses the contagiousness of Willis's leadership.

TEACHER GROWTH: MEANING IN EVERY EXPERIENCE

To return to the first question of this chapter: Can educators live successful, gratifying careers while remaining in the classroom? Margaret Willis did. She knew that if teachers stayed in the classroom throughout their careers, they must be intellectually nourished. This consisted not merely of reading or traveling. Learning became as important to the teacher as it was to students; but, equally important, searching for meanings became the teachers' quest. Willis fashioned this quest as a theme in the sole published "philosophy" of the University School. Although authored by many names, Paul Klohr, former director of the school from 1952 to 1957, refers to the document as "Margaret's."[55] The statement explicates the meaning of a democratic society and a democratic personality, and stresses "that teaching is a calling which demands professional preparation appropriate to the functions of education in a democracy."[56] The preparation and continuing education of the teacher was placed at the center of the University School's program.

Willis had voiced strong views on the preparation of teachers for the Eight Year Study: "Selection should require that all successful (teacher) applicants demonstrate qualities of resourcefulness, personal balance, open-mindedness, ability to take criticism from 'inferiors' as well as 'superiors,' and a kind of intellectual curiosity and awareness which squeezes meaning out of every experience. One of the poorest criteria would be that an individual has been so happy and secure in a stereotyped learning situation that he wants to go on perpetuating it."[57] Inscribed above a doorway to the school, "The Old Order Changeth," captures Willis's view of teacher development as a way to break the traditional, the stereotypical, and the monotonous. In fact, Willis seemed to flourish in the most problematic of situations in which she was engaged in creative problem solving, and she sought to help and nurture others. She expected and demanded more of educators than a well-managed classroom. Management and discipline were neither the goal of teachers nor a way to build a life in the classroom. In 1936, during the midst of the Eight Year Study, Willis wrote, "There are so many interesting things that need doing, that a person owes it to herself to work at something she likes. I enjoy doing things that present new problems or old problems in forms which are new enough to be challenging."[58] These challenges found their way into Willis's pedagogical life. She described the satisfying yet exhausting nature of teaching at the laboratory school, yet "it was not the weariness of meaningless routine. There was routine, of course, but unless it served some educational purpose, it could be discarded. It was not the fatigue which comes from frustration [and] it was not the nervous drain of petty bickering and backbiting."[59] There would be no meaningless routine, and Willis criticized, rather strongly, those teachers who were having "one year's experience repeated fifteen times."[60]

Teacher development emerged as Willis's practice to draw meaning from every experience. She seemed unperplexed by what would arise in class; the weeks and months offered "new occasions and new duties."

> One female student told Willis she wanted to study sexual intercourse so she would be properly prepared when the time for such knowledge would matter. Willis was unperturbed, instead offering several viewpoints from which such a study might be pursued. One can only imagine the enthusiasm with which male students looked forward to the young woman's forthcoming report, but when the day came her report was treated with the same degree of academic analysis as others. Few teachers might have been able to see such an investigation as worthy, or appropriate for a high school senior. But Willis saw it as only one of millions of subjects which students might study, evaluate, think about and, best of all, to write about. Willis saw every experience as worthy of sharing, and all material worthy for writing where ideas are expressed and evaluated.[61]

On another occasion, in 1965, Butche described when he was sponsoring higher education experiences for children of families that he had met in Thailand. The first student to arrive, one who had not completed secondary education in Bangkok, was in need of further evaluation. Butche called Margaret Willis for advice. Willis told him to bring the youth to the University School.

> Margaret did not see this as a burden, nor as an obligation to a former student. She saw the Thai student as a resource for her senior class. Upon arrival, Willis began discussing the flora and fauna of Ohio and how different it must be to that of Thailand. Flora and fauna was not of much interest to the city kid from Bangkok, but it proved to be a topic that provoked discussions among students and guests just long enough for Willis to determine his suitability for inclusion in her class. She made her decision with typical speed and self-assurance: Willis offered to take the Thai student into her senior class for the remainder of

> the school year. After a period of time Willis concluded that the Thai youth was not ready for university-level work and suggested that he enroll in an English-as-a-second language program. But that was Margaret—always searching for interesting occasions for her cluster of students and helping those who needed help.[62]

Willis's long pedagogical life was a result of squeezing meaning out of every experience and embracing problems—not fleeing or trying to end them. Perhaps her independence, her confidence, or other unrecognized traits came into play to establish her leadership. But Willis, with her unbridled curiosity, lived the academic chase and knew that all experiences had legitimate and intellectual meaning. Development became a form of teaching that was more than providing experiences for her students—she was nourishing herself. Willis would not allow her students to be dictated by a rigid curriculum, nor would she allow herself to become trapped by the mundane. And teacher growth could arise in this setting, as the staff engaged in many collaborative projects. In fact, this became a defining aspect for Willis as she described teacher growth: "Of great significance to teacher growth at the University School was the wide variety of situations in which teachers worked together, learned from each other, and refined their techniques and tested their understandings of general principles; they extended their knowledge of psychology, child development, and the practical implications of philosophy as they shared their insights about child behavior or worked out plans of cooperative action."[63] From such a community, questions of philosophy dominated discourse, as teachers sought to explore the meaning of "democracy as a way of life" in the everyday circumstances of an active school and placed themselves, continually, in the role of learner and among learners—namely, their students. Teachers were living intellectual lives alongside their students. And through independence and cooperation, teachers and students were making decisions. This was a teacher's life and one carefully and slowly constructed by Margaret Willis. And this was one reason for a long life of teacher growth and fulfillment. Willis constructed a community with interesting companions. She enjoyed her setting and allowed students and colleagues to create an intellectually alive space. As she stayed at the University School year after year—viewed as leader from the beginning—she ultimately grew into the role of "tribal elder." When asked why Willis never became the director of the school, Butche cracked, "because she didn't want to take a demotion."[64]

THE HONORABLE FIGHT

My description thus far must not imply that Ohio State University was a utopian setting. The school was exceptional. But, as we know, schools are part of communities, and the Ohio State University community proved ultimately to undermine the school. In 1953, Margaret Willis wrote (to a friend), "The causes I fight for? Better schools, better teaching, free minds for free men."[65] Little did she know the fight that would occur a decade later. Detailed in the recent publication by Butche, a brilliant example of local school research, and summarized by Robert Bullough, insurmountable forces came together to end this community.

> . . . a dominating trustee, former United States Senator, one-time presidential candidate, arch conservative and McCarthyite John W. Bricker, who saw in the College of Education a breeding ground of liberalism and of dangerous dissenters who needed silencing in order to protect the university he served so long and loved so well . . . a dean, Donald P. Cottrell,

> who was punished by his president, Novice G. Fawcett, for, among other reasons, not encouraging the graduate school to waive residency requirements so he might receive a Ph.D. The president punished the College to punish the dean, whose salary was frozen for a dozen years in the apparent hope he would leave . . . a university president (Fawcett) who came to office from the superintendency of the public schools of Columbus, Ohio, and who wanted the University School closed, Butche argues, because he believed it diverted resources and attention away from the public schools . . . two ambitious faculty members representing a new breed of education professors, one hoping to be dean, David Clark, and the other seeking to make his mark in empirical research, Egon Guba, who supported closing of the School and took on the "old guard" at Ohio State, professors who cherished the traditions represented by the University School and who fiercely believed in its mission.[66]

The University School community came together to fight. In the thick of it was Margaret Willis. Her presence, her voice, and her strength are still apparent from the worn, musty "Save University School" file stashed away in her professional papers. From that file appears an apology from Guba, then Director of the Ohio State University Bureau of Educational Research, "I feel that I owe you an apology for my testy remarks at the Faculty Committee."[67] In response to this memo (as well as to an altercation a few days before), Willis, a "mere" high school teacher, responded to Guba, the director of a nationally renowned research center, "As you see, my reactions on April 9 were not just personal pique, but rather stemmed from bafflement over how anybody as smart as you could come to conclusions which seemed to me contrary to the data!"[68] Willis was ready for the fight but, alas, the powers were too great. Margaret Willis's "defense of the University School before its enemies, both open and secret, was grounded in a more fundamental issue, the place of democratic decision making on American campuses."[69] The University School, a setting that fostered "democracy as a way of life" for its students and staff, was closed. A school that believed in open discourse was terminated in the most devious ways. Willis was in attendance for the final high school commencement and commented to others with great pride on the final student address—a statement criticizing the closing of the school. She kept a copy of the address in its own special file among her professional papers.

INTERESTED IN EVERYTHING AND APPARENTLY SUFFERING FROM NOTHING

> To many we were a strange pair, two women tourists traveling in summer heat in a land whose beauties and marvels aren't suggested for summer visiting. We were a rarity, two women tourists traveling alone, interested in everything and apparently suffering from nothing—heat, inconveniences, flies, cats, nor contaminated water. We endured them all, let none be deceived. But this was our vacation, and we were going to enjoy Iran. We were not going to pass up this opportunity to see as much as we could—weather or no.[70]

Willis represents the common, unnamed teacher whose entire life is spent in the classroom. Yet, what a life; "interested in everything" while engaged in adventures with students, staff, colleagues, and friends. Willis's story is also uncommon and remarkably unique as she emerged as leader for this democratic community. Butche's comments remain all too true. Margaret Willis is a "learned and complex person not so easily labeled."

Within a "strictly-defined" child-centered classroom, feminist pedagogy and teacher authority prove complex. "Progressive and child-centered teachers do exercise authority, but that authority is a vexing problem for many, associated in their minds with the more traditional classroom settings they want to avoid."[71] This topic, treated by Frances Maher, Ellen Lagemann, Kathleen Weiler, and others, offers many fresh insights into those educational settings, including Dewey's Laboratory School, where male professors generated theory and female teachers applied that knowledge.[72] I wish to step aside, however, from many progressive schools and hold this one, the Ohio State University School, as somewhat different. How different is a question that will be open to future researchers and biographers who wish to explore this venue. For now, the Ohio State University School suggests different forms of power, classroom autonomy, and intellectual pursuits and calls for seeing the "progressive" classroom in other ways.

If progressive education seeks not primarily to determine the needs of students nor to revolve the classroom curricula around the interests of the child, that is, to place the woman teacher in a "relatively passive, nurturing, enabling female 'other' of the 'active' masculinized child,"[73] but, instead, to construct community, to foster independent thinking, and to balance personal and social needs as a model for democracy, then progressive education becomes a much different order. "Teacher as guide" becomes not a mere slogan but a meaningful motif and instructional style. And when we learn of a teacher such as Margaret Willis, actively engaged in the adventure of ideas, and the Ohio State University School, a setting for the experimentation of progressive practices, "learning communities" becomes not a cliche but, instead, a concept for countless insights.

Margaret Willis gladly taught at the University School for 36 years. Throughout this period "gladly would she learn." Progressive education has taken many hard hits—then as well as now—for focusing on the needs of students, for frivolous life adjustment programs, and for blatant signs of anti-intellectualism. These criticisms are no doubt accurate in certain settings that called themselves progressive. Yet, Margaret Willis was so intellectually alive and the Ohio State University School placed such importance on knowledge that such traditional criticisms seem superfluous.

Researchers of the Eight Year Study, the project Willis first found herself immersed upon becoming a teacher at the Ohio State University School, noticed the striking changes in attitudes of hundreds of teachers. The research staff attributed the differences to "the discovery of the possibilities of personal growth through teaching, the discovery of a new faith in the democratic ideal and the place of education in achieving it, and the assurance of a modicum of freedom and security."[74] Willis and her colleagues discovered possibilities of personal growth through teaching, and this growth became the world of teachers who engaged in a life of learning so that they could bestow the same joy to their students. Progressive education and the Ohio State University School permitted the building of such communities, and teachers and staff were given "the greatest stimulus of all . . . the sense of belonging to an adventurous company."[75] Willis's *adventurous company* included all learners, namely, students, colleagues, and friends; and within this company, she emerged as a daring, independent leader. And gladly learn and gladly teach—this is what is overlooked as we reexamine many of these remarkable progressive schools. Teachers taking pride in their friendships with students, teachers establishing learning communities with their students and evolving as school leaders without ever taking administrative reins, and teachers being learners and their students being teachers. This is what is overlooked as we examine leadership styles, and this is leadership that resulted from working together in a spirit of cooperation, independence, curiosity, adventure, and democracy. Such was the grand experiment at the Ohio State University School, and such is the testimony of the life of Margaret Willis.

NOTES

1. I thank my mentors, Paul R. Klohr, professor emeritus at Ohio State University, and Robert V. Bullough, Jr., professor emeritus at University of Utah, for their advice on this vignette. I thank my teachers, Robert Butche and Samantha Maddox, for their encouragement, and I thank my colleagues, Mary Bull, Vicky Newman, and Alan Wieder, for their comments. Finally, I wish to thank two very special high school teachers, Edwin Epps and Thomas Horton, for living Margaret Willis lives.
2. Kenneth Winetrout, "Boyd H. Bode: The Professor and Social Responsibility," in *Teachers and Mentors* ed. C. Kridel et al. (New York: Garland, 1996), pp. 71–79; R. V. Bullough, Jr., *Democracy in Education—Boyd H. Bode* (Bayside, N.H.: General Hall, 1981).
3. Robert Butche, *Image of Excellence: The Ohio State University School* (New York: Peter Lang, 2000), p. 134.
4. *The Record of the University School* (Columbus: Ohio State University, 1934), p. 5.
5. Wilford M. Aikin, *The Story of the Eight-Year Study* (New York: Harper and Brothers, 1942).
6. Boyd H. Bode, *Democracy as a Way of Life* (New York: Macmillan, 1937); Boyd H. Bode, *Progressive Education at the Crossroads* (New York: Newson, 1938). "Democracy as a way of life," according to Bode, provided opportunities for "the common man" to develop capacities and to liberate intelligence for the good of society. The goal of schooling thereby became the development of individual growth that would foster a democratic way of living.
7. Margaret Willis, "Democracy in the Formulation of School Policies," *Educational Method* 19, no. 4 (January 1940): 220.
8. Robert V. Bullough, Jr., and the author are addressing this topic in *With Adventurous Company: The History of the Eight Year Study*, forthcoming.
9. Robert V. Bullough, Jr., "Musing on Life Writing," in *Writing Educational Biography* ed. C. Kridel (New York: Garland, 1998), p. 20.
10. Phyllis Rose, *Parallel Lives* (New York: Knopf, 1984), p. 5.
11. I presume Willis would have become somewhat impatient with my efforts to depict her life and would much prefer to read about the University School and the lives of her students than about herself. These are my mere impressions from the few telephone conversations we had from 1984 to 1986. Self-effacing would not be the first descriptor of Willis; however, neither did she seek out the spotlight.

    With the telling of the story, however, the power of biography goes to the biographer to shape the narrative. Any portrayal of Margaret Willis brings to the forefront unique research issues, certainly for the biographer and especially for a male biographer. "Disclosing certain kinds of details about a woman's life is sometimes described as an arbitrary choice the biographer makes, rather than the duty of any biographer searching for truth." (Linda Wagner-Martin, *Telling Women's Lives* [New Brunswick, N.J.: Rutgers University Press, 1988], p. 13) I recognize the many arbitrary choices that have come to shape my view of Willis, and this brief vignette is quite selective. There are many silences caused, in part, by limited space, and some silences will not go unnoticed. Important aspects of Willis's life will be unspoken today; however, through the efforts of Samantha Maddox, archivist of the Margaret Willis Professional Papers, these dimensions will ultimately be told. Willis's salons, her friendship with Eleanor Roosevelt, and her many feminist stances with university administration are just a few of the many important aspects of a career that must be further explored.

12. Margaret Willis, "Teaching Abroad," *Wellesley College Alumnae Magazine* (24 January 1953), Museum of Education, University of South Carolina, Margaret Willis Professional Papers (hereafter called Margaret Willis Papers), box 5, folder 3.
13. Willis, "Teaching Abroad."
14. Margaret Willis, memorandum to H. Alberty, n.d., Margaret Willis Papers box 1, folder 1
15. Margaret Willis, letter to R. Lindquist, 5 April 1936; Margaret Willis Papers, box 5, folder 2.
16. Willis, memorandum to H. Alberty, Margaret Willis Papers.
17. Willis, "Teaching Abroad."
18. Ibid.
19. Margaret Willis, letter to David Clark, 14 April 1964; Margaret Willis Papers, box 5, folder 2.
20. Willis returned to Skagit County, Washington, in her retirement years (after a two-month, 8,000 mile motoring tour of Turkey and Iran in the fall of 1967, described by Willis in the 30 June 1968 issue of The *New York Times* Travel Section). Willis proceeded to work with a newly formed historical society and ultimately wrote and edited eight books about Skagit County.
21. Willis, "Teaching Abroad."
22. Margaret Willis, letter to Wilford Aikin, 3 September, 1940, Margaret Willis Papers box 2, folder 2.
23. H. Harry Giles et al., *Exploring the Curriculum* (New York: Harper & Brothers, 1942), p. 307.
24. George S. Counts, *Dare the School Build a New Social Order* (New York: John Day, 1932).
25. "The fatuous business of needs" was argued in Boyd H. Bode, "Needs and the Curriculum," *Progressive Education* 17, no. 8 (1940): 532–536, and V. T. Thayer, AV. T. Thayer Replies," *Progressive Education* 17, no, 8 (1940): 537–540. "Stanwood Cobb, the founder of the PEA, lamented, 'The Association was an enthusiastic well-coordinated working organization during the first decade, but then something happened.' When asked what happened, Cobb replied, 'Well, they took it away from us.' Cobb identified 'they' as the group from Teachers College." Patricia A. Graham, *Progressive Education: From Arcady to Academe* (New York: Teacher College Press, 1967), p. 57.
26. Eugene R. Smith. "A Message from the President of the Progressive Education Association," *Progressive Education* 1, no. 2 (1924): 99.
27. Another overlooked feature of this era is the different forms of educational research and experimentation. Bureaus of Educational Research existed throughout the country during the 1930s; yet, their research would most often be defined today as numerical tabulations—surveys—and not examples of social science research designs as we now expect. The Eight Year Study did not embody a pretest/post-test format. Students were tested, and a form of control and experimental groups was devised; yet, classroom experimentation often involved teachers trying out new practices. Today, with our more sophisticated social science perspective, we may dismiss "trying out new practices." Such experimentation, however, proved more adventurous than many scientifically controlled experiments of today.
28. H. Gordon Hullfish, *Toward a Democratic Education* (Columbus: Ohio State University, 1960), pp. 77–78.
29. Willis, "Democracy," p. 218.
30. Margaret Willis, *Three Dozen Years* (Columbus: Ohio State University, 1968) pp. 8–9.
31. Margaret Willis, "Three Dozen Years," Report of the Committee on the University School, 1 November 1954; Margaret Willis Papers, box 1, folder 9.
32. See *Thirty Schools Tell Their Story* (New York: Harper, 1942), pp. 718–757; Harold Alberty (and Elsie Alberty), *Reorganizing the High School Curriculum* (New York: Macmillan, 1947, 1953, 1962); and Harold Fawcett, *The Nature of Proof* (New York: Teachers College, 1938).

33. Margaret Willis, lecture, 28 November 1959; Margaret Willis Papers, box 5, folder 9.
34. Willis, *Three Dozen Years,* p. 30.
35. *The Record of the University School,* p. 11.
36. Margaret Willis, letter to M. Stickney, 5 April 1931, Margaret Willis Papers, box 5, folder 2.
37. Robert Butche, interview with the author, Columbus, Ohio, 3 July 1997.
38. Willis, "Democracy," p. 216.
39. Ibid., pp. 216–217.
40. Paul Klohr, interview with the author, Columbus, Ohio, 8 January 2000.
41. Butche, *Image of Excellence,* p. 134.
42. Klohr, interview with author.
43. P. Thomas, "The Paradoxes of Lou LaBrant," Museum of Education's Writing Educational Biography keynote address, 20 July 1999; later published as "The Paradoxes of Lou LaBrant: Choreographer of the Learner's Mind," *Vitae Scholasticae* 18, no. 2 (Fall 1999): 35–54.
44. Klohr, interview with the author.
45. Robert Butche, correspondence with the author, 31 August 2000.
46. Butche, correspondence with the author.
47. Class of 1938, *Were We Guinea Pigs?* (New York: Henry Holt), p. 295.
48. Ibid., p. 1.
49. Ibid., pp. 298–299.
50. Willis, letter to R. Lindquist.
51. Margaret Willis, *The Guinea Pigs After 20 Years* (Columbus: Ohio State University Press), p. 11.
52. Ibid., p. vii.
53. Class of 1938, interviews with the author, Class of 1938 class reunion, Columbus, Ohio, 20–21 May, 1988.
54. Willis, *The Guinea Pigs After 20 Years,* p. ix.
55. Klohr, interview with the author.
56. Faculty of the University School, *The Philosophy and Purposes of the University School* (Columbus: Ohio State University), p. 2.
57. Margaret Willis, Eight Year Study Questionnaire, 3 September 1940, Margaret Willis Papers, box 2, folder 2.
58. Willis, letter to R. Lindquist.
59. Willis, *Three Dozen Years,* p. 77.
60. Ibid., p. 72.
61. Butche, correspondence with the author.
62. Butche, interview with the author.
63. Willis, *Three Dozen Years,* p. 76.
64. Butche, interview with the author.
65. Margaret Willis, letter to M. C. Lyons, 24 January 1953, Margaret Willis Papers, box 5, folder 3.
66. Robert V. Bullough, Jr., review of Robert Butche's "Image of Excellence," unpublished manuscript.
67. Eugene Guba, letter to Margaret Willis, 13 April 1964; Margaret Willis Papers, box 5, folder 2.
68. Margaret Willis, letter to Eugene Guba, 13 April 1964; Margaret Willis Papers, box 5, folder 2.
69. Bullough, Jr., review.
70. Margaret Willis, "To many" manuscript, n.d., p. 1; Margaret Willis Papers, box 5, folder 9.
71. Frances A. Maher, "Progressive Education and Feminist Pedagogies: Issues in Gender, Power, and Authority," *Teachers College Record* 101, no. 1 (Fall 1999): 45.

72. Frances A. Maher, "Toward a Richer Theory of Feminist Pedagogy," *Journal of Education* 169, no. 3 (1987): 91–99; Ellen C. Lagemann, "Experimenting with Education: John Dewey and Ella Flagg Young at the University of Chicago," *American Journal of Education* 104 (May): 171–185; and Kathleen Weiler and Sue Middleton, *Telling Women's Lives* (London: Open University Press, 1994).
73. Frances A. Maher, "Progressive Education and Feminist Pedagogies," p. 45.
74. Giles et al., *Exploring the Curriculum,* p. 308.
75. Ibid., p. 308.

CHAPTER FOURTEEN

FROM SUSAN ISAACS TO LILLIAN WEBER AND DEBORAH MEIER: A PROGRESSIVE LEGACY IN ENGLAND AND THE UNITED STATES

JODY HALL

> . . . the unicellular organism, Amoeba . . . a naked, jelly-like speck of protoplasm, lives in the debris at the bottom of fresh-water ponds. . . . If it finds itself suspended in water, completely free of any surface, the organism may throw out "pseudopodia" in every direction until it seems to be nothing but a number of slowly moving filaments of protoplasm . . . which reach out until contact with some surface is made. Then the pseudopodium in touch with the surface applies itself, the others being gradually withdrawn until the normal shape and movements of the creature are resumed.
>
> —Susan Brierley, *An Introduction to Psychology*, Methuen, 1921

Susan Fairhurst Brierley Isaacs (1885–1948), English child psychologist, psychoanalyst and school reformer, saw in this instance of an amoeba's behavior a life energy by which all organisms seek equilibrium between inner processes and environmental changes. Such an analogy is indicative of the use of evolutionary biology to inform psychology in its beginning years. From 1924 to 1927, as director of the experimental Malting House School in Cambridge, for children aged 2 to 10, Isaacs gathered evidence of children "reaching out" to materials, events, plants, and animals, and to peers and adults, much in the way that the amoeba's pseudopodium, on touching a surface, changed its behavior—an act, Isaacs wrote, of "positive activity towards its environment."[1] The legacy of Susan Isaacs is not so much the founding of a school, as it is the synthesis of a school of thinking about schooling that undergirded reform. Although the Malting House School was short-lived, Isaacs drew from the data to create a persuasive set of guidelines and exemplars for practice. In so doing, she blended ideas from

Darwinian biology, philosopher John Dewey, and psychoanalysts Sigmund Freud and Melanie Klein. Isaacs lived the theory that she espoused, through immersion as teacher and researcher in the day-to-day life of the Malting House School, in her own psychoanalyses with J. C. Flugel, Otto Rank, and Joan Riviere, in teaching graduate students in education and psychology, in psychoanalytic practice, and as a training analyst.

Her passion was to improve conditions for children. To that end, Isaacs conducted research and interpreted the results for audiences of parents, childcare providers, nursery and primary teachers, school inspectors, educational researchers, policy makers, psychologists, teacher educators, psychoanalysts, and training analysts.[2] In the field of schooling, Isaacs advocated providing children with a range of environments in which to pursue their interests, with the support of adults, and she contributed to the schooling community a rationale for children's use of fantasy in play as a bridge to thought. She elevated the value of children's interests, questions, and ideas to a position of utmost importance in the endeavor of schooling, and she provided powerful exemplars for the teaching community to follow. As head of the Department of Child Development at the Institute of Education, the University of London (1933–1943), Isaacs inspired confidence and conviction in teachers, school officials, and future education leaders. Her former student and successor as head of department, Dorothy Gardner, compiled a richly detailed biography[3] that drew from public records, interviews with her friends, relatives, colleagues, former students and psychoanalysts, memorabilia collected by her husband Nathan Isaacs, and correspondence in private collections—the originals of which are, for the most part, no longer available. This essay contributes a reading of Isaacs in the context of the early twentieth century intertwining of evolutionary biology and psychology in school practice and explores how Isaacs's vision subsequently helped Americans to renew their commitment to pragmatic practices of education in the 1960s.

JOHN DEWEY IN THE ENGLISH SCHOOLING COMMUNITY

At the end of the nineteenth century, philosopher John Dewey raised educational issues about the emergence of urban, industrial communities with worldwide markets and easy communication and distribution among all parts.[4] Change in school practice was regarded as a necessary corollary to change in society, and it was Dewey, according to, University Professor of Education John Adams at the University of London, who called the attention of the schooling community to this relationship.[5] The shift from proprietary capitalism to corporate capitalism was accompanied by increased agitation for franchising the working class and women, new concern for social life, and belief in progress. In schooling, the shift led to the addition of vocational education, increased requirements for school attendance, and the use of scientific analogies to inform practice. In creating a naturalistic argument for human psychology, Dewey drew largely from William James's perspective on evolutionary biology in *Principles of Psychology* (1890): Survival for humans depended on "conscious deliberation and experimentation."[6] He proposed a model for school reform aimed to psychologize content within the "range and scope of the child's life" and promote adaptability for participation in society, which was accomplished by continuous reconstruction of children's experiences out into organized bodies of knowledge and by cultivation of inquiry in the context of real problems.[7]

What accounts for the positive response to Dewey's work in sectors of the English schooling world, in this case exemplified by Susan Isaacs, in the first four decades of the twentieth century? Recognizing England's deeply entrenched class system, the long-standing classicist tradition in elite schools, and the emerging disposition to regard intelligence as fixed, we might not expect Dewey's ideas to register in the English schooling community. Indeed, there is much evidence of their misinterpretation.[8] Yet his theory was represented in two seminal references for English school reform: the Board of Education's Hadow report, published as *The Report of the Consultative Committee on the Primary School* (1931), and Susan Isaacs's *Intellectual Growth of Young Children* (1930).[9]

To understand the affinity of some in the English school reform community to Dewey in the first part of the twentieth century, we turn first to the late 1880s and early 1890s when his thinking focused on ethics and politics "in which psychology became the handmaiden of moral argument."[10] Dewey directed his attention to Oxford don T. H. Green, about whom he wrote: "Both theoretically and personally, the deepest interests of his times were the deepest interest of Professor Green. The most abstruse and critical of his writings are . . . attempts to solve the problems of his times. . . . He saw in what is called philosophy only a systematic search for and justification of the conviction by which men should live."[11] From about 1880 to 1910, Green persuaded social reformers in England to view progress as the widening of the spectrum of people among whom there is a common good and duty.[12] Green's liberalism was a development that ran parallel to Fabian political theory,[13] to which Isaacs subscribed as a young adult, and to the idealist philosophy of Edward Caird, whose views inspired R. H. Tawney, a Fabian educational reformer and powerful member of the Board of Education Consultative Committee in the years following World War I. One explanation for the positive response to Dewey's educational philosophy, therefore, is a common intellectual tradition and social action agenda. English educationalists aligned with these positions were committed to solve problems of the times by opening up society to include the working classes and women and to move in the direction of a liberating pedagogy.

J. J. Findlay, professor of education at Manchester, introduced Dewey to English readers in 1906,[14] three years before he would support Susan Isaacs's entrance to the university. When Findlay studied at Jena in Germany in the 1890s, he encountered "a band of eager young [American] teachers from western states who were transplanting the Herbartian banner to Illinois." After a period of discipleship, they "challenged the strict shibboleths imported from Germany" with "independent research and reconstruction." Findlay regarded Dewey as foremost among American school reformers, because of his firsthand work in "conducting a school for young children . . . in the presence of teachers and students who were seeking to reconcile theory and practice." Of particular interest to English teachers in 1906, Findlay suggested, was Dewey's exposition of child development—"of children's minds, as changing in aspect between the ages of four and twelve." As for universities and departments of education, Dewey had shown how universities could "combine the obligation to teach and 'train' with the obligation to investigate and advance the boundaries of knowledge."

Six years later, Findlay reported[15] that English schooling was "being slowly readjusted to a new social order" in which "the child of poverty" could advance in society, where movements for technical or vocational education existed, where a new aim was emerging as to "the purpose and method of the school"—that of establishing "initiative, independence and freedom for the child" in "the spirit of democracy, of equality," and in which freedom from convention permitted the growth of character and virtue. Yet, according to Findlay, there was much to do: The study of Greek among nongifted teachers and scholars was made a fetish, newly established science was taught but without relation to experience, and the newly recognized "phenomenon of child nature" continued to be subjugated to established conventions.

Although World War I dampened the general belief in progress, the onus remained to realign institutions to a new political, social, and economic order. Social action in England took the form of the Education Act of 1918, the expansion of departments of state (1919), the raising of the leaving age from state school from 12 to 14 (1921), the institution of the Unemployment Acts (1920–1922), and the school reform work of the Board of Education's Hadow Committee.

In 1930, Findlay wrote with mild alarm about the trend of reform.[16] Instead of educating a people elevated to democracy by science and a pragmatic initiation to the accumulated wisdom of the ages—the vision held by Dewey and Findlay—society had given birth to a people mostly eager to acquire a roadster and a wireless. Reform-minded educationalists had roused popular enthusiasm with slogans that were narrow and uninformed by research. In a more hopeful vein, Findlay looked to the Dewey-inspired movement of "Activity, or Self-Activity"—reforms prevalent since 1920 that were currently "looking for support in the newer expositions of psychology" in which "a solid contribution is being made to reform in Method"—quite likely an allusion to his former student Susan Isaacs's work at the Malting House School. Activity had emerged as the key to a variety of reforms: "The pupil himself, as an active being, can be set free to display his own energy, and thus progress." Susan Isaacs described and explained specific conditions for this vague term "activity" and qualified the use of the term "freedom" in her work on schooling.

SUSAN ISAACS: HER LIFE AND WORK

Susan Isaacs was well suited to accept the challenge that her university sponsor J. J. Findlay had made to the schooling community to take up the ideas of Dewey. Isaacs came of age at the turn of the century in the industrialized, Lancashire cotton town of Bromley Cross near Bolton.[17] An admirer of suffragist Emmeline Pankhurst, the Brontës and George Eliot in her youth, Isaacs was well aware of women's issues. As a young adult at the turn of the century, she joined the Fabian Society and supported the women's suffrage movement.

Isaacs was the seventh of eight surviving children—two died in infancy. Her Lancashire father, William Fairhurst, the editor of the *Bolton Journal* and *Guardian,* and a part-time Methodist lay preacher, placed a high value on intellectual activity, clear expression of thought, and correct use of grammar. Her Scottish mother Miriam Sutherland Fairhurst, who died when Isaacs was ten, was remarkable for her efficient, energetic management of household affairs on limited means. Their home was full of books, and discussion of social, political, and literary matters occurred on a regular basis. Darwin's *The Origin of Species,* for example, was much discussed. According to a friend from her youth, Isaacs and her sisters "declaimed themselves socialists, vegetarians, unitarians, and finally agnostics, to the horror of their father. They wore arty dresses of sage green which fascinated us." When Isaacs renounced religion, her father did not speak to her for two years and removed her from school, saying, "If education makes women Godless, they are better without it."

After the interruption in her schooling from age 14 to 23—the years partly filled with private teaching and child-care posts—Isaacs entered an infant teacher training program at Manchester University. There, she first learned of John Dewey from Miss Grace Owen, a leader for progressive education of young children. Owen recognized Isaacs's academic promise and suggested to Professor J. J. Findlay that Isaacs should leave the teacher training course to pursue a bachelor's degree at the university. Findlay wrote personally to Isaacs's father, William Fairhurst, prevailing upon him to let Isaacs undertake a degree. From Manchester University, where she took a first in philosophy in 1912, she went on to graduate work in psychology at Cambridge University and then to lecture for a year in infant school education at Darlington Training College and for a year in logic at Manchester. Following her marriage to botanist William Brierley, whom she had known at Manchester, she moved to Richmond in 1916, as Brierley had received an appointment in research into plant diseases at Kew Gardens. In London, she gave lectures in psychology and began her lifelong interest in psychoanalysis.

Isaacs's life soon took major turns as she remarried and took a position as head of the Malting House School. Her second husband, Nathan Isaacs, was a metallurgist ten years her junior, who had attended her lectures. To one of these lectures, Nathan Isaacs responded with an essay "so far outside the usual students' calibre that she realized immediately she had caught a whale in her herring net."[18] In 1922, they traveled to Austria, where she obtained a divorce and married Nathan.

Her first book, *An Introduction to Psychology*, appeared in 1921, and she became secretary of the British Psychological Society's newly formed committee for research in education. After three years, she reported, "The amount of research being done is small."[19] In accepting the position at the Malting House School, Isaacs took up psychological research into the emotional life and intellectual processes of childhood.

In the spring of 1924, a full-page advertisement appeared in the *New Statesman* and the *British Journal of Psychology:* "Wanted–an Educated Young Woman with honours degree–preferably first class–or equivalent, to conduct education of a small group of children aged 2 1/2–7, as a piece of scientific work and research. . . . Hence a training in any of the natural sciences is a distinct advantage." The anonymous advertisers were Margaret and Geoffrey Pyke. Geoffrey Pyke had accumulated enough money in the purchase and sale of metals to found a school, and when their son was born in 1921, Pyke, having "read some psychoanalytic writings, decided characteristically that the infant must be educated free from neurosis."[20]

Susan Isaacs was more than qualified to conduct such research. She accepted the position, provided that she would be fully in charge. Pyke agreed. After many meetings, they agreed upon a plan of action and made their views public: "The provision of the right environment . . . requires the most careful scientific method, expert knowledge of child psychology and wide educational experience. . . . each child shall be free . . . to develop his natural interests, individual powers, and means of expression, while living in a happy children's community, the conditions of which lead to normal social development."[21]

In 1926, Pyke persuaded Nathan Isaacs to leave his job as a metal merchant in the City (London) and join the staff to write "a number of books on the theory of knowledge."[22] Nathan Isaacs (1895–1966) would go on to write, lecture and consult in the schooling community; he was particularly active in the primary science education reform of the 1960s.[23] The Isaacses remained at the school until 1927; it closed in 1929 because of Pyke's business losses.

During the next 21 years, until her death from cancer in 1948 at age 63, Susan Isaacs devoted herself to the education and psychoanalytic communities. Returning to London, Isaacs spent much of her time writing. Drawing from her work at the Malting

House School, she wrote *The Nursery Years* (1929) and *The Children We Teach* (1932) for childcare providers, parents, and teachers. The American edition of *The Nursery Years* (1937) was awarded a medal by the Parents' Institute in America. For professional educators oriented to more technical discussion of theory and practice, she wrote *Intellectual Growth in Young Children* (1930) and *Social Development of Young Children* (1933), for which she received a doctorate from Manchester University in 1933. Nathan Isaacs's "Children's 'Why' Questions" appeared as an appendix to *Intellectual Growth in Young Children.* In addition to these books, Isaacs wrote a column for parents—"Ursula Wise" in *The Nursery World* (1929–1936)—and many scholarly articles and reviews;[24] she also gave lectures in England and elsewhere, most notably in Australia and New Zealand.

In 1933 Isaacs was appointed the first head of the newly founded Department of Child Development at the Institute of Education at the University of London. In setting up the department, Sir Percy Nunn argued that an earlier Statement of Policy for the institute had made provision for setting up "a department whose aim would be to enlarge and improve the scientific foundations upon which the education of young children should be based." Appealing to nationalist sentiment, the statement pointed out that England needed its own research center comparable to those of other countries, for example, at Yale University and the Institute Rousseau in Geneva. As head, Isaacs carried out her responsibilities on a part-time basis. When World War II started, she moved to Cambridge, where she conducted a study of the problems of evacuation for children and their foster parents. Returning to London, Isaacs brokered the transfer of leadership of the department to Dorothy Gardner in 1943 and continued her work as a writer, consultant, lecturer, analyst, and training analyst until her death in 1948.

In addition to her interest in schooling, Isaacs contributed to the field of psychoanalysis with distinction until her death.[25] In 1922, she became qualified as an analyst to take patients and continued to do so part-time on a daily basis until the end of her life. Melanie Klein came to London from Berlin in 1925, where she continued to relate Freud's theories to the emotional development of children. After Isaacs's departure from the Malting House School in 1927, she entered an extended analysis with Kleinian analyst Mrs. Joan Riviere, because Isaacs was convinced of the value of Klein's work and characteristically wished to learn about it firsthand. In the 1940s Isaacs qualified to work as a training analyst.

INTELLECTUAL HISTORY OF SUSAN ISAACS

Isaacs thrived on argument. Her writing shows clear exposition and a critical reading of a wide range of professional discourse. Her synthesis of contemporary beliefs from the fields of biology, psychoanalysis, and philosophy of education provided the schooling community with a powerful case for reconceptualizing the conditions of children's development.

Like other psychologists of the period, including Jean Piaget, Isaacs used Darwinian biology as a springboard for theorizing human growth. She was skeptical about "regarding 'mind' as some mysterious independent entity distinct from 'body,' yet existing in equally mysterious relation with 'body.'"[26] From Isaacs's point of view, rationalists had lost credibility with the arrival of the evolutionary perspective: Human behavior, like that of creatures other than humans, "is very clearly the manifestation of native tendencies which are released by environmental stimuli."[27] Psychologists had much to learn from biology in studying change—a basic characteristic of life. Change rested on "interests" in the behavior of organisms: "What actually happens is that an organism adjusts its bodily and mental relation with significant stimuli in its environment; and when this occurs on the conscious level of behavior, we say that the organism "attends to" these stimuli; that the object is "interesting" to the creature."[28]

As an undergraduate at Manchester, Dewey caught Isaacs's attention with his ideas of schooling based on the pursuit of interests in experimental activity. His *Psychology* (1887) and *How We Think* (1909) were particularly influential.[29] Dewey argued that the "interests of the self" break up "the hard rigidity of psychical life" and introduce "flexibility" and "perspective."[30] He believed that experiences in the real world provided the vehicle for the pursuit of interests. Of particular importance to educators was the distinction between two basic kinds of experience: Empirically, experience was "dominated by the past, by custom and routine . . . often opposed to the reasonable"; experimentally, experience "includes reflection which sets us free."[31]

Whereas Dewey spoke in terms of interests in events and problems out in the world, psychoanalyst Melanie Klein helped Isaacs to expand her ideas about interests to take into account young children's unconscious fantasies, wishes, and anxieties. The "deepest sources" of intellectual growth, Isaacs argued[32] in accord with Klein, lay in the "first 'symbol formation' of infantile mental life." The young child's interest in physical objects was "certainly derivative" of and drew "its impetus from early infantile wishes and fears in relation to its parents." Engines, motors, fires, lights, water, mud, and animals have symbolic meaning, and children's ability to concern themselves with real objects and happenings could be used, Isaacs said, for intellectual growth. Similarly, in their dramatic play, children "work out their inner conflicts in an external field, thus lessening the pressure of the conflict, and diminishing guilt and anxiety," making it "easier for the child to control his real behaviour, and to accept the limitations of the real world." The physical world was a canvas "upon which to project his personal wishes and anxieties, and his first form of interest in it is one of dramatic representation." The job of educators was "to counter-act the effect of these dramatic tensions in children by bringing in, wherever possible, the real world." Elaborated make-believe had "cognitive value" as a "bridge" from fantasy to thought.

Isaacs conceptualized children's intellectual and social development as "biological in its groundwork" and affected by "social influences."[33] In 1935, Isaacs was skeptical about dividing development "into a series of distinct 'stages'" as "too simple to fit the facts, although it contains elements of truth." She saw broad phases, one of which was early childhood (roughly from age 2 to 5) and later childhood (roughly from age 5 to 12). The early phase was marked by "instinctual drives of bodily love and aggression, with intense and conflicting emotions . . . directed mainly towards the parents." This phase was further characterized by the use of dramatic play to deal with stress, growth of bodily skills, of knowledge of the physical world and people, and of language—all of which "brings a certain measure of control and of adaptation to the real world." The phase of later childhood was marked by "massive repression" of "early emotional conflict with regard to the parents," with other children providing "emotional satisfaction." There was a "turning away from phantasy [*sic*] to real achievement and knowledge of the real world." Of particular note in this phase was the increase in "verbal formulation of experience and verbal reasoning in concrete terms."

ISAACS IN PERSON

At first glance, Susan Isaacs could appear shy and reserved. But when lecturing or discussing matters of importance, she was highly engaging, using her "lightening mind" and reasoning and speaking abilities to present her perspective. One of her former students recalls a meeting at Dartington Hall in which Isaacs kept the audience of students, staff, parents, teachers, and workers on the estate "spellbound" with a talk on intellectual growth in young children. New Zealand Director of Education Arnold Campbell recalled her ability, during a 1937 lecture tour, to diffuse hostile questions by treating them seriously and clearly explaining evidence for her position. A sense of humor and playfulness underscored her appreciation of children's inventiveness with language and thought: Five-year old "butcher" James, for example, fetched his "customer" Mrs. I, a small amount of meat with a long pole and asked, "Will that be enough—how many children have you?' She replied, 'Fourteen'; he laughed and said promptly, 'Well, that won't be enough meat.'"

As a teacher of graduate students, she was sensitive and generous, especially to those from other countries. Dorothy Gardner wrote, "My most vivid recollections are of the many ways in which she led us to do our own thinking; of superb teaching, but also of wise silences until we had worked out a problem to a point where her help became essential and we could really assimilate it." She expected much of her students, who if unprepared for class would not repeat that mistake. She once commented about her relationship with her students to a colleague that she was "like a dog on the scent of a hare."

Isaacs was highly organized. It is hard to believe that her job as head of department at London's Institute of Education was technically part-time. Her 1939 report[34] about her performance in that position lists a staggering number of duties: She gave one lecture and three seminars each week, engaged a handful of adjunct faculty, did tutorial work for all education students, met socially with overseas students once a week, worked with two to four graduate psychology students, arranged field work, organized and supervised a play center for children, interviewed prospective students, conducted advisory correspondence, met with many visiting distinguished psychologists, educators, and child guidance workers from overseas, served on six committees and four editorial boards of journals, edited a series of handbooks and pamphlets, and replied to letters from parents and teachers. The list did not include her writing, her paid "examinerships" for universities, or her work at the London Clinic of Psycho-Analysis. Her efficiency enabled her to undertake an immense amount of work: She sometimes met for a tutorial at a public place just before a meeting or lecture, and used the odd few minutes to add a paragraph or two to a manuscript or take a quick nap.

Nathan Isaacs was perhaps her greatest fan and collaborator. Over the course of their 26-year marriage, they sometimes had fiery exchanges. D. W. Winnicott, a psychoanalyst colleague and friend of Susan Isaacs, wrote, "When I heard Nathan ruthlessly criticize her ideas and formulations I felt maddened, but I found that she valued exactly this from him and that she made positive use of his ruthlessness, as of his terrific intellect."[35] His success in the business world must have enabled Susan to pursue her professional interests at the low salaries that she received, though there is little reference to this factor in any of the literature about her life.

Her combative temperament led her to challenge those whose view might be damaging, in her opinion, to the treatment of children. In the course of her life, she challenged "inadequate psychology" in teacher training colleges, for example, separate faculties of the mind, separate instincts, "dreary old theories of play," and sense training. "Bewildered lecturers in training colleges were tremendously grateful for her clarity of exposition" on these subjects. She criticized Maria Montessori for the contradiction inherent in her recommended practices: "[the belief] that if she gave 'freedom' to children they would respond in precisely the ways she expected, and indeed required." She opposed those who attempted to "pigeonhole human personalities into types. . . . And in fact [these] typologists constantly warn us that their 'types' are never found pure. Why not, then, express the facts in terms of process and tendency?" In a time of great interest in Freudian terminology, she spoke out against labeling people. She felt it "prevented real thought about the people concerned and led to slick judgments." In line with her opposition to "typologists" and labels was her concern about Piaget's classification system: "He formalizes the differences thus revealed by calling them 'isms.' By substantifying [*sic*] them and thus sharpening their contours, he dispenses himself from looking for likeness and continuities." In these reactions to others' ideas of children, Isaacs sought always to avoid reductive thinking.

ISAACS IN PRACTICE

"[A boy of six years of age] experimentally opened the case of a piano, and spent hours examining its structure and watching the action of the hammer on the wire, and the relation of the striking of the key to the movement of the hammer. At no point did he ask why, but made continual observations in the form of how and what. A day or two later he told me that his mother had said to him that 'when the hammer struck the wire a little fairy that was in the wire came out and sang.'"[36] This is a classic Isaacs-type example of the merging of fantasy and realism in an educational learning situation. Intellectual growth is liberally sprinkled with many such instances of what she called "experiment, observation, and discovery."

Isaacs explained conditions for provoking children's responses to events in the real world—those conditions included a certain degree of freedom to explore a range of stimulating environments around the school and grounds. The concept of freedom, Isaacs cautioned, was "extremely naïve and misleading." The role of teacher was not to be neutrally passive. Instead, educators should engage in a "discriminative technique of interaction" with children: capable of coinvestigating with children, joining in—not taking the lead but following theirs, helping them to pass over from fantasy life to the real world, allowing a certain degree of untidiness for the sake of experimenting, conditioning children to the outcomes of careless or angry acts, avoiding lengthy verbal instruction and explanations, and throwing questions back to children, such as "What do you think?" and "Let's find out." Children have a strong "desire to touch and handle, to pull to pieces, to 'look inside.' . . ." Fostering children's interests and questions was paramount:

> . . . our technique was to meet the spontaneous inquiries of the children, as they were shown day by day, and to give them the means of following these inquiries out in sustained and progressive action. So the facts of their behavior with fire and water and ice, with pulleys and see-saw and pendulum, and later with drilling machine and Bunsen burner, can be taken as immediate evidence of the spontaneous direction of their interests. We did not 'teach' our children about these things, nor try to create an interest in them, nor introduce any experiments or apparatus until the need for them had actually arisen. . . . It was . . . their eager questions . . . that led me gradually to give them material that would allow of these interests being followed out for their own sake.[37]

The Malting House School achieved celebrity, both good and bad, over the course of its brief tenure in the world. A description of the school appeared in the progressive education journal, the *New Era,* characterizing Isaacs's work at the school as "very valuable experimental work under definitely scientific conditions," with teachers "carefully selected for their experience in research and psychology,"[38] but a few in nearby Cambridge gossiped about it as a "pre-genital brothel."[39] The responses of children were recorded frankly and explained in the context of intellectual and social development, for example, children's commentary on the mechanics of bicycles, the innards of dissected animals (including Mrs. I's dead cat), and boys' erections. At a time when the health hazards of cigarettes were unknown, children were allowed to smoke. They could make fires on the grounds of the school. However, after burning down a shack while a movie of the school was being made, children were limited to three matches a day and were required to carry a bucket of water. Most occurrences at the Malting House School, however, over the course of the Isaacs's three-year involvement were more ordinary: discussing relations between heavy, dark clouds and rain, and between air blown through a rubber tube and the rising and falling of bubbles, and exploring the melting of materials on a hot-water pipe. In these exemplars, Isaacs set forth clear evidence for ways that children could pursue their interests in the environment.

The question remains as to how elements of Deweyan pragmatism retained their power into the early thirties and beyond, reflected as they were in government policy reports and the infiltration of Susan Isaacs's perspective in state school reform in practices that reached maturity in the 1950s and 1960s.[40] There are three possible explanations. The first explanation has already been discussed: a common intellectual tradition and social action agenda. Second, it is widely acknowledged that unlike Americans, the English resisted the shift to widespread scientific management of society until mid-century.[41] Whereas Dewey's pragmatism is a version of a scientific approach, its tentativeness, fluidity, and emphasis on process disqualified it in America, where more mechanistic approaches to public school organization and instruction were in favor;[42] yet, it was embraced by some in England, where reformers who had worked their way into the state sector remained focused on a more gradual humanistic-pragmatic approach to change. This gradual approach can be seen in the Fabian socialist movement. Although the power of the Fabian movement was largely spent by 1918, its approach of "permeation"—of Fabians' joining organizations to make change gradually[43]—left its mark on the generation who came of age at the turn of the century and went on to powerful positions in education, most notably Susan Isaacs and G. H. Tawney. Third, the entrenched English class system supported the placement of a civil service, well-educated in the humanities, in positions of power—the more liberal of whom could sympathize with a pragmatic approach for restructuring schooling. Those allied with social reformers in their formative years would have felt comfortable with Dewey's ideas about schooling.

AMERICAN RESPONSE TO SUSAN ISAACS'S LEGACY

In 1961, American educational historian Lawrence Cremin chronicled the apparent collapse of American progressive education in the 1950s in *The Transformation of the American School: Progressivism in American Education 1876–1957* (1961).[44] Soon after its publication, word of a large number of English primary schoolrooms organized around children's open-ended exploration of centers began to circulate in America. In particular, infant schools for children aged four to seven[45] and the English primary science education reform movement of the 1960s show strong evidence of Isaacs's impact on school practice.[46] Many Americans made their way to England to view this phenomenon first-hand and returned to change practice. According to American school reformer Lillian Weber (1917–1994), in *The English Infant School and Informal Education* (1971)—her 1965–1966 study of 47 state schools in England—the English experience "should encourage us to search for possibilities and steel us against a polemic that rejects these premises as inapplicable or as failures without permitting any real trial."[47] Weber founded the Workshop Center for Open Education at New York City College in 1967 largely on the grounds of what she had learned in her 18-month study of English schooling.

The English Infant School is important for its candid explanation of the resulting shift in Weber's thinking about schooling. In 1963, there was an "avalanche of public pressure" for preschool education leading to the federal funding of Head Start. At that time, Weber, concerned about quick fixes, heard about preschool education in England at a conference at Vassar College and visited England in 1965. At first, Weber interpreted curriculum and instruction in English state schools as comparable to American private preschools, and she was particularly interested in extending private school practices to public education. Weber conceptualized private school practices as "the kind of teacher-child relationship and classroom organization stressing greater individualization." Gradually, however, her notions about informal schooling changed to a theoretical perspective, stemming from Susan Isaacs. Weber reported, "Informal, as I understand it, refers to the setting, the arrangements, the teacher-child relations that maintain, restimulate if necessary, and extend what is considered to be the most intense form of learning, the already existing child's way of learning through play and through the experiences he seeks out for himself."[48] Interpreting this italicized definition, Weber underscored a child's curiosity as a driving force: "The active force . . . is . . . curiosity, interest, and the needs of the child's own search for definition and relevance." Susan Isaacs had provided the rationale for informal education. Weber commented, "She gave an objective basis, in the context of genetic psychology, to previous generalities on natural development, on the deep connections linking inner and outer reality, emotional and intellectual life."[49]

Other Americans heard about English school reform in the 1960s and 1970s, and some went to see for themselves. Scientists and teachers involved with the Boston area Elementary Science Study, for example, learned about it at their 1963 summer conference from staff members Bill Hull and Leonard Sealey—the latter, an advisor for junior schools in Leicestershire, England. Some observed English schools and workshops while participating in the Nuffield Junior Science Project. These American scientists and their collaborators, although unaware of the historical context and largely uninterested in a theoretical, psychological perspective,[50] returned with a vision of schooling more compatible with their emerging ideas about science instruction than what they saw in American schools.[51] Education professor Lydia Smith, of Simmons College, made her own pilgrimage in 1972 to interview elderly "pioneer teachers," after a conversation with Ted Sizer, dean of the Harvard Graduate School of Education, about the value of their oral histories. Smith called what she saw "activity and experience" education. Many teachers whom she interviewed referred directly to the work of Susan Isaacs: "As I visited the remote corners of the lovely English countryside the name Susan Isaacs came up often."[52] This led Smith to contribute her valuable 1985 biography of Isaacs, *To Understand and to Help: The Life and Work of Susan Isaacs (1885–1948).*

Although Isaacs's work is no longer circulating actively in the American and English schooling communities, it fostered such contemporary educators as American school reformer Deborah Meier. Around 1967, soon after moving back to New York City from

Chicago, Meier[53] heard Lillian Weber speak. In turn, Weber helped half-time kindergarten teacher Meier and her Public School (P.S.) 144 colleagues in Harlem to start a cluster of classes that ranged from pre-K to grade-two. In 1970, Meier made her own five-week study tour of English schools. After leaving P.S. 144, Meier worked at Weber's Workshop Center at City College for a year and then opened a program connected to City College in District 2, helping schools to create clusters of open-style schoolrooms. In 1974, Meier started a small elementary school at Central Park East that grew to encompass four schools—elementary and secondary—about which she wrote in *The Power of Their Ideas: Lessons for America about a Small School in Harlem.*[54] At the outset, Meier and her colleagues regarded their children as having been "driven into dumbness by a failure to challenge their curiosity." Meier explained that the teachers' own kind of classroom was "literally full of stuff: books of every sort, paints as well as paintings, plants, animals, radios to repair—things. The curriculum we sought was both conceptual and tangible. We wanted children to fall in love as we had with stories of the past, including their own; we wanted schools that would evoke a sense of wonder."

Asked how Weber influenced her work, Meier replied, "I think Weber affected my work in so many ways that it would be hard to describe briefly. For one thing, she was one of the first tough, politically savvy, well-educated Renaissance women I had met with an interest in early childhood education and a commitment to public education. She was an inspiration and probably sealed it for me, that this was where I would devote my life's work." As for what she had absorbed from Isaacs and the example of English schooling, Meier reported that their work had "that wonderful combination of detail—actual children appeared in ways that captured both my imagination and my attention—along with a theoretical curiosity about how such stories and anecdotes could be understood." In addition to their way of seeing and responding to children, English educators, according to Meier, were "intellectually curious people who found young children's minds of serious interest." Meier had previously found herself similarly curious, but in the American context had "felt in many ways isolated with that idea—like a freak. Few serious people in my larger world of politics and academia could imagine taking the ideas of children so seriously, and finding these ideas so provocative and fascinating." In the tradition of Dewey, Isaacs, Gardner, and Weber, Meier acts as an intellectual and wants the same for her students. In *The Power of Their Ideas,* she wrote, "We might even want all our young people to be intellectuals—all of them. That's where my vote would go. . . . If we agree that what we want are citizens with a lively curiosity—who ask, How come? and, Why? and, Is it truly so?—we'll have the start of a new definition of 'well-educated.'"

REFLECTIONS

Evolutionary biology provided analogies for the early days of psychology from which both John Dewey and Susan Isaacs drew in sketching out a basis for reform in schooling. Susan Isaacs created substantive common ground in tune with her time. She generated theory and practice that stood at the crossroads of early twentieth century biology, philosophy, and psychology. Isaacs led with her pen. In her writing and in her speaking, she engaged a wide range of audiences: parents in her "Ursula Wise" columns; practically minded child-care workers and teachers; research-oriented and psychologically knowledgeable professional educators and policy makers; and the psychoanalytic community.

By the late 1960s, large segments of the English schooling community had taken up progressive practices. The maturation of these practices in the 1960s coincided with the widespread circulation in England of Jean Piaget's ideas about the development of reason in children—a perspective steeped in evolutionary biology. Although Isaacs had a more generous view of young children's ability to use reason than Piaget did and took a dim view of Piaget's highly defined, fixed stages, they had much in common—especially their belief in the necessity of children's using reason in relation to experience to develop mature thought.

The connections that Dewey and Isaacs had made among evolutionary biology, psychology, and school practice were largely out of sight by the 1960s—historical artifacts. The need to replace divinity and absolute explanation with organic human dispositions to intellectual and social development—a need that drove early psychologists to psychology and social reform—had dissipated. Yet, an interest in practices resulting from this union appeared from several different sources. Beginning in the mid-1950s, American psychologists[55] involved with the newly founded cognitive sciences turned away from the dominant, mechanistic learning theory approaches toward a psychology in which powers of the mind—hypothesizing, high-level problem solving, strategies—replaced a theory of associative learning. Some looked to Swiss psychologists Jean Piaget and Barbel Inhelder, who long had been studying the development of logic. Psychologist Jerome Bruner introduced their work to a massive American audience.[56] In another quarter, scientists emerging from the laboratory communities of World War II wanted to create similar conditions in America's public schools, when they were enlisted to reform science education. They traveled to England to observe approvingly the open-ended, experiential-learning conditions of English schools. The American Civil Rights movement called attention to inequities in American society and led other American educators, like Lillian Weber, to England to observe a model of publicly funded schooling that regarded all children as harboring curiosity when exposed to a stimulating environment. Many people involved in the expansion of preschool education and Head Start programs looked to the example of English infant school practices.

At the time of Isaacs's most developed writing about schooling in the early 1930s, she was somewhat vague when it came to curriculum, except in the area of dramatic play and physical causality. Subjects were to be avoided, because the real world did not break knowledge into categories. Instead, children were to follow up on interests, and learning of the subjects would follow. Considering this kind of curriculum from the perspective of the turn of the twenty-first century—characterized by exponential growth of knowledge, highly professionalized career ladders, new technologies to access and produce knowledge, the publication of a vast literature for children in various fields, awareness of gender, multicultural and environmental issues, and questions about government of and by the people when corporate interests have so much power—one must ask the question, Is this too simple?

Some have responded to the growth of knowledge and the need for knowledgeable, skilled workers by regulating children's learning in the categories and hierarchies of state and national standards. It does seem prudent to expose children and young people to a certain degree of breadth and depth of knowledge, and skill. When standards are viewed as guidelines suggestive of possibilities, they allow for dynamic interaction. The large number of content categories and hierarchies in curriculum documents, however, can strip the learning enterprise of sense-making exchanges among teachers and their students. Inquiry takes time, and yet the development of inquiry is necessary if we are to use knowledge and technology to respond to pressing social issues and to contribute new knowledge. If the large number of content standards are used as ends or outcomes to be cemented as the accumulated wisdom of the ages in the minds of children, they prevent the investigation of the new and shifting realities of the present and interpretations of the past. It is interesting that in a time of technological revolution in which literate inquirers of all ages have access to a range of databases, the balance in schooling has shifted to cultivating foundational content.

Does Susan Isaacs have anything of value to say to us at this point in our history? Perhaps more than anything else, the life, work, and legacy of Isaacs serve as a witness to the value of the curiosity of children and observant teachers leading curriculum in the day-to-day life of a significantly enriched school setting. Responding to the challenges of our time requires the cultivation of homely acts of curiosity. This happened when the children built a wormery at the Malting House School. "Mrs. I" had noticed their interest in worms while digging in the garden, and she suggested that they fill a box with soil and a number of worms and put some decayed leaves on the surface. A few days later, the children made an important discovery: They noticed that some of the leaves had been pulled down into the soil. Following her early insight about the amoeba's pseudopodium reaching out to make contact with some surface, Susan Isaacs set up a rich school environment in which children could observe everyday things, like worms, and learn from them with the help of their teacher. Isaacs reminds us of a timeless lesson about education: Exciting discoveries about seemingly mundane events drive learning and motivate children and teachers to raise questions. Their experience makes them want to learn more.

NOTES

1. Susan Isaacs, *An Introduction to Psychology* (London: Methuen, 1921).
2. For a discussion of her writings, especially her educational contributions, and a complete bibliography, see Lydia A. H. Smith, *To Understand and To Help: The Life and Work of Susan Isaacs (1885–1948)* (Cranberry, N.J.: Associated University Presses, 1985).
3. Dorothy E. M. Gardner, *Susan Isaacs* (London: Methuen Educational, 1969).
4. John Dewey, *The School and Society* (Chicago: University of Chicago Press, 1900).
5. John Adams, *Modern Developments in Educational Practice* (London: University of London Press, 1922), p. 171.
6. For a discussion of Dewey and James, see Robert Westbrook, *John Dewey and American Democracy* (Ithaca: Cornell University Press, 1991), pp. 65–71. Quoted in Westbrook from John Dewey, "Evolution and Ethics" *The Early Works of John Dewey, 1882–1898* (Carbondale: Southern Illinois University Press, 1973).
7. John Dewey, "The Child and the Curriculum" in *The School and the Curriculum* ed. J. J. Findlay (London: Blackie, 1906).
8. Jody S. Hall, "John Dewey and Pragmatism in the Primary School: a Thing of the Past?" *Curriculum Studies* 4, no. 1 (1996): 5–23; idem, "The Reception of John Dewey in the English Schooling Community, 1906–1930," Paper presented at the annual meeting of the International Standing Conference for the History of Education, Berlin, Germany, September 1995; Kevin J. Brehony, "An 'Undeniable' and 'Disastrous' Influence? Dewey and English Education (1895–1939)" *Oxford Review of Education* 23, no. 4 (1997): 427–445.
9. Board of Education, *Report of the Consultative Committee on the Primary School* (London: Her Majesty's Stationary Office, 1931); Susan Isaacs, *Intellectual Growth in Young Children, with an Appendix on "Children's 'Why' Questions" by Nathan Isaacs* (London: Routledge and Kegan Paul, 1930); Hall, "John Dewey and Pragmatism," p. 96.

10. Westbrook, *John Dewey and American Democracy,* pp. 34–38.
11. Ibid., p. 37. Quoted in Westbrook from "The Philosophy of Thomas Hill Green" (1889), *The Early Works of John Dewey, 1882–1898* (Carbondale: Southern Illinois University Press, 1973).
12. Peter Gordon and John White, *Philosophers as Educational Reformers: The Influence of Idealism on British Educational Thought and Practice* (London: Routledge and Kegan Paul, 1979), pp. 9–11.
13. A. M. McBriar, *Fabian Socialism and English Politics 1884–1918* (Cambridge: University Press, 1962), p. 74.
14. Findlay, ed., *The School and the Curriculum.*
15. J. J. Findlay, *The School: An Introduction to the Study of Education* (London: Williams and Norgate, 1912).
16. J. J. Findlay, "The Practice of Education" *The Foundations of Education: A Survey of Principles and Projects* (London: University of London, 1930).
17. Unless noted, the details describing the life of Susan Isaacs come from Gardner, *Susan Isaacs.*
18. Evelyn Lawrence, "Foreward" to *A Brief Introduction to Piaget* by Nathan Isaacs (New York: Agathon Press, 1960), p. 7.
19. R. W. Selleck, *English Primary Education and the Progressives, 1914–1939* (London: Routledge and Kegan Paul, 1972), p. 190.
20. W.A.C. Stewart, *The Educational Innovators: Volume II: Progressive Schools 1881–1967* (London: Macmillan, 1968), p. 119.
21. Willem van der Eyken and Barry Turner, *Adventures in Education* (London: Allan Lane, Penguin Press, 1969), pp. 22–23.
22. Ibid., p. 35.
23. Jody S. Hall, "Psychology and Schooling: the Impact of Susan Isaacs and Jean Piaget on 1960s Science Education Reform," *History of Education* 29, no. 2 (1999): 153–170.
24. See Smith, "Works by Susan Isaacs," in *To Understand and to Help,* pp. 334–336.
25. The following essay currently circulates in the psychiatric community: Susan Isaacs, "The Nature and Function of Phantasy," in *Developments in Psycho-Analysis* by Melanie Klein et al. (New York: Da Capo Press, 1983).
26. Isaacs, *Introduction to Psychology,* p. 8
27. Ibid., p. 49.
28. Ibid., p. 55.
29. Smith, *To Understand and to Help,* p. 161.
30. John Dewey, *Psychology* (New York: Harper, 1887), p. 108.
31. John Dewey, *How We Think* (London: D. C. Heath, 1909), p. 228.
32. Susan Isaacs, "The Function of the School for the Young Child," *The Forum of Education and Journal of Experimental Psychology* (June 1927), pp. 122–125; idem, *Intellectual Growth,* pp. 101–102; and idem, "The Nature and Function of Phantasy," pp. 107–111.
33. Susan Isaacs, *The Psychological Aspects of Child Development* (London: Evans Brothers, 1935), pp. 7–14.
34. Gardner, "Report made by Susan Isaacs to Sir Fred Clarke on the Department of Child Development (February 1939)," Appendix II in *Susan Isaacs,* pp. 179–184.
35. Gardner, "Foreward" by D. W. Winnicott in *Susan Isaacs,* pp. 5–6.
36. Isaacs, "The Function of the School for the Young Child," p. 125.
37. Isaacs, *Intellectual Growth,* pp. 80–81.
38. "The Malting House School," *New Era* 9 (1928): 72.
39. van der Eyken and Turner, *Adventures in Education,* p. 37.
40. W. Boyd and W. Rawson, *The Story of the New Education* (London: Heinemann, 1965), p. 84; Gardner, *Susan Isaacs,* p. 73; Lillian Weber, *The English Infant School and Informal Education* (Englewood Cliffs, N.J.: Prentice Hall, 1971), p. 168; Lydia Smith, *Activity*

and Experience: Sources of English Informal Education (New York: Agathon Press, 1976), p. 10; Peter Cunningham, *Curriculum Change in the Primary School Since 1945: Dissemination of the Progressive Ideal* (London: Falmer Press, 1988); Brian Simon, *Education and the Social Order 1940–1990* (London: Lawrence and Wishart, 1991), p. 362; Joanna S. Hall, "Experience, Experiment and Reason: English Primary Science 1959–1967" (Ph.D. diss., University of Liverpool, 1993).

41. J. G. Crowther, *Science in Modern Society* (New York: Harper, 1968); Hilary Rose and Steven Rose, *Science and Society* (London: Alan Lane, Penguin Press, 1969).
42. Herbert M. Kliebard, *The Struggle for the American Curriculum 1893–1958* (New York: Routledge, 1986).
43. A. McBriar, *Fabian Socialism,* pp. 95–97.
44. Lawrence Cremin, *The Transformation of the School: Progressivism in American Education 1876–1957* (New York: Vintage, 1961), p. ix.
45. Weber, *The English Infant School;* Smith, *Activity and Experience.*
46. Hall, "Psychology and Schooling," pp. 153–170.
47. Weber, *The English Infant School,* p. 274.
48. Ibid., p. ll.
49. Ibid., p. 171.
50. Jody S. Hall, "Changing Views of Children as Young Scientists: The Elementary Science Study and Developmental Psychology 1956–1966." Paper presented at the annual meeting of the History of Education Society, Atlanta, Georgia, October 1999; idem, "Physicists Discovered Inhelder and Piaget in 1959." Paper presented at the Annual Meeting of the Jean Piaget Society, Montreal, Canada, June 2000.
51. Emily Romney and Mary Jane Neuendorffer, *The Elementary Science Study: A History* (Newton, Mass.: Educational Development Center, 1973), pp. 36–39.
52. Smith, *Activity and Experience,* p. 10.
53. Interview of Deborah Meier by Jody S. Hall in August 2000 and e-mail correspondence between Meier and Hall in October 2000.
54. Deborah Meier, *The Power of Their Ideas: Lessons for America from a Small School in Harlem* (Boston: Beacon Press, 1995).
55. Jerome S. Bruner, J. J. Goodnow, and G. A. Austin, *A Study of Thinking* (New York: John Wiley, 1956).
56. Jerome S. Bruner, *The Process of Education* (New York: Vintage Books, 1960).

CONCLUSION

ALAN R. SADOVNIK AND SUSAN F. SEMEL

Historians of progressive education have often overlooked the contributions of women to the movement. Although some of the women chronicled in this book are mentioned in various histories,[1] this is more often than not in the context of brief discussions of the schools that they founded, rather than in a discussion of their lives and careers. More importantly, histories of progressive education tend to be histories of great men, including John Dewey, William Heard Kilpatrick, Harold Rugg, and George Counts, or their opponents, including Robert Hutchins, Arthur Bestor, Hyman Rickover, and Isaac Kandel.[2] In *Left Back: A Century of Failed School Reform,* a historical critique of progressive education, Diane Ravitch continues this trend by attributing the founding of many of the child-centered schools of the early twentieth century to upper middle-class parents, rather than the women founders of these schools.[3]

Historians can debate the reasons for these omissions. Perhaps because the dominant figures in progressive education were men, like Dewey and Kilpatrick, the attention given to women has been lessened. Another explanation is that because the women discussed in this book were practitioners and almost all worked outside of the academy, their lives and careers have received considerably less attention from scholars of progressive education. This supports critical and feminist theories, which argue that women's contributions to educational history have been devalued due to the dominance of theory over practice and to women's marginalized position in schools as teachers rather than administrators.[4] Of those included here, Margaret Haley and Ella Flagg Young are probably the best known. Both spent the majority of their careers in public education and both were more directly involved in policy making. Many of the others spent their lives in schools—most of them private—and like the schools they founded, have often wound up ignored, lumped together, or as historical footnotes.[5]

One of the main contributions of feminist scholarship over the past three decades has been to correct what historian of education, Clarence Karier, termed "inexcusable omissions."[6] The first, and perhaps most important, contribution of the chapters in this book is to present, in one volume, the life histories and contributions of a group of women founders and leaders during the Progressive Era. Their contributions to educational philosophy and curriculum and practice need to be positioned directly within the historical record so that women's contributions to the history of progressive education are not marginalized.

Collectively, the chapters in this book highlight a number of important themes in the history of education, including female leadership and feminist theory, progressive education and democratic education, and the importance of race, class, gender, religion, geography, and networks to understanding people's lives. Additionally, the lives of these women provide important lessons for contemporary school reform, especially reforms that place significant emphasis on leadership.

FEMALE LEADERSHIP AND FEMINIST THEORY

Feminist theories of leadership have stressed the ways in which women leaders differ from male leaders. Based on the works of Carol Gilligan and Nel Noddings, these theories argue that women are more likely to lead through cooperation, inclusion, caring, and connectedness.[7] Using the concept of a "web of inclusion," or the more recent term, "relational leadership," many feminist theories and research suggest that women are more democratic, less hierarchical, and more relational in their approach.[8]

These approaches have been criticized by some feminist researchers. For example, sociologist Cynthia Fuchs Epstein has argued that these types of feminist approaches are simple-minded and reduce women's behavior to a narrow range of human behaviors and to stereotypical female gender roles.[9] A recent Wellesley Center for Women study cautions against using notions of democratic, caring female leadership simplistically, as women leadership is often far more complex. In citing Helen Parkhurst, Caroline Pratt, and Carmelita Hinton in her previous study of female leadership, Semel cautions against using contemporary feminist theories of leadership to explain the past.[10] The examination of the leadership styles of these three school founders suggested what was termed a progressive paradox—democratic education autocratically delivered. The majority of women discussed in this book support this caveat.

All of the women leaders portrayed here were strong, driven, intense, tenacious, visionary, and charismatic and possessed a strong sense of social mission. Many were less than democratic and often difficult to work with, including Charlotte Hawkins Brown, Pratt, Parkhurst, Hinton, Haley, and Laura Bragg. Some were more collaborative, including Elsie Ripley Clapp, Flora Cooke, Young, and Margaret Willis. The lives and careers of these women indicate that female leadership during the Progressive Era was just as complex as contemporary female leadership. There is little evidence to suggest that these leaders, as a group, were more caring, connected, and democratic than their male counterparts. School founders, especially, Pratt, Parkhurst, and Hinton, were far more autocratic than democratic, with the challenges of founding, funding, and running a school with a particular philosophy, which made consensus decision making difficult, if not impossible (and for some undesirable).

The charisma of the founders often proved to be a double-edged sword, especially for their schools. On the one hand, their charisma was essential to the founding, funding, and development of the school. On the other hand, upon their retirement or death, it became difficult, if not impossible, to find these qualities in the new leaders. Thus, many of the schools had great difficulty reproducing charismatic leadership (e.g., the Organic School after Marietta Johnson) and never fully recovered. Those that remained successful managed to create rational or traditional forms of leadership based on the works of the founders, but not based exclusively on charisma or personality (such as the Dalton School and the City and Country School).

PROGRESSIVE EDUCATION AND DEMOCRATIC EDUCATION

In *"Schools of Tomorrow," Schools of Today: What Happened to Progressive Education*[11] we discussed a second paradox of progressive education, especially related to child-centered private schools: the paradox of democratic education for the elite. Most of the private, child-centered schools discussed there and in this book were founded on democratic principles, but attracted student bodies that were largely white and economically privileged. Although we argued that this paradox should not preclude the study of these schools, we suggested that their practices needed to be applied to public school settings, with more diverse student populations. This study of women leaders indicates that the issues of democratic education and diversity were as complex as those of female leadership.

Not surprisingly, the founders of private schools attracted and educated nondiverse populations, often despite their best efforts to recruit a more diverse student body. Some of this was due to resistance on the part of working class parents to progressive education; some of it was due to the legal and social forces of segregation; much of it was due to financial constraints. For example, the schools founded by Margaret Naumburg, Pratt, Parkhurst, and Hinton attracted, for the most part, privileged students. Despite attempts to provide scholarships, these schools reflected both their geographic locations and private school tuitions. The schools founded by Johnson, Clapp, and Brown were racially segregated, reflecting the legal and social constraints of the times in the South. Of the private schools, Cooke's Parker School, although made up of a majority of privileged children, paid the most attention to issues of diversity. Part of this may be attributed to its founder, Colonel Parker, and its benefactress, Anita Blaine; part of it to Cooke's steadfast devotion to the their democratic principles. Johnson's School of Organic Education, although private, was actually a quasi-public school, as Johnson used the tuition and fees of boarding students to provide a tuition-free education for local children. There, significant attention was paid to economic diversity, but given its location in the segregated South, it did not have racial diversity.

The nonfounding leaders tended to serve the interests of democratic education more fully. Because they worked in public education, Haley, Young, and Charl Williams were in a better position to champion principles of diversity and democratic education, especially for their constituencies, including women teachers. Less attention was placed on race, especially for Haley, as definitions of democratic education reflected the racism of the times. Bragg, on the other hand, a Northerner transplanted in the South, used the Charleston Museum as a vehicle for inclusion and democratic access, bringing the museum to the people, especially African Americans, in the segregated South. Because the Ohio State University School was public and drew from a university community, Willis worked within a more diverse setting, although not significantly more than the private schools. Although committed to democratic education, like her private school counterparts, Willis continued to be limited by the student population attracted to her school.

The works of these women indicate that the progressive principles of democratic education and inclusion often represented ideology rather than reality. This was determined by time and place. Like the issue of female leadership style, educational reform during the Progressive Era and the leaders of these reforms should not be judged by contemporary standards. Rather, whereas many of their democratic principles were not easily realized, they set the foundation for continuing democratic reform that, although still not fully realized, continues to be a central focus of educational debates today.

SOCIOLOGICAL FACTORS AND BIOGRAPHY

The chapters in this book suggest that a number of background variables were important in the women's personal lives and workplaces. These included, but were not limited to, social class, race, gender and family relations, religion, geography, and networks.

A majority of the women came from middle to upper middle-class backgrounds, including Brown, Johnson, Naumburg, Pratt, Parkhurst, Clapp, Hinton, Cooke, Bragg, Susan Isaacs, and Willis; only a few came from lower middle-class or working-class backgrounds, including Haley, Young, and Williams. The majority were Protestant (Haley was Irish Catholic; Naumburg was Jewish). All except Brown were white. Many came from religious backgrounds (e.g., Bragg and Johnson). Although they were born and raised in numerous places, including the South (Williams and Brown), New York City (Naumburg and Clapp), upstate New York (Pratt and Young), the Midwest (Johnson, Parkhurst, Hinton, and Cooke), Massachusetts (Bragg), and England (Isaacs), many settled or lived for a number of years in urban areas, especially Chicago (Cooke, Hinton, Haley, and Young) and New York (Naumburg, Pratt, Parkhurst, and Clapp). Some of the women were married (Brown, Naumburg, Hinton, Johnson, Isaacs, and Young)—some more than once—others were never married (Pratt, Parkhurst, Clapp, Haley, Bragg, and Willis); and some were in women-committed relationships (Pratt and Young).[12]

For the most part, these women leaders were middle and upper middle class, white and Protestant. Because many came from religious backgrounds, perhaps they came to see progressive education as a form of social missionary work and were committed to the principles of child-centered and/or democratic education. Although not all of them grew up in urban areas, a majority settled in cities for significant parts of their lives, especially Chicago and New York. Within this context, they were part of a network of progressive educators, including John Dewey, Jane Addams, Colonel Francis Parker, and the Progressive Education Association (PEA), established originally by many of the founders of the child-centered schools discussed in this book. Many of the women knew each other and worked together in a variety of projects. For example, in Chicago, Haley, Young, Cooke, and Hinton were part of networks that included Parker, Dewey and Addams. In New York, Pratt, Parkhurst, Naumburg, and Clapp were part of networks that included Dewey, when he moved to Columbia, other private school heads, including Lucy Sprague Mitchell at City and Country and then Bank Street, Elizabeth Irwin at Little Red School House, and Abraham Flexner at the Lincoln School, as well as the progressives at Teachers College, including George Counts, William Heard Kilpatrick, Harold Rugg, and Ann Shumaker. The chapters in this book indicate that many of the women worked together, sometimes collaboratively and sometimes conflictually, as was the case with Haley and the circle in Chicago around Jane Addams. The important point is that many of these leaders were part of progressive movements, often centered in cities, dedicated to democratic school and societal reform.

The careers of these women raise issues concerning the differences between public and private education. The majority worked for some or much of their careers in the private sector (Brown, Johnson, Naumburg, Pratt, Parkhurst, Clapp at City and Country, Hinton, and Cooke). Others worked primarily in the public sphere (Clapp at Ballard and Arthurdale, Haley, Young, Williams, Bragg, and Willis). Clapp and Willis worked in public schools, albeit experimental ones; Johnson started her career in rural public schools, worked for some years in public normal schools as a critic teacher and used these experiences to make the Organic School a quasi-public school. Haley, Young, and Williams were the only ones to work in public school systems, either in teacher unions or as a superintendent. The question is why? Although there are examples of women superintendents, as Jackie Blount indicates, they have had little real power in public education.[13] During the Progressive Era, at a time when administrative progressivism was becoming the dominant strain in public education, it was understandable that most of the female school founders worked in the private sector.[14] As Blount asks, "Why would large numbers of women suddenly desire a position that was created largely by and for a few men?"[15] The fact is that the private sector gave these driven, charismatic, and visionary leaders the space and freedom to do things that would not have been possible in the increasingly bureaucratic and male-dominated public sector. Those who worked in the public sector, like Haley, Young, and Williams, devoted much of their careers to improving the conditions of women teachers. Although Willis worked in a public school, the Ohio State University School had far more in common with private schools than public schools.[16]

Feminist history has provided an analysis of the relationship between the public and private spheres of life, arguing that differences between the public and private lives of men and women are necessary for understanding male domination.[17] Within this context, Blount examines the personal lives of women educational leaders and indicates that some were never married and/or had long-term relationships with other women.[18] The women chronicled here often merged the public and private spheres of their lives, with their work often becoming their lives. Although some were married at some point, many of them were divorced at least once (including Brown, Naumburg, and Isaacs) or widowed (Hinton, Johnson, and Young). Some had short- or long-term relationships with other women (Pratt with Helen Marot, and Young with Laura Brayton); others had children (including Naumburg, Johnson, and Hinton). The chapters in this book do not provide sufficient data on their personal lives to generalize about the relationship between the public and private spheres of life for women leaders, other than the vast majority did not assume the traditional gender roles of their time. Whether married, unmarried, or with or without children, these women were married to their workplaces. Clearly, in terms of gender roles, they were women ahead of their times. These chapters indicate the need for further biographical work in this area.

LESSONS FROM THE PAST

In *"Schools of Tomorrow," Schools of Today,* we argued that one of the reasons for exploring the history of education is to apply lessons from the past to contemporary educational reform. Although we caution against using the present to analyze the past, we found a number of important lessons in the history of progressive schools for contemporary policy and practice.[19] The lives and careers of the women in this book provide similar lessons.

Research on school effectiveness has focused on the essential role of leadership, particularly in turning around low-performing urban schools.[20] This book chronicles the leadership qualities of a number of remarkable women, all of whom were singularly dedicated to their visions and to their schools or workplaces. Their leadership styles, however, are not easily categorized, with many of them being authoritarian, and a smaller number, democratic and collaborative. All were driven, tenacious, individualistic and charismatic: qualities that are not easily replicated or taught. These qualities also were more apt to flourish in private settings (or today, in small, alternative public ones) rather than in large, bureaucratic public institutions. This trend supports research on leadership that suggests that private schools often provide more room for creative, innovative, and nontraditional leadership than public schools, as they are more often free from bureaucratic constraints.[21]

The increasing popularity of alternative schools and public charter schools, especially in urban areas, is a reflection of the concern about bureaucracy limiting successful school reforms.[22] Alternative and charter school advocates argue that freed from the bureaucratic constraints of large public school districts, these schools can create innovative and democratic practices to raise achievement levels of low-income children. Although the jury is still out regarding their democratic accomplishments, it is clear that alternative and charter school heads have a great deal to learn from these women. Starting and running a school is extremely difficult, requires significant private sources of funding (even in the public sector), and the school is too often dependent on the superhuman efforts of the leader and her dedicated staff. Most important, school heads do not stay forever, and unless schools can institutionalize leadership and processes that transcend the cult of personality, schools may not be able to persist over time. Some schools were able to do this; others were not. When Parkhurst left Dalton, her successor, Charlotte Durham, was able to codify the Dalton Plan and the school continued successfully. When Pratt left City and Country, her successor, Jean Murray, continued successfully the school's practices in the tradition of its founder. After Johnson retired from the Organic School, the school was never the same, despite the best efforts of the directors and teachers who followed her. The Parker School had great difficulty successfully replacing Cooke. An important lesson is that although leadership is vitally important, schools cannot become totally dependent on the vision, drive, and charisma of a single individual. It is clear that alternative and charter schools need to prepare and train successors to replace their founding heads and so ensure smooth transitions.

The connection between past and present is exemplified in the work and life of Susan Isaacs. The only European in the group—and the only one whose major contributions were at the university level—Isaacs's work had an influence on progressive practices in both the United Kingdom and the United States. Through City College of New York, Professor Lillian Weber's importation of her ideas and their implementation in the English Infant School, the child-centered strand of progressive education was rediscovered in the United States in the 1960s and 1970s. Weber affected the work of Deborah Meier in New York City, who went on to found Central Park East in East Harlem in the 1970s, based partly on her own childhood experiences at a New York City's private, progressive school. CPESS has become a model for alternative urban schools nationwide in providing progressive education for low-income students and students of color. Like the women in this book, Meier took advantage of progressive networks to foster a growing school reform movement. Today, she works with Theodore Sizer and the Coalition of Essential Schools (CES), which comprises a national school reform network reflecting many of the progressive practices of the women in this book, and is principal of the Mission Hill Charter School in the Boston Public Schools. Moreover, the CES comprises a formal network to advance progressive education in public and private schools, in some ways similar to the network developed by some of these women through the PEA during the Progressive Era.

From Haley, Williams, and Young, we learn the importance of changing organizational structure and power relations, as well as working within the public education system to change it. These women worked to change the public schools, as union leaders (Haley, Young, and Williams) and as a superintendent (Young). For any long-term democratic change to occur, especially with respect to improving the educational opportunities of low-income students, it must occur in large urban systems, not just in small alternative and charter schools. Large urban districts are in desperate need of innovative, visionary leaders, as principals and superintendents, to transform failing schools and districts.

Finally, the lives and works of these women illustrate the importance of teaching in school reform. Research on school improvement points to the necessity of qualified, competent, and knowledgeable teachers in every classroom as a prerequisite for successful reform.[23] The majority started their careers as teachers before moving to their work as founders and leaders. The one exception, Laura Bragg, received her degree in library science and did not teach in the schools. As a museum educator, however, she was a teacher and educator, using Lawrence Cremin's broad definition of institutions that educate, which includes museums. Margaret Willis was the one women who remained in the classroom for her entire career. An example of what today is called a teacher-leader, Willis felt her most important contributions could be made in the classroom. Although the other women illustrate the importance of leadership, Willis reminds us that without excellent teachers who remain in the classroom, educational reform cannot succeed. Thus, educational reform must strengthen both leadership and teaching, but not at the expense of each other.

Although these lessons are important, the significance of this book is its contribution to the history of women and education. All of the women led remarkable lives, and their legacies are embedded in education today. Their names should be more than historical footnotes or asides in the history of progressive education.

NOTES

1. See Lawrence A. Cremin, *The Transformation of the School: Progressivism in American Education, 1876–1957* (New York: Vintage Books, 1961).
2. See Cremin, *The Transformation of the School;* Diane Ravitch, *The Troubled Crusade* (New York: Basic Books, 1985).
3. Diane Ravitch, *Left Back: A Century of Failed School Reform* (New York: Simon and Schuster, 2000), p. 177.
4. For discussions of women's place in schools and the devaluing of practice, see Michael Apple, "Teaching and Women's Work: A Comparative Historical and Ideological Analysis" *Teachers College Record* 86 (Spring 1985): 455–473; Jesse Goodman, "The Disenfranchisement of Elementary Teachers and Strategies for Resistance" *Journal of Curriculum and Supervision* 3 (Spring 1988): 201–220; and Frances A. Maher, Progressive Education and Feminist Pedagogies: Issues in Gender, Power, and Authority," *Teachers College Record* 101, no. 1 (Fall 1999): 45. For a discussion of the omission of women in the history of education, see Julie K. Teel, *Laura Zirbes (1884–1967): An American Progressive in an Era of Education Conservatism* (Ph.D. diss., University of Indiana, 2001).
5. For example, in the classic, *The Transformation of the School,* although acknowledging the importance of the private, child-centered schools, Cremin treats them as if they were mostly the same and gives scant attention to their women founders. The exception to this are the schools and founders portrayed in John and Evelyn Dewey's *Schools of To-Morrow* (New York: Dutton, 1962; original 1915), including Marietta Johnson and Caroline Pratt.

6. See Karen Graves, Timothy Glander, and Christine Shea, *Inexcusable Omissions: Clarence Karier and the Critical Tradition in History of Education Scholarship* (New York: Peter Lang, 2001), for a detailed discussion of Karier's critique of Cremin's liberal democratic approach. Although Karier's perspective was more far reaching and included the omissions of race, social class, and gender, critical histories of education included feminist perspectives that focused on the omission of women, especially teachers in the history of schooling.
7. Carol Gilligan, *In a Different Voice* (Cambridge, Mass.: Harvard, 1982); Nell Noddings, *Caring: A Feminist Approach to Ethics & Moral Education* (Berkeley, Calif.: University of California Press, 1984); idem, *The Challenge to Care in School: An Alternative Approach to Education* (New York: Teachers College Press, 1992).
8. See Judy B. Rosener, "Ways Women Lead," *Harvard Business Review* (November–December 1990): 119–125, for a discussion of webs of inclusion. See Joyce Fletcher, *Disappearing Acts: Gender, Power, and Relational Practice at Work* (Cambridge, Mass.: MIT Press, 1999); idem, *Inside Women's Power: Learning from Leaders* (Wellesley, Mass.: Wellesley Centers for Women Research Report, 2001), for a full discussion of this research. For a critical discussion of the tensions between feminist pedagogy and progressive education, see Maher, "Progressive Education and Feminist Pedagogies" p. 45.
9. Cynthia Fuchs Epstein, *Deceptive Distinctions: Sex, Gender, and the Social Order* (New Haven, Conn.: Yale University Press, 1988). See also, Jaclyn Fierman, "Do Women Manage Differently" *Fortune* (17 December 1990), p. 116.
10. Susan F. Semel, "Female Founders and the Progressive Paradox," In *Social Reconstruction Through Education,* ed. Michael E. James (Norwood, N.J.: 1995), pp. 89–108.
11. Susan F. Semel and Alan R. Sadovnik, *"Schools of Tomorrow," Schools of Today: What Happened to Progressive Education* (New York: Peter Lang, 1999), pp. 353–376.
12. For a discussion of woman committed relationships, see Patricia Palmieri, *In Adamless Eden* (New Haven, Conn.: Yale University Press, 1995).
13. Jackie M. Blount, *Destined to Rule the Schools: Women and the Superintendency, 1873–1995* (Albany, N.Y.: SUNY Press, 1998). For a biography of another woman school superintendent, see Clinton Allison, ed., *Kellie McGarrh's Hangin' in Tough: The Life of Mildred Doyle* (New York: Peter Lang, 1999). Mildred Doyle was superintendent of the Knox County, Tennessee, schools from 1946 to 1976.
14. For a discussion of male dominance of administrative progressivism during the Progressive Era, see David Tyack and Elisabeth Hansot, *Managers of Virtue* (New York: Basic Books, 1982).
15. Blount, *Destined to Rule the Schools,* p. 162.
16. See Robert Butche, *Image of Excellence: The Ohio State University School* (New York: Peter Lang, 2000).
17. Linda K. Kerber, "Separate Spheres, Female Worlds, Woman's Place: The Rhetoric of Women's History," *Journal of American History* 75 (1988): 30–37.
18. Relationships between women and lesbianism has been the subject of considerable discussion. A number of terms, including Boston marriages and women-committed relationships, have been used, especially where no evidence of sexual relationships exist in the historical record. For detailed discussions, see Blount, *Destined to Rule;* Allison, *Kellie McGarr's Hangin' in Tough;* and Palmieri, *In Adamless Eden.*
19. Semel and Sadovnik, *"Schools of Tomorrow," Schools of Today,* pp. 353–376.
20. See U.S. Department of Education, National Center for Educational Statistics, *School Quality: An Indicators Report* (Washington, D.C.: National Center for Educational Statistics, 2000).
21. See Anthony Bryk, Valerie Lee, and Peter Holland, *Catholic Schools and the Common Good* (Cambridge, Mass.: Harvard University Press, 1993).

22. For discussions of charter schools, bureaucracy, and accountability and charter schools and democratic education, see Bruce Fuller, ed., *Inside Charter Schools: The Paradox of Radical Decentralization* (Cambridge, Mass.: Harvard University Press, 2000); Amy Stuart Wells, et al. *Beyond the Rhetoric of Charter School Reform: A Study of Ten California Districts* (Los Angeles: UCLA, 1998). For a discussion of alternative schools, including charter schools, see Evans Clinchy ed., *Creating New Schools: How Small Schools are Changing American Education* (New York: Teachers College Press, 2000); Deborah Meier, *The Power of Their Ideas* (Boston: Beacon, 1995); Mary Ann Raywid, "Urban Academy: A School That Really Works" in *"Schools of Tomorrow," Schools of Today,* eds. Semel and Sadovnik, pp. 289–312.
23. "School Quality: An Indicators Report," *What Matters Most: Teaching for America's Future* (New York: National Commission on Teaching and America's Future, 1996).

INDEX

The Pursuit of God

The Pursuit *of* God

A. W. Tozer

The Pursuit of God – A. W. Tozer

First edition published 1948, by Christian Publications, Inc.
Harrisburg, Pa.

Cover Design: Amber Burger

Cover Photography: Amanda Carden/Shutterstock

Editors: Ruth Zetek, Jeremiah Zeiset

Printed in the United States of America

Aneko Press

www.anekopress.com

Life Sentence Publishing, Inc.
203 E. Birch Street
P.O. Box 652
Abbotsford, WI 54405

RELIGION / Christian Life / Spiritual Growth

Hardcover ISBN: 978-1-62245-356-6

Paperback ISBN: 978-1-62245-296-5

eBook ISBN: 978-1-62245-297-2

MP3 CD ISBN: 978-1-62245-361-0

10 9

Available where books are sold.

Contents

Recommended Reading:

God's Pursuit of Man

by A. W. Tozer

Introduction

Here is a masterly study of the inner life by a heart thirsting after God, eager to grasp at least the outskirts of His ways, the abyss of His love for sinners, and the height of His unapproachable majesty – and it was written by a busy pastor in Chicago!

Who could imagine David writing the twenty-third Psalm on South Halsted Street, or a medieval mystic finding inspiration in a small study on the second floor of a frame house on that vast, flat checkerboard of endless streets.

Where cross the crowded ways of life
Where sound the cries of race and clan,
In haunts of wretchedness and need,
On shadowed threshold dark with fears,
And paths where hide the lures of greed . . .

But even as Dr. Frank Mason North of New York says in his immortal poem, so Mr. Tozer says in this book:

Above the noise of selfish strife
We hear Thy voice, O Son of Man.

My acquaintance with the author is limited to brief visits and loving fellowship in his church. There I discovered a self-made scholar, an omnivorous reader with a remarkable library of theological and devotional books, and one who seemed to burn the midnight oil in pursuit of God. His book is the result of long meditation and much prayer. It is not a collection of sermons. It does not deal with the pulpit and the pew but with the soul thirsty for God. The chapters could be summarized in Moses' prayer, *Show me thy glory*, or Paul's exclamation, *O the depth of the riches both of the wisdom and of the knowledge of God!* It is theology not of the head but of the heart.

There is deep insight, sobriety of style, and a universality of outlook that is refreshing. The author has few quotations but he knows the saints and mystics of the centuries – Saint Augustine, Nicholas of Cusa, Thomas à Kempis, Friedrich von Hügel, Charles Finney, John Wesley, and many more. The ten chapters are heart searching and the prayers at the close of each are for the closet, not the pulpit. *I felt the nearness of God while reading them.*

Here is a book for every pastor, missionary, and devout Christian. It deals with the deep things of God and the riches of His grace. Above all, it has the keynote of sincerity and humility.

Samuel M. Zwemer
New York City

Preface

In this hour of all-but-universal darkness, one cheering gleam appears: Within the fold of conservative Christianity, there are to be found increasing numbers of people whose religious lives are marked by a growing hunger after God Himself. They are eager for spiritual realities and will not be put off with words, nor will they be content with correct "interpretations" of truth. They are thirsty for God, and they will not be satisfied until they have drunk deep at the fountain of living water.

This is the only real harbinger of revival which I have been able to detect anywhere on the religious horizon. It may be the cloud the size of a man's hand for which a few saints here and there have been looking. It can result in a resurrection of life for many souls and a recapture of that radiant wonder which should accompany faith in Christ, that wonder which has all but fled the church in our day.

But this hunger must be recognized by our religious leaders. Current evangelicalism has (to change the figure) laid the altar and divided the sacrifice into parts, but now seems satisfied to count the stones and rearrange the pieces with never a care that there is not a sign of fire upon the top of lofty Mount Carmel. But God be thanked that there are a few who care. They are those who, while they love the altar and delight in the sacrifice, are yet unable to reconcile themselves to the continued absence of fire. They desire God above all. They are thirsty to taste for themselves the "piercing sweetness" of the love of Christ about whom all the holy prophets did write and the psalmists did sing.

There is today no lack of Bible teachers to set forth correctly the principles of the doctrines of Christ, but too many of these seem satisfied to teach the fundamentals of the faith year after year, strangely unaware that there is in their ministry no manifest presence, nor anything unusual in their personal lives. They minister constantly to believers who feel within their breasts a longing which their teaching simply does not satisfy.

I trust I speak in love, but the lack in our pulpits is real. Milton's terrible sentence applies to our day as accurately as it did to his: "The hungry sheep look up, and are not fed." It is a solemn thing, and no small scandal in the kingdom, to see God's children starving while actually seated at the Father's table. The truth of Wesley's words is established before our eyes: "Orthodoxy, or right opinion, is, at best, a very slender part of religion. Though right tempers cannot subsist without right opinions, yet right opinions may subsist without right tempers. There may be a right opinion of God without either love or one right temper toward Him. Satan is a proof of this."

Thanks to our splendid Bible societies and to other effective agencies for the dissemination of the Word, there are today many millions of people who hold "right opinions," probably more than ever before in the history of the church. Yet I wonder if there was ever a time when true spiritual worship was at a lower ebb. To great sections of the church, the art of worship has been lost entirely, and in its place has come that strange and foreign thing called the "program." This word has been borrowed from the stage and applied with sad wisdom to the type of public service which now passes for worship among us.

Sound Bible exposition is an imperative *must* in the church of the living God. Without it, no church can be a New Testament church in any strict meaning of that term. But exposition may be carried on in such way as to leave the hearers devoid of any true spiritual nourishment whatsoever. For it is not mere words that nourish the soul, but God Himself; and unless and until the hearers find God in personal experience, they are not the better for having heard the truth. The Bible is not an end in itself, but a means to bring men to an intimate and satisfying knowledge of God, that they may enter into Him, that they may delight in His presence, may taste and know the inner sweetness of the very God Himself in the core and center of their being, their spirit.

The Bible is not an end in itself, but a means to bring men to an intimate and satisfying knowledge of God.

This book is a modest attempt to aid God's hungry

children so to find Him. Nothing here is new except in the sense that it is a discovery which my own heart has made of spiritual realities most delightful and wonderful to me. Others before me have gone much farther into these holy mysteries than I have done, but if my fire is not large, it is yet real, and there may be those who can light their candle at its flame.

A. W. Tozer
Chicago, Ill.
June 16, 1948

Chapter 1

Following Hard after God

My soul clings to You; Your right hand upholds me. (Psalm 63:8)

Christian theology teaches the doctrine of prevenient grace, which briefly stated means this: that before a man can seek God, God must first have sought the man.

Before a sinful man can think a right thought of God, there must have been a work of enlightenment done within him; it may be imperfect, but it is a true work nonetheless, and is the secret cause of all desiring and seeking and praying which may follow.

We pursue God because, and only because, He has first put an urge within us that spurs us to the pursuit. *No one can come to me*, said our Lord, *unless the Father who has sent me draws him,* and it is by this very prevenient *drawing* that God takes from us every vestige of credit for the act of coming. The impulse to pursue God originates with God, but the outworking of that impulse is our following hard after Him; and all the time we are pursuing Him we are already in His hand: *Your right hand upholds me.*

In this divine "upholding" and human "following" there is no contradiction. All is of God, for as von Hügel teaches, *God is always previous.* In practice, however (that is, where God's previous working meets man's present response), man must pursue God. On our part, there must be positive reciprocation if this secret drawing of God is to eventuate in identifiable experience of the Divine. In the warm language of personal feeling, this is stated in the forty-second Psalm: *As the deer pants for the water brooks, so my soul pants for You, O God. My soul thirsts for God, for the living God; When shall I come and appear before God?* This is deep calling unto deep, and the longing heart will understand it.

The doctrine of justification by faith – a biblical truth, and a blessed relief from sterile legalism and unavailing self-effort – has in our time fallen into evil company and been interpreted by many in such manner as actually to bar men from the knowledge of God. The whole transaction of religious conversion has been made mechanical and spiritless. Faith may now be exercised without a jar to the moral life and without embarrassment to the Adamic ego. Christ may be "received" without creating any special love for Him in the spirit of the receiver. The man is "saved," but he is not hungry or thirsty after God. In fact, he is specifically taught to be satisfied and encouraged to be content with little.

The modern scientist has lost God amid the wonders

of His world; we Christians are in real danger of losing God amid the wonders of His Word. We have almost forgotten that God is a spirit and, as such, can be cultivated as any person can, for we are spirit (*God created man in His own image, in the image of God He created him;* Genesis 1:27a). It is inherent in personality to be able to know other personalities, but full knowledge of one personality by another cannot be achieved in one encounter. It is only after long and loving mental interchange that the full possibilities of both can be explored.

All social interchange between human beings is a response of personality to personality, grading upward from the most casual brush between man and man to the fullest, most intimate communion of which the human spirit is capable. Religion, so far as it is genuine, is in essence the response of created personalities to the creating personality, God. *This is eternal life, that they may know You, the only true God, and Jesus Christ whom You have sent.* In the deep of His mighty nature God thinks, wills, enjoys, feels, loves, desires, and suffers as any other person may. In making Himself known to us, He stays by the familiar pattern of personality. He communicates with us through the avenues of our minds, our wills, and our emotions. The continuous and unembarrassed interchange of love and thought between God and the spirit of the redeemed man is the throbbing heart of New Testament religion.

This interchange between God and our spirit is known to us in conscious personal awareness. It is personal; that is, it does not come through the body of believers, as such, but is known to the individual, and to the body through the individuals which compose it. And it is conscious; that is, it does not stay below the threshold of consciousness and work there unknown to the soul (as, for instance, infant baptism is thought by some to do), but comes within the field of awareness where the man can "know" it as he knows any other fact of experience.

You and I are in little degree (except for our sins) what God is in large degree. Being made in His image, we have within us the capacity to know Him. In our sins we lack only the power. The moment the Spirit has quickened us to life in regeneration, our whole being senses its kinship to God and leaps up in joyous recognition. That is the heavenly birth without which we cannot see the kingdom of God. It is, however, not an end but an inception, for now begins the glorious pursuit, the heart's happy exploration of the infinite riches of the Godhead. That is where we begin, I say; but where we stop, no man has yet discovered, for there is neither limit nor end in the awful and mysterious depths of the triune God.

Shoreless Ocean, who can sound Thee?
Thine own eternity is round Thee,
Majesty divine!

To have found God and still to pursue Him is a paradox of love, scorned indeed by the too-easily-satisfied religious person, but justified in happy experience by the children of the burning heart. Saint Bernard of Clairvaux stated this holy paradox in a musical four-line poem that will be instantly understood by every worshipping soul:

> *We taste Thee, O Thou Living Bread,*
> *And long to feast upon Thee still:*
> *We drink of Thee, the Fountainhead*
> *And thirst our souls from Thee to fill.*

Come near to the holy men and women of the past and you will soon feel the heat of their desire after God. They mourned for Him, they prayed and wrestled and sought for Him day and night, in season and out, and when they had found Him, the finding was all the sweeter for the long seeking. Moses used the fact that he knew God as an argument for knowing Him better. *Now therefore, I pray You, if I have found favor in Your sight, let me know Your ways that I may know You, so that I may find favor in Your sight*; and from there he rose to make the daring request: *I pray You, show me Your glory.* God was frankly pleased by this display of ardor, and the next day called Moses into the mount, and there in solemn procession made all His glory pass before him.

Come near to the holy men and women of the past and you will soon feel the heat of their desire after God.

David's life was a torrent of spiritual desire, and his psalms ring with the cry of the seeker and the glad shout of the finder. Paul confessed the mainspring of his life to be his burning desire after Christ. *That I may know Him* was the goal of his heart, and to this he sacrificed everything. *More than that, I count all things to be loss in view of the surpassing value of knowing Christ Jesus my Lord, for whom I have suffered the loss of all things, and count them but rubbish so that I may gain Christ.*

The hymns are sweet with the longing after God, the God whom, while the singer seeks, he knows he has already found. "His track I see and I'll pursue," sang our fathers only a short generation ago, but that song is heard no more in the great congregation. How tragic that we in this dark day have had our seeking done for us by our teachers. Everything is made to center upon the initial act of "accepting" Christ (a term, incidentally, which is not found in the Bible), and we are not expected thereafter to crave any further revelation of God to our spirit. We have been snared in the coils of a false logic which insists that if we have found Him, we need no more seek Him. This is set before us as the last word in orthodoxy, and it is taken for granted that no Bible-taught Christian ever believed otherwise; thus, the whole testimony of the worshipping, seeking, singing church on that subject is crisply set aside. The experiential heart-theology of a grand army of fragrant saints is rejected in favor of a smug interpretation of Scripture, which would certainly have sounded strange to a Saint Augustine, a Samuel Rutherford, or a David Brainerd.

In the midst of this great chill there are some, I rejoice to acknowledge, who will not be content with shallow logic. They will admit the force of the argument, and then turn away with tears to hunt some lonely place and pray, "O God, show me Your glory." They want to taste, to touch with their hearts, to see with their eyes the wonder that is God.

I want deliberately to encourage this mighty longing after God. The lack of it has brought us to our present low estate. The stiff and wooden quality about our religious lives is a result of our lack of holy desire. Complacency is a deadly foe of all spiritual growth. Acute desire must be present or there will be no manifestation of Christ to His people. He waits to be wanted. Too bad that with many of us He waits so long, so very long, in vain.

Every age has its own characteristics. Right now we are in an age of religious complexity. The simplicity which is in Christ is rarely found among us. In its stead are programs, methods, organizations, and a world of nervous activities which occupy time and attention but can never satisfy the longing of the heart. The shallowness of our inner experience, the hollowness of our worship, and that servile imitation of the world which marks our promotional methods all testify that we, in this day, know God only imperfectly, and the peace of God scarcely at all.

I want deliberately to encourage this mighty longing after God.

If we would find God amid all the religious externals, we must first determine to find Him, and then proceed in the way of simplicity. Now as always God reveals Himself to "babes" and hides Himself in thick darkness from the wise and the prudent. We must simplify our approach to Him. We must strip down to essentials (and they will be found to be blessedly few). We must put away all effort to impress, and come with the guileless candor of childhood. If we do this, without doubt God will quickly respond.

When religion has said its last word, there is little that we need other than God Himself. The evil habit of seeking *God-and* effectively prevents us from finding God in full revelation. In the "and" lies our great woe. If we omit the "and" we shall soon find God, and in Him we shall find that for which we have all our lives been secretly longing.

We need not fear that in seeking God only we may narrow our lives or restrict the motions of our expanding hearts. The opposite is true. We can well afford to make God our all, to concentrate, to sacrifice the many for the One.

The author of the quaint, old English classic, *The Cloud of Unknowing*, teaches us how to do this. "Lift up thine heart unto God with a meek stirring of love; and mean Himself, and none of His goods. And thereto, look thee loath to think on aught but God Himself. So that nought work in thy wit, nor in thy will, but only God Himself. This is the work of the soul that most pleaseth God."

When religion has said its last word, there is little that we need other than God Himself.

Again, he recommends that in prayer we practice a further stripping down of everything, even of our theology. "For it sufficeth enough, a naked intent direct unto God without any other cause than Himself." Yet underneath all his thinking lay the broad foundation of New Testament truth, for he explains that by "Himself" he means "God that made thee, and bought thee, and that graciously called thee to thy degree." And he is all for simplicity: If we would have religion "lapped and folden in one word, for that thou shouldst have better hold thereupon, take thee but a little word of one syllable: for so it is better than of two, for even the shorter it is the better it accordeth with the work of the Spirit. And such a word is this word GOD or this word LOVE."

When the Lord divided Canaan among the tribes of Israel, Levi received no share of the land. God said to him simply, *I am your portion and your inheritance*, and by those words made him richer than all his brethren, richer than all the kings and princes who have ever lived in the world. And there is a spiritual principle here, a principle still valid for every priest of the Most High God.

The man who has God for his treasure has all things in One. Many ordinary treasures may be denied him, or if he is allowed to have them, the enjoyment of them will be so tempered that they will never be necessary to his happiness. Or if he must see them go, one after one, he will scarcely feel a sense of loss, for having the Source of all things he has in One all satisfaction, all pleasure, all delight. Whatever he may lose, he has actually lost nothing, for he now has it all in One, and he has it purely, legitimately, and forever.

> *O God, I have tasted thy goodness, and it has both satisfied me and made me thirsty for more. I am painfully conscious of my need of further grace. I am ashamed of my lack of desire. O God, the triune God, I*

want to want thee; I long to be filled with longing; I thirst to be made more thirsty still. Show me thy glory, I pray thee, that so I may know thee indeed. Begin in mercy a new work of love within me. Say to my soul, "Rise up, my love, my fair one, and come away." Then give me grace to rise and follow thee up from this misty lowland where I have wandered so long. In Jesus' name, Amen.

Chapter 2

The Blessedness of Possessing Nothing

Blessed are the poor in spirit, for theirs is the kingdom of heaven. (Matthew 5:3)

Before the Lord God made man upon the earth, He first prepared for him by creating a world of useful and pleasant things for his sustenance and delight. In the Genesis account of the creation, these are called simply "things." They were made for man's uses, but they were meant always to be external to the man and subservient to him. In the deep heart of the man was a shrine where none but God was worthy to come. Within him was God; without, a thousand gifts which God had showered upon him.

But sin has introduced complications and has made those very gifts of God a potential source of ruin to the soul.

Our woes began when God was forced out of His central shrine and "things" were allowed to enter.

Within the human heart, "things" have taken over. Men have now, by nature, no peace within their hearts, for God is crowned there no longer, but there in the moral dusk, stubborn and aggressive usurpers fight among themselves for first place on the throne.

This is not a mere metaphor, but an accurate analysis of our real spiritual trouble. There is within the human heart a tough fibrous root of fallen life whose nature is to possess, always to possess. It covets "things" with a deep and fierce passion. The pronouns "my" and "mine" look innocent enough in print, but their constant and universal use is significant. They express the real nature of the old Adamic man better than a thousand volumes of theology could do. They are verbal symptoms of our deep disease. The roots of our hearts have grown down into *things*, and we dare not pull up one rootlet lest we die. Things have become necessary to us, a development never originally intended. God's gifts now take the place of God, and the whole course of nature is upset by the monstrous substitution.

Our Lord referred to this tyranny of *things* when He said to His disciples, *If anyone wishes to come after Me, he must deny himself, and take up his cross and follow Me. For whoever wishes to save his life will lose it; but whoever loses his life for My sake will find it.*

Breaking this truth into fragments for our better understanding, it would seem that there is within each of us an enemy which we tolerate at our peril. Jesus called it "life" and "self," or as we would say, the *self-life.* Its chief characteristic is its possessiveness; the words "gain" and "profit" suggest this. To allow this enemy to live is in the end to lose everything. To repudiate it and give up all for Christ's sake is to lose nothing at last, but to preserve everything unto life eternal. And possibly also a hint is given here as to the only effective way to destroy this foe: It is by the Cross. *Let him . . . take up his cross and follow me.*

The way to deeper knowledge of God is through the lonely valleys of soul poverty and giving up of all things. The blessed ones who possess the kingdom are they who have repudiated every external thing and have rooted from their hearts all sense of possessing. These are the "poor in spirit." They have reached an inward state paralleling the outward circumstances of the common beggar in the streets of Jerusalem; that is what the word "poor" as Christ used it actually means. These blessed poor are no longer slaves to the tyranny of *things*. They have broken the yoke of the oppressor; and this they have done not by fighting but by surrendering. Though free from all sense of possessing, they yet possess all things. *Theirs is the kingdom of heaven.*

Let me exhort you to take this seriously.

Let me exhort you to take this seriously. It is not to be understood as mere Bible teaching to be stored away in the mind along with an inert mass of other doctrines. It is a marker on the road to greener pastures, a path chiseled against the steep sides of the mount of God. We dare not try to bypass it if we would follow on in this holy pursuit. We must ascend a step at a time. If we refuse one step, we bring our progress to an end.

As is frequently true, this New Testament principle

of spiritual life finds its best illustration in the Old Testament. In the story of Abraham and Isaac, we have a dramatic picture of the surrendered life as well as an excellent commentary on the first beatitude.

Abraham was old when Isaac was born, old enough indeed to have been his grandfather, and the child became at once the delight of his heart. From that moment when he first stooped to take the tiny form in his arms, he was an eager love slave of his son. God went out of His way to comment on the strength of this affection. And it is not hard to understand. The baby represented everything sacred to his father's heart: the promises of God, the covenants, the hopes of the years, and the long messianic dream. As he watched him grow from babyhood to young manhood, the heart of the old man was knit closer and closer with the life of his son, until at last the relationship bordered upon the perilous. It was then that God stepped in to save both father and son from the consequences of an uncleansed love.[1]

Take now your son, said God to Abraham, *your only son, whom you love, Isaac, and go to the land of Moriah, and offer him there as a burnt offering on one of the mountains of which I will tell you.* The sacred writer spares us a closeup of the agony that night on the slopes near Beersheba when the aged man had it out with his God, but respectful imagination may view in awe the bent form and convulsive wrestling alone

1 *Publisher's Note:* We are unaware of this thought being supported in scripture. Rather, Genesis 22:1 simply says *And it came to pass after these things that God proved Abraham*. This verse seems to indicates a test rather than a reprimand.

under the stars. Possibly not again until a "Greater than Abraham" wrestled in the garden of Gethsemane did such mortal pain visit a human soul. If only the man himself might have been allowed to die. That would have been easier a thousand times, for he was old now, and to die would have been no great ordeal for one who had walked so long with God. Besides, it would have been a last sweet pleasure to let his dimming vision rest upon the figure of his stalwart son who would live to carry on the Abrahamic line and fulfill in himself the promises of God made long before in Ur of the Chaldees.

How could he slay the lad? Even if he could get the consent of his wounded and protesting heart, how could he reconcile the act with the promise: *For through Isaac your descendants shall be named*? This was Abraham's trial by fire, and he did not fail in the crucible. While the stars still shone like sharp white points above the tent where the sleeping Isaac lay, and long before the gray dawn had begun to lighten the east, the old saint had made up his mind. He would offer his son as God had directed him to do, and *then trust God to raise him from the dead*. This, says the writer to the Hebrews, was the solution his aching heart found sometime in the dark night, and he rose *early in the morning* to carry out the plan. It is beautiful to see that, while he erred as to God's method, he had correctly sensed the secret of His great heart. And the solution accords well with the New Testament Scripture, *Whoever loses . . . for my sake shall find*.

God let the suffering old man go through with it up to the point where He knew there would be no retreat, and then forbade him to lay a hand upon the boy. To the wondering patriarch He now says in effect, "It's all right, Abraham. I never intended that you should actually slay the lad. I only wanted to remove him from the temple of your heart that I might reign unchallenged there. I wanted to correct the perversion that existed in your love. Now you may have the boy, sound and well. Take him and go back to your tent. Now I know that thou fearest God, seeing that thou hast not withheld thy son, thine only son, from me."

Then heaven opened and a voice was heard saying to him, *By Myself I have sworn, declares the LORD, because you have done this thing and have not withheld your son, your only son, indeed I will greatly bless you, and I will greatly multiply your seed as the stars of the heavens and as the sand which is on the seashore; and your seed shall possess the gate of their enemies. In your seed all the nations of the earth shall be blessed, because you have obeyed My voice.*

The old man of God lifted his head to respond to the voice, and stood there on the mount strong and pure and grand, a man marked out by the Lord for special treatment, a friend and favorite of the Most High. Now he was a man wholly surrendered, a man utterly obedient, a man who possessed nothing. He had concentrated his all in the person of his dear son, and God had taken it from him. God could have begun out on the margin of Abraham's life and worked inward to the center; He chose rather to cut quickly to the heart and have it over in one sharp act of separation. In dealing thus, He practiced an economy of means and time. It hurt cruelly, but it was effective.

I have said that Abraham possessed nothing. Yet was not this poor man rich? Everything he had owned before was his still to enjoy: sheep, camels, herds, and goods of every sort. He had also his wife and his friends, and best of all he had his son Isaac safe by his side. He had everything, but *he possessed nothing.* There is the spiritual secret. There is the sweet theology of the heart, which can be learned only in the school of renunciation. The books on systematic theology overlook this, but the wise will understand.

After that bitter and blessed experience, I think the words "my" and "mine" never again had the same meaning for Abraham. The sense of possession which they connote was gone from his heart. *Things* had been cast out forever. They had now become external to the man. His inner heart was free from them. The world said, "Abraham is rich," but the aged patriarch only smiled. He could not explain it to them, but he knew that he owned nothing, that his real treasures were inward and eternal.

There can be no doubt that possessive clinging to things is one of the most harmful habits in the life.

There can be no doubt that this possessive clinging to things is one of the most harmful habits in the life. Because it is so natural, it is rarely recognized for the evil that it is; but its outworkings are tragic.

We are often hindered from giving up our treasures to the Lord out of fear for their safety; this is especially true when those treasures are loved relatives and friends. But we need have no such fears. Our Lord came not to destroy but to save. Everything is safe which we commit to Him, and nothing is really safe which is not so committed.

Our gifts and talents should also be turned over to Him. They should be recognized for what they are, God's loan to us, and should never be considered in any sense our own. We have no more right to claim credit for special abilities than for blue eyes or strong muscles. *For who regards you as superior? What do you have that you did not receive?*

The Christian who is alive enough to know himself even slightly will recognize the symptoms of this possession malady, and will grieve to find them in his own heart. If the longing after God is strong enough within him, he will want to do something about the matter. Now, what should he do?

First of all, he should put away all defense and make no attempt to excuse himself either in his own eyes or before the Lord. Whoever defends himself will have himself for his defense, and he will have no other; but let him come defenseless before the Lord and he will have for his defender no less than God Himself. Let the inquiring Christian trample under foot every slippery trick of his deceitful heart and insist upon frank and open relations with the Lord.

Then he should remember that this is holy business. No careless or casual dealings will suffice. Let him come to God in full determination to be heard. Let him insist that God accept his all, that He take *things* out of his heart and Himself reign there in power. It may be he will need to become specific, to name things and people by their names one by one. If he will become drastic enough, he can shorten the time of his toil from years to minutes, and enter the good land long before his slower brethren who coddle their feelings and insist upon caution in their dealings with God.

Let us never forget that such a truth as this cannot be learned by rote, as one would learn the facts of physical science. They must be *experienced* before we can really know them. We must in our hearts live through Abraham's harsh and bitter experiences if we would know the blessedness which follows them. The ancient curse will not go out painlessly; the tough, old miser within us will not lie down and die obedient to our command. He must be torn out of our heart like a plant from the soil; he must be extracted in agony and blood like a tooth from the jaw. He must be expelled from our soul by violence as Christ expelled the moneychangers from the temple. And we shall need to steel ourselves against his piteous begging, and to recognize it as springing out of self-pity, one of the most reprehensible sins of the human heart.

If we would indeed know God in growing intimacy, we must go this way of renunciation. And if we are set upon the pursuit of God, He will sooner or later bring us to this test. Abraham's testing was, at the time, not known to him as such, yet if he had taken some course other than the one he did, the whole history of the Old Testament would have been different. God would have found His man, no doubt, but the loss to Abraham would have been tragic beyond the telling. So we will be brought one by one to the testing place, and we may never know when we are there. At that testing place there will be no dozen possible choices for us; there will be just one and an alternative, but our whole future will be conditioned by the choice we make.

> *Father, I want to know thee, but my coward heart fears to give up its toys. I cannot part with them without inward bleeding, and I do not try to hide from thee the terror of the parting. I come trembling, but I do come. Please root from my heart all those things which I have cherished so long and which have become a very part of my living self, so that thou mayest enter and dwell there without a rival. Then shalt thou make the place of thy feet glorious. Then shall my heart have no need of the sun to shine in it, for thyself wilt be the light of it, and there shall be no night there. In Jesus' name, Amen.*

Chapter 3

Removing the Veil

Therefore, brethren, since we have confidence to enter the holy place by the blood of Jesus. (Hebrews 10:19)

Among the famous sayings of the church fathers, none is better known than Augustine's "Thou hast formed us for Thyself, and our hearts are restless till they find rest in Thee."

The great saint states here in few words the origin and interior history of the human race. God made us for Himself; that is the only explanation that satisfies the *heart* of a thinking man, whatever his wild reason may say. Should faulty education and perverse reasoning lead a man to conclude otherwise, there is little that any Christian can do for him. For such a man I have no message. My appeal is addressed to those who have been previously taught in secret by the wisdom of God; I speak to thirsty hearts whose longings have been wakened by the touch of God within them, and such as they need no reasoned proof. Their restless hearts furnish all the proof they need.

God formed us for Himself. The *Shorter Catechism*, "Agreed upon by the Reverend Assembly of Divines at Westminster," as the old *New-England Primer* has it, asks the ancient questions *what* and *why* and answers them in one short sentence hardly matched in any uninspired work. "Question: What is the chief end of man? Answer: Man's chief end is to glorify God, and to enjoy Him forever." With this agree the four and twenty elders who fall on their faces to worship Him that liveth forever and ever, saying, *Worthy are You, our Lord and our God, to receive glory and honor and power; for You created all things, and because of Your will they existed, and were created.*

Man's chief end is to glorify God, and to enjoy Him forever.

God formed us for His pleasure, and so formed us that we, as well as He, can in divine communion enjoy the sweet and mysterious mingling of kindred personalities. He meant us to see Him and live with Him and draw our life from His smile. But we have been guilty of that "foul revolt" of which Milton speaks when describing the rebellion of Satan and his hosts. We have broken with God. We have ceased to obey Him or love Him, and in guilt and fear have fled as far as possible from His presence.

Yet who can flee from His presence when the heaven and the heaven of heavens cannot contain Him? when the wisdom of Solomon testifies that the Spirit of the Lord fills the world? The omnipresence of the Lord is one thing, and is a solemn fact necessary to His perfection. His *manifest* presence is another thing altogether, and from that presence we have fled, like Adam, to hide among the trees of the garden, or like Peter to shrink away crying, *Go away from me Lord, for I am a sinful man.*

So the life of man upon the earth is a life away from His presence, wrenched loose from that "blissful center" which is our right and proper dwelling place, our first estate, which we kept not, the loss of which is the cause of our unceasing restlessness.

The whole work of God in redemption is to undo the tragic effects of that foul revolt, and to bring us back again into right and eternal relationship with Himself. This required that our sins be disposed of satisfactorily, that a full reconciliation be effected and the way opened for us to return again into conscious communion with God and to live again in His presence as before. Then by His prevenient working within us, He moves us to return. This first comes to our notice when our restless hearts feel a yearning for the presence of God and we say within ourselves, "I will arise and go to my Father." That is the first step, and as the Chinese sage Lao-Tze has said, "The journey of a thousand miles begins with a single step."

The interior journey of the soul from the wilds of sin into the enjoyed presence of God is beautifully illustrated in the Old Testament tabernacle. The returning sinner first entered the outer court where he offered a blood sacrifice on the brazen altar and washed himself in the laver that stood near it. Then through a veil he passed into the Holy Place where no natural light could come, but the golden candlestick, which spoke of Jesus the Light of the World, threw its soft glow over all. There also was the shewbread to tell of Jesus, the Bread of Life, and the altar of incense, a figure of unceasing prayer.

Though the worshipper had enjoyed so much, still he had not yet entered the presence of God. Another veil separated from the Holy of Holies where above the mercy seat dwelt the very God Himself in reverential and glorious manifestation. While the tabernacle stood, only the high priest could enter there, and that but once a year, with blood which he offered for his sins and the sins of the people. It was this last veil which was *rent* when our Lord gave up the ghost on Calvary, and the sacred writer explains that this rending of the veil opened the way for every worshipper in the world to come by the new and living way straight into the divine presence.

Everything in the New Testament accords with this Old Testament picture. Ransomed men need no longer pause in fear to enter the Holy of Holies. *God wills that we should push on into His presence and live our whole life there.* This is to be known to us in conscious experience. It is more than a doctrine to be held; it is a life to be enjoyed every moment of every day.

This flame of His presence was the beating heart of the Levitical order. Without it, all the appointments of the tabernacle were characters of some unknown language; they had no meaning for Israel or for us. The greatest fact of the tabernacle was that *Jehovah was there*; a presence was waiting within the veil. Similarly, the presence of God is the central fact of Christianity. At the heart of the Christian message is God Himself waiting for His redeemed children to push in to conscious awareness of His presence. That type of Christianity which happens now to be the vogue knows this presence only in theory. It fails to stress the Christian's privilege of present realization. According to its teachings, we are in the presence of God positionally, and nothing is said about the need to experience that presence actually. The fiery urge that drove men like Robert Murray McCheyne is wholly missing. And the present generation of Christians measures itself by this imperfect rule. Lowly contentment takes the place of burning zeal. We are satisfied to rest in our *judicial* possessions and, for the most part, we bother ourselves very little about the absence of personal experience.

Who is this within the veil who dwells in fiery manifestations? It is none other than God Himself, "One God the Father Almighty, Maker of heaven and earth, and of all things visible and invisible," and "One Lord Jesus Christ, the only begotten Son of God; begotten of His Father before all worlds, God of God, Light of Light, Very God of Very God; begotten, not made; being of one substance with the Father," and "the Holy Ghost, the Lord and Giver of life, Who proceedeth from the Father and the Son, Who with the Father and the Son together is worshipped and glorified." Yet this Holy Trinity is One God, for "we worship one God in Trinity, and Trinity in Unity; neither confounding the Persons, nor dividing the Substance. For there is one Person of the Father, another of the Son, and another of the Holy Ghost. But the Godhead of the Father, of the Son, and of the Holy Ghost, is all one: the glory equal and the majesty co-eternal." So in part run the ancient creeds, and so the inspired Word declares.

Behind the veil is God, that God after whom the world, with strange inconsistency, has felt, "if haply they might find Him." He has revealed Himself to some extent in nature, but more perfectly in the incarnation; now He waits to show Himself in ravishing fullness to the humble of soul and the pure in heart.

The world is perishing for lack of the knowledge of God, and the church is famishing for want of His presence. The instant cure of most of our religious ills would be to enter His presence in spiritual experience, to become suddenly aware that we are in God and that God is in us. This would lift us out of our pitiful narrowness and cause our hearts to be enlarged. This would burn away the impurities from our lives as the bugs and fungi were burned away by the fire that dwelt in the bush.

What a broad world to roam in, what a sea to swim in is this God and Father of our Lord Jesus Christ. He is *eternal*, which means that He predates time and is wholly independent of it. Time began in Him and will end in Him. To it, He pays no tribute, and from it, He suffers no change. He is *immutable*, which means that He has never changed and can never change in any smallest measure. To change, He would need to go from better to worse or from worse to better. He cannot do either, for being perfect He cannot become more perfect, and if He were to become less perfect He would be less than God. He is *omniscient*, which means that He knows in one free and effortless act all matter, all spirit, all relationships, all events. He has no past and He has no future. He *is*, and none of the limiting and qualifying terms used of creatures can apply to Him. *Love* and *mercy* and *righteousness* are His, and *holiness* so ineffable that no comparisons or figures will avail to express it. Only fire can give even a remote conception of it. In fire He appeared at the burning bush; in the pillar of fire He dwelt through all the long wilderness journey. The fire that glowed between the wings of the cherubim in the Holy Place was called the *shekinah*, the presence through the years of Israel's glory. And when the old had given place to the new, He came at Pentecost as a fiery flame and rested upon each disciple.

The great ones of the kingdom have been those who loved God more than others did.

Baruch Spinoza wrote of the intellectual love of God, and he had a measure of truth there; but the highest love of God is not intellectual, it is spiritual. God is spirit and only the spirit of man can know Him really. In the deep spirit of a man, the fire must glow, or his love is not the true love of God. The great ones of the kingdom have been those who loved God more than others did. We all know who they have been and gladly pay tribute to the depths and sincerity of their devotion. We have but to pause for a moment and their names come trooping past us smelling of myrrh and aloes and cassia out of the ivory palaces.

Frederick Faber was one whose soul panted after God as the roe pants after the water brook, and the measure in which God revealed Himself to his seeking heart set the good man's whole life afire with a burning adoration rivaling that of the seraphim before the throne. His love for God extended to the three persons of the Godhead equally, yet he seemed to feel for each One a special kind of love reserved for Him alone. Of God the Father he sings:

Only to sit and think of God,
Oh what a joy it is!
To think the thought, to breathe the Name;
Earth has no higher bliss.

Father of Jesus, love's reward!
What rapture will it be,
Prostrate before Thy throne to lie,
And gaze and gaze on Thee!

His love for the person of Christ was so intense that it threatened to consume him; it burned within him as a sweet and holy madness and flowed from his lips like molten gold. In one of his sermons he says, "Wherever we turn in the church of God, there is Jesus. He is the beginning, middle, and end of everything to us. . . . There is nothing good, nothing holy, nothing beautiful, nothing joyous which He is not to His servants. No one need be poor, because, if he chooses, he can have Jesus for his own property and possession. No one need be downcast, for Jesus is the joy of heaven, and it is His joy to enter into sorrowful hearts. We can exaggerate about many things; but we can never exaggerate our obligation to Jesus, or the compassionate abundance of the love of Jesus to us. All our lives long we might talk of Jesus, and yet we should never come to an end of the sweet things that might be said of Him. Eternity will not be long enough to learn all He is, or to worship Him for all He has done, but then, that matters not; for we shall be always with Him, and we desire nothing more." And addressing our Lord directly he says to Him:

I love Thee so, I know not how
My transports to control;
Thy love is like a burning fire
Within my very soul.

Faber's blazing love extended also to the Holy Spirit. Not only in his theology did he acknowledge His deity and full equality with the Father and the Son, but he also celebrated it constantly in his songs and in his prayers. He literally pressed his forehead to the ground in his eager, fervid worship of the third person of the Godhead. In one of his great hymns to the Holy Spirit he sums up his burning devotion thus:

O Spirit, beautiful and dread!
My heart is fit to break
With love of all Thy tenderness
For us poor sinners' sake.

I have risked the tedium of quotation that I might show by pointed example what I have set out to say, namely, that God is so vastly wonderful, so utterly and completely delightful that He can, without anything other than Himself, meet and overflow the deepest demands of our total nature, mysterious and deep as that nature is. Such worship as Faber knew (and he is but one of a great company which no man can number) can never come from a mere doctrinal knowledge of God. Hearts that are "fit to break" with love for the Godhead are those who have been in His presence and have looked with opened eye upon the majesty of the Deity. Men of the breaking hearts had a quality about them not known to or understood by common men. They habitually spoke with spiritual authority. They had been in the presence of God and they reported what they saw there. They were prophets, not scribes, for the scribe tells us what he has read, and the prophet tells us what he has seen.

The distinction is not an imaginary one. Between the scribe who has read and the prophet who has seen, there is a difference as wide as the sea. We are today overrun with orthodox scribes, but the prophets, where are they? The hard voice of the scribe sounds over evangelicalism, but the church waits for the tender voice of the saint who has penetrated the veil and has gazed with inward eye upon the wonder that is God. And yet, thus to penetrate, to push in sensitive living experience into the holy presence, is a privilege open to every child of God.

With the veil removed by the rending of Jesus' flesh, with nothing on God's side to prevent us from entering, why do we tarry without? Why do we consent to abide all our days just outside the Holy of Holies and never enter at all to look upon God? We hear the Bridegroom say, *Let me see your form, let me hear your voice; for your voice is sweet, and your form is lovely.* We sense that the call is for us, but still we fail to draw near, and the years pass and we grow old and tired in the outer courts of the tabernacle. What hinders us?

The answer usually given, simply that we are "cold," will not explain all the facts. There is something more serious than coldness of heart, something that may be behind that coldness and be the cause of its existence. What is it? What could it be but the presence of *a veil in our hearts*? It is a veil not taken away as the first veil was, but which remains there still shutting out the light and hiding the face of God from us. It is the veil of our fleshly fallen nature that lives on, unjudged within us, uncrucified, and unrepudiated. It is the close-woven veil of the self-life which we have never truly acknowledged, of which we have been secretly ashamed, and which for these reasons we have never brought to the judgment of the cross. It is not too mysterious, this opaque veil, nor is it hard to identify. We have but to look in our own hearts and we shall see it there, sewn and patched and repaired it may be, but there nevertheless, an enemy to our lives and an effective block to our spiritual progress.

This veil is not a beautiful thing and it is not a thing about which we commonly care to talk, but I am addressing the thirsting souls who are determined to follow God, and I know they will not turn back even though the way leads temporarily through the blackened hills. The urge of God within them will assure their continuing the pursuit. They will face the facts however unpleasant, and endure the cross for the joy set before them. So I am bold to name the threads out of which this inner veil is woven.

It is woven of the fine threads of the self-life, the "hyphenated" sins of the human spirit. They are not something we do; they are something we *are*, and therein lies both their subtlety and their power.

To be specific, the self-sins are these: self-righteousness, self-pity, self-confidence, self-sufficiency, self-admiration, self-love, and a host of others like them. They dwell too deep within us and are too much a part of our natures to come to our attention until the light of God is focused upon them. The grosser manifestations of these sins – egotism, exhibitionism, self-promotion – are strangely tolerated in Christian leaders, even in circles of impeccable orthodoxy. They are so much in evidence as actually, for many people, to become identified with the gospel. I trust it is not a cynical observation to say that they appear these days to be a requisite for popularity in some sections of the church visible. Promoting self under the guise of promoting Christ is currently so common as to excite little notice.

One should suppose that proper instruction in the doctrines of man's depravity and the necessity for justification through the righteousness of Christ alone would deliver us from the power of the self-sins; but it does not work out that way. Self can live unrebuked at the very altar. It can watch the bleeding victim die and not be in the least affected by what it sees. It can fight for the faith of the Reformers and preach eloquently the creed of salvation by grace, and gain strength by its efforts. To tell all the truth, it seems actually to feed upon orthodoxy and is more at home in a Bible conference than in a tavern. Our very state of longing after God may afford it an excellent condition under which to thrive and grow.

Self is the opaque veil that hides the face of God from us. It can be removed only in spiritual experience, never by mere instruction. We may as well try to instruct leprosy out of our system. There must be a work of God in destruction before we are free. We must invite the cross to do its deadly work within us. We must bring our self-sins to the cross for judgment. We must prepare ourselves for an ordeal of suffering in some measure like that through which our Savior passed when He suffered under Pontius Pilate.

Self is the opaque veil that hides the face of God from us.

Let us remember: When we talk of the rending of the veil, we are speaking figuratively, and the thought of it is poetical, almost pleasant; but in actuality, there is nothing pleasant about it. In human experience, that veil is made of living spiritual tissue; it is composed of the conscious, quivering stuff of which our whole beings consist, and to touch it is to touch us where we feel pain. To tear it away is to injure us, to hurt us and make us bleed. To say otherwise is to make the cross no cross and death no death at all. It is never fun to die. To rip through the dear and tender stuff of which life is made can never be anything but deeply painful. Yet that is what the cross did to Jesus and it is what the cross would do to every man to set him free.

Let us beware of tinkering with our inner life in the hope of rending the veil ourselves. God must do everything for us. Our part is to yield and trust. We must confess, forsake, repudiate the self-life, and then reckon it crucified. But we must be careful to distinguish lazy "acceptance" from the real work of God. We must insist upon the work being done. We dare not rest content with a neat doctrine of self-crucifixion. That is to imitate King Saul and spare the best of the sheep and the oxen.

Insist that the work be done in very truth and it will be done. The cross is rough, and it is deadly, but it is effective. It does not keep its victim hanging there forever. There comes a moment when its work is finished and the suffering victim dies. After that is resurrection glory and power, and the pain is forgotten for the joy that the veil is taken away and we have entered in actual, spiritual experience the presence of the living God.

Lord, how excellent are thy ways, and how devious and dark are the ways of man. Show us how to die to our selfish desires, that we may rise again to newness of life. Rend the veil of our self-life from the top down as thou didst rend the veil of the temple. We would draw near in full assurance of faith. We would dwell with thee in daily experience here on this earth, so that we may be accustomed to the glory when we enter thy heaven to dwell with thee there. In Jesus' name, Amen.

CHAPTER 4

Apprehending God

O taste and see. (Psalm 34:8)

It was Canon Holmes of India who more than twenty-five years ago called attention to the inferential character of the average man's faith in God. To most people God is an inference, not a reality. He is a deduction from evidence which they consider adequate; but He remains personally unknown to the individual. "He *must* be," they say, "therefore, we believe He is." Others do not go even so far as this; they know of Him only by hearsay. They have never bothered to think the matter out for themselves, but have heard about Him from others, and have put belief in Him into the back of their minds, along with the various odds and ends that make up their total creed. To many others God is but an ideal, another name for goodness, or beauty, or truth; or He is law, or life, or the creative impulse behind the phenomena of existence.

These notions about God are many and varied, but

those who hold them have one thing in common: They do not know God in personal experience. The possibility of intimate acquaintance with Him has not entered their minds. While admitting His existence, they do not think of Him as knowable in the sense that we know things or people.

Christians, to be sure, go further than this, at least in theory. Their creed requires them to believe in the personality of God, and they have been taught to pray, *Our Father who is in heaven*. Now personality and fatherhood carry with them the idea of the possibility of personal acquaintance. This is admitted, I say, in theory; but for millions of Christians, nevertheless, God is no more real than He is to the non-Christian. They go through life trying to love an ideal and be loyal to a mere principle.

Over against all this cloudy vagueness stands the clear scriptural doctrine that God can be known in personal experience. A loving personality dominates the Bible, walking among the trees of the garden and breathing fragrance over every scene. Always a living person is present, speaking, pleading, loving, working, and manifesting Himself whenever and wherever His people have the receptivity necessary to receive the manifestation.

The Bible assumes as a self-evident fact that men can know God with at least the same degree of immediacy as they know any other person or thing that comes within the field of their experience. The same terms are used to express the knowledge of God as are used to express the knowledge of physical things. *O **taste*** *and see that the LORD is good. All Your garments are* ***fragrant*** *with myrrh and aloes and cassia; out of ivory palaces. My sheep* **hear** *my voice. Blessed are the pure in heart, for they shall* **see** *God.* (Emphases added.) These are but four of countless such passages from the Word of God. And more important than any proof text is the fact that the whole import of the Scripture is toward this belief.

What can all this mean except that we have in our hearts organs by means of which we can know God as certainly as we know material things through our familiar five senses. We apprehend the physical world by exercising the faculties given us for the purpose, and we possess spiritual faculties by means of which we can know God and the spiritual world if we will obey the Spirit's urge and begin to use them.

Where faith is defective, the result will be inward insensibility and numbness toward spiritual things.

That a saving work must first be done in the heart is taken for granted here. The spiritual faculties of the unregenerate man lie asleep in his nature, unused and for every purpose dead; that is the stroke which has fallen upon us by sin. They may be quickened to active life again by the operation of the Holy Spirit in regeneration; that is one of the immeasurable benefits which come to us through Christ's atoning work on the cross.

But for the very ransomed children of God themselves, why do they know so little of that habitual conscious communion with God which the Scriptures seem to offer? The answer is our chronic unbelief. Faith enables our spiritual sense to function. Where faith is defective, the result will be inward insensibility and numbness toward spiritual things. This is the condition of vast numbers of Christians today. No proof is necessary to support that statement. We have but to converse with the first Christian we meet or enter the first church we find open to acquire all the proof we need.

A spiritual kingdom lies all about us, enclosing us, embracing us, altogether within reach of our inner selves, waiting for us to recognize it. God Himself is here waiting for our response to His presence. This eternal world will come alive to us the moment we begin to reckon upon its reality.

I have just now used two words which demand definition; or if definition is impossible, I must at least make clear what I mean when I use them. They are "reality" and "reckon."

What do I mean by *reality*? I mean that which has existence apart from any idea any mind may have of it, and which would exist if there were no mind anywhere to entertain a thought of it. That which is real has being in itself. It does not depend upon the observer for its validity.

I am aware that there are those who love to poke fun at the plain man's idea of reality. They are the idealists who spin endless proofs that nothing is real outside of the mind. They are the relativists who like to show that there are no fixed points in the universe from which we can measure anything. They smile down upon us from their lofty intellectual peaks and settle us to their own satisfaction by fastening upon us the reproachful term "absolutist." The Christian is not put out of countenance by this show of contempt. He can smile right back at them, for he knows that there is only One who is absolute, and that is God. But he knows also that the absolute One has made this world for man's uses, and, while there is nothing fixed or real in the last meaning of the words (the meaning as applied to God), *for every purpose of human life we are permitted to act as if there were.* And every man does act thus except the mentally ill. These unfortunates also have trouble with reality, but they are consistent; they insist upon living in accordance with their ideas of things. They are honest, and it is their very honesty that constitutes a problem.

The idealists and relativists are not mentally ill. They prove their soundness by living their lives according to the very notions of reality which they in theory repudiate, and by counting upon the very fixed points which they prove are not there. They could earn a lot more respect for their notions if they were willing to live by them; but this they are careful not to do. Their ideas are brain-deep, not life-deep. Wherever life touches them, they repudiate their theories and live like other men.

The Christian is too sincere to play with ideas for their own sake. He takes no pleasure in the mere spinning of gossamer webs for display. All his beliefs are practical. They are geared into his life. By them he lives or dies, stands or falls for this world and for all time to come. From the insincere man he turns away.

The sincere plain man knows that the world is real. He finds it here when he wakes to consciousness, and he knows that he did not think it into being. It was here waiting for him when he came, and he knows that when he prepares to leave this earthly scene it will be here still to bid him good-bye as he departs. By the deep wisdom of life he is wiser than a thousand men who doubt. He stands upon the earth and feels the wind and rain in his face and he knows that they are real. He sees the sun by day and the stars by night. He sees the hot lightning play out of the dark thundercloud. He hears the sounds of nature and the cries of human joy and pain. These he knows are real. He lies down on the cool earth at night and has no fear that it will prove illusory or fail him while he sleeps. In the morning, the firm ground will be under him, the blue sky above him, and the rocks and trees around him as when he closed his eyes the night before. So he lives and rejoices in a world of reality.

With his five senses he engages this real world. All things necessary to his physical existence he apprehends by the faculties with which he has been equipped by the God who created him and placed him in such a world as this.

Now, by our definition God is also real. He is real in the absolute and final sense that nothing else is. All other reality is contingent upon His. The great reality is God, who is the author of that lower and dependent reality which makes up the sum of created things, including ourselves. God has objective existence independent of and apart from any notions which we may have concerning Him. The worshipping heart does not create its object. It finds Him here when it wakes from its moral slumber in the morning of its regeneration.

Another word that must be cleared up is the word *reckon*. This does not mean to visualize or imagine. Imagination is not faith. The two are not only different from, but also stand in sharp opposition to each other. Imagination projects unreal images out of the mind and seeks to attach reality to them. Faith creates nothing; it simply reckons upon that which is already *there*.

God and the spiritual world are real. We can reckon upon them with as much assurance as we reckon upon the familiar world around us. Spiritual things are there (or rather we should say *here*), inviting our attention and challenging our trust.

God and the spiritual world are real.

Our trouble is that we have established bad thought habits. We habitually think of the visible world as real and doubt the reality of any other. We do not deny the existence of the spiritual world, but we doubt that it is real in the accepted meaning of the word.

The world of sense intrudes upon our attention day and night for the whole of our lifetime. It is clamorous, insistent, and self-demonstrating. It does not appeal to our faith; it is here, assaulting our five senses, demanding to be accepted as real and final. But sin has so clouded the lenses of our hearts that we cannot see that other reality, the city of God, shining around us. The world of sense triumphs. The visible becomes the enemy of the invisible, and the temporal, of the eternal. That is the curse inherited by every member of Adam's tragic race.

At the root of the Christian life lies belief in the invisible. The object of the Christian's faith is unseen reality.

Our uncorrected thinking, influenced by the blindness of our natural hearts and the intrusive ubiquity of visible things, tends to draw a contrast between the spiritual and the real; but actually, no such contrast exists. The antithesis lies elsewhere: between the real and the imaginary, between the spiritual and the material, between the temporal and the eternal; but between the spiritual and the real, never. The spiritual *is* real.

If we would rise into that region of light and power plainly beckoning us through the Scriptures of truth, we must break the evil habit of ignoring the spiritual. We must shift our interest from the seen to the unseen. For the great unseen reality is God. *He who comes to God must believe that He is and that He is a rewarder of those who seek Him.* This is basic in the life of faith. From there we can rise to unlimited heights. *Believe in God,* said our Lord Jesus Christ, *believe also in Me.* Without the first, there can be no second.

If we truly want to follow God, we must seek to be otherworldly. This I say knowing well that that word has been used with scorn by the sons of this world and applied to the Christian as a badge of reproach. So be it. Every man must choose his world. If we who follow Christ, with all the facts before us and knowing what we are about, deliberately choose the kingdom of God as our sphere of interest, I see no reason why anyone should object. If we lose by it, the loss is our own; if we gain, we rob no one by so doing. The "other world," which is the object of this world's disdain and the subject of the drunkard's mocking song, is our carefully chosen goal and the object of our holiest longing.

But we must avoid the common fault of pushing the "other world" into the future. It is not future, but present. It parallels our familiar physical world, and the doors between the two worlds are open. *But you have come,* says the writer to the Hebrews (and the tense is plainly present), *to Mount Zion and to the city of the living God, the heavenly Jerusalem, and to myriads of angels, to the general assembly and church of the firstborn who are enrolled in heaven, and to God, the Judge of all, and to the spirits of the righteous made perfect, and to Jesus, the mediator of a new covenant, and to the sprinkled blood, which speaks better than the blood of Abel.* All these things are contrasted with *the mount that might be touched* and *the sound of a trumpet and the voice of words* that might be heard. May we not safely conclude that, as the realities of Mount Sinai were apprehended by the senses, so the realities of Mount Zion are to be grasped by the soul? And this not by any trick of the imagination, but in downright actuality. The soul has eyes with which to see and ears with which to hear. Feeble they may be from long disuse, but by the life-giving touch of Christ alive now and capable of sharpest sight and most sensitive hearing.

As we begin to focus upon God, the things of the spirit will take shape before our inner eyes. Obedience to the Word of Christ will bring an inward revelation of the Godhead (John 14:21-23). It will give acute perception, enabling us to see God even as is promised to the pure in heart. A new God-consciousness will seize upon us and we shall begin to taste and hear and inwardly feel the God who is our life and our all. There will be seen the constant shining of the light that lights up every man that comes into the world. More and more, as our faculties grow sharper and more sure, God will become to us the great "all," and His presence the glory and wonder of our lives.

O God, quicken to life every power within me, that I may lay hold on eternal things. Open my eyes that I may see; give me acute spiritual perception; enable me to taste thee and know that thou art good. Make heaven more real to me than any earthly thing has ever been. In Jesus' name, Amen.

Chapter 5

The Universal Presence

Where can I go from Your Spirit? Or where can I flee from Your presence? (Psalm 139:7)

In all Christian teaching, certain basic truths are found, hidden at times, and rather assumed than asserted, but necessary to all truth as the primary colors are found in and necessary to the finished painting. Such a truth is the divine immanence.

God dwells in His creation and is everywhere indivisibly present in all His works. This is boldly taught by prophet and apostle and is accepted by Christian theology generally; that is, it appears in the books, but for some reason it has not sunk into the average Christian's heart so as to become a part of his believing self. Christian teachers shy away from its full implications, and, if they mention it at all, mute it down until it has little meaning. I would guess the reason for this is the fear of being charged with pantheism; but the doctrine of the divine presence is definitely not pantheism.

Pantheism's error is too palpable to deceive anyone. It is that God is the sum of all created things. Nature and God are one, so that whoever touches a leaf or a stone touches God. That is of course to degrade the glory of the incorruptible Deity and, in an effort to make all things divine, banish all divinity from the world entirely.

The truth is that while God dwells in His world, He is separated from it by a gulf forever impassable. However closely He may be identified with the work of His hands, *they* are and must eternally be *other than He*, and He is and must be antecedent to and independent of them. He is transcendent above all His works even while He is immanent within them.

What now does the divine immanence mean in direct Christian experience? It means simply that *God is here*. Wherever we are, God is here. There is no place, there can be no place, where He is not. Ten million intelligences standing at as many points in space and separated by incomprehensible distances can each one say with equal truth, God is here. No point is nearer to God than any other point. It is exactly as near to God from any place as it is from any other place. No one is in mere distance any farther from or any nearer to God than any other person is.

The truth is that while God dwells in His world, He is separated from it by a gulf forever impassable.

These are truths believed by every instructed Christian. It remains for us to think on them and pray over them until they begin to glow within us.

In the beginning God. Not "In the beginning *matter*," for matter is not self-causing. It requires an antecedent cause, and God is that cause. Not "In the beginning *law*," for law is but a name for the course which all creation follows. That course had to be planned, and the planner is God. Not "In the beginning *mind*," for mind also is a created thing and must have a creator behind it. But *In the beginning God*, the uncaused cause of matter, mind, and law. There we must begin.

Adam sinned and, in his panic, frantically tried to do the impossible: He tried to hide from the presence of God. David also must have had wild thoughts of trying to escape from God's presence, for he wrote: *Where can I go from Your Spirit? Or where can I flee from Your presence?* Then he proceeded through one of his most beautiful psalms to celebrate the glory of the divine immanence. *If I ascend to heaven, You are there; if I make my bed in Sheol, behold, You are there. If I take the wings of the dawn, if I dwell in the remotest part of the sea, even there Your hand will lead me, and Your right hand will lay hold of me.* And he knew that God's *being* and God's *seeing* are the same, that the seeing "Presence" had been with him even before he was born, watching the mystery of unfolding life. Solomon exclaimed, *But will God indeed dwell on the earth? Behold, heaven and the highest heaven cannot contain You, how much less this house which I have built?* Paul assured the Athenians that God *is not far from each one of us; for in him we live and move and exist.*

If God is present at every point in space, if we cannot go where He is not, cannot even conceive of a place where He is not, why then has not that presence become the one universally celebrated fact of the world? The patriarch Jacob, *in the howling waste of a wilderness*, gave the answer to that question. He saw a vision of God and cried out in wonder, *Surely the LORD is in this place, and I did not know it.* Jacob had never been for one small division of a moment outside the circle of that all-pervading presence. But he knew it not. That was his trouble, and it is ours. Men do not know that God is here. What a difference it would make if they knew.

The presence and the manifestation of the presence are not the same. There can be the one without the other. God is here when we are wholly unaware of it. He is *manifest* only when and as we are aware of His presence. On our part there must be surrender to the Spirit of God, for His work it is to show us the Father and the Son. If we cooperate with Him in loving obedience, God will manifest Himself to us, and that manifestation will be the difference between a nominal Christian life and a life radiant with the light of His face.

Always, everywhere, God is present, and always He seeks to reveal Himself. To each one he would reveal not only that He is, but *what* He is as well. He did not have to be persuaded to reveal Himself to Moses. *The LORD descended in the cloud and stood there with him as he called upon the name of the LORD.* He not only made a verbal proclamation of His nature, but He also revealed His very self to Moses so that the skin of Moses' face shone with the supernatural light. It will be a great moment for some of us when we begin to believe that God's promise of self-revelation is literally true: that

He promised much, but He promised no more than He intends to fulfill.

Our pursuit of God is successful just because He is forever seeking to manifest Himself to us. The revelation of God to any man is not God coming from a distance upon a time to pay a brief and momentous visit to the man's soul. Thus to think of it is to misunderstand it all. The approach of God to the soul or of the soul to God is not to be thought of in spatial terms at all. There is no idea of physical distance involved in the concept. It is not a matter of miles but of experience.

God promised much, but He promised no more than He intends to fulfill.

To speak of being near to or far from God is to use language in a sense always understood when applied to our ordinary human relationships. A man may say, "I feel that my son is coming nearer to me as he gets older," and yet that son has lived by his father's side since he was born and has never been away from home more than a day or so in his entire life. What then can the father mean? Obviously, he is speaking of *experience*. He means that the boy is coming to know him more intimately and with deeper understanding, that the barriers of thought and feeling between the two are disappearing, that father and son are becoming more closely united in mind and heart.

So when we sing, "Draw me nearer, nearer blessed Lord," we are not thinking of the nearness of place, but of the nearness of relationship. It is for increasing degrees of awareness that we pray, for a more perfect consciousness of the divine presence. We need never shout across the spaces to an absent God. He is nearer than our own soul, closer than our most secret thoughts.

Why do some people "find" God in a way that others do not? Why does God manifest His presence to some and let multitudes of others struggle along in the half-light of imperfect Christian experience? Of course, the will of God is the same for all. He has no favorites within His household. All He has ever done for any of His children He will do for all of His children. The difference lies not with God but with us.

Pick at random a score of great saints whose lives and testimonies are widely known. Let them be Bible characters or well-known Christians of post-biblical times. You will be struck instantly with the fact that the saints were not alike. Sometimes the unlikenesses were so great as to be positively glaring. How different, for example, was Moses from Isaiah; how different was Elijah from David; how unlike each other were John and Paul, Saint Francis and Luther, Finney and Thomas à Kempis. The differences are as wide as human life itself: differences of race, nationality, education, temperament, habit, and personal qualities. Yet they all walked, each in his day, upon a high road of spiritual living far above the common way.

Their differences must have been incidental and in the eyes of God of no significance. In some vital quality they must have been alike. What was it?

I venture to suggest that the one vital quality which they had in common was *spiritual receptivity.* Something in them was open to heaven, something which urged them Godward. Without attempting anything like a profound analysis, I shall say simply that they had spiritual awareness and that they went on to cultivate it until it became the biggest thing in their lives. They differed from the average person in that when they felt the inward longing, they *did something about it.* They acquired the lifelong habit of spiritual response. They were not disobedient to the heavenly vision. As David put it neatly, *When You said, Seek My face, my heart said to You, Your face, O LORD, I shall seek.*

As with everything good in human life, behind this receptivity is God. The sovereignty of God is here, and is felt even by those who have not placed particular stress upon it theologically. The pious Michelangelo confessed this in a sonnet:

> *My unassisted heart is barren clay,*
> *That of its native self can nothing feed:*
> *Of good and pious works Thou art the seed,*
> *That quickens only where Thou sayest it may:*
> *Unless Thou show to us Thine own true way*
> *No man can find it: Father! Thou must lead.*

These words will repay study as the deep and serious testimony of a great Christian.

Important as it is that we recognize God working in us, I would yet warn against a too-great preoccupation with the thought. It is a sure road to sterile passivity. God will not hold us responsible to understand the mysteries of election, predestination, and the divine sovereignty. The best and safest way to deal with these truths is to raise our eyes to God and in deepest reverence say, *O*

LORD, thou knowest. Those things belong to the deep and mysterious profoundness of God's omniscience. Prying into them may make theologians, but it will rarely make saints.

Receptivity is not a single thing; rather, it is a compound, a blending of several elements within the soul. It is an affinity for, a bent toward, a sympathetic response to, a desire to have. From this it may be gathered that it can be present in degrees, that we may have little or more or less, depending upon the individual. It may be increased by exercise or destroyed by neglect. It is not a sovereign and irresistible force which comes upon us as a seizure from above. It is a gift of God, indeed, but one which must be recognized and cultivated as any other gift if it is to realize the purpose for which it was given.

Failure to see this is the cause of a very serious breakdown in modern evangelicalism. The idea of cultivation and exercise, so dear to the saints of old, has now no place in our total religious picture. It is too slow, too common. We now demand glamour and fast-flowing dramatic action. A generation of Christians reared among push buttons and automatic machines is impatient with slower and less-direct methods of reaching their goals. We have been trying to apply machine-age methods to our relationships with God. We read our chapter, have our short devotions, and rush away, hoping to make up for our deep inward bankruptcy by attending another gospel meeting or listening to another thrilling story told by a religious adventurer lately returned from afar.

The tragic results of this spirit are all about us. Shallow lives, hollow religious philosophies, the preponderance of the element of fun in gospel meetings, the glorification of men, trust in religious externalities, quasi-religious fellowships, salesmanship methods, the mistaking of dynamic personality for the power of the Spirit: These and such as these are the symptoms of an evil disease, a deep and serious malady of the soul.

For this great sickness that is upon us no one person is responsible, and no Christian is wholly free from blame. We have all contributed, directly or indirectly, to this sad state of affairs. We have been too blind to see, or too timid to speak out, or too self-satisfied to desire anything better than the poor, average diet with which others appear satisfied. To put it differently, we have accepted one another's notions, copied one another's lives, and made one another's experiences the model for our own. And for a generation the trend has been downward. Now we have reached a low place of sand and burnt-wire grass, and, worst of all, we have made the Word of Truth conform to our experience and accepted this low plane as the very pasture of the blessed.

It will require a determined heart and more than a little courage to wrench ourselves loose from the grip of our times and return to biblical ways.

It will require a determined heart and more than a little courage to wrench ourselves loose from the grip of our times and return to biblical ways. But it can be done. Every now and then in the past, Christians have had to do it. History has recorded several large-scale returns led by such men as Saint Francis, Martin Luther, and George Fox. Unfortunately, there seems to be no Luther or Fox on the horizon at present. Whether or not another such return may be expected before the coming of Christ is a question upon which Christians are not fully agreed, but that is not of too great importance to us now.

What God in His sovereignty may yet do on a worldwide scale I do not claim to know; but what He will do for the plain man or woman who seeks His face I believe I do know and can tell others. Let any man turn to God in earnest, let him begin to exercise himself unto godliness, let him seek to develop his powers of spiritual receptivity by trust and obedience and humility, and the results will exceed anything he may have hoped in his leaner and weaker days.

Any man who by repentance and a sincere return to God will break himself out of the mold in which he has been held, and will go to the Bible itself for his spiritual standards, will be delighted with what he finds there.

Let us say it again: The universal presence is a fact. God is here. The whole universe is alive with His life. And He is no strange or foreign God, but the familiar Father of our Lord Jesus Christ whose love has for these thousands of years enfolded the sinful race of men. And always He is trying to get our attention, to reveal Himself to us, to communicate with us. We have within us the ability to know Him if we will but respond to His overtures. (And this we call pursuing God!) We will know Him in increasing degree as our receptivity becomes more perfect by faith and love and practice.

O God and Father, I repent of my sinful preoccupation with visible things. The world has been too much with me. Thou hast been here and I knew it not. I have been blind to thy presence. Open my eyes that I may behold thee in and around me. In Jesus' name, Amen.

Chapter 6

The Speaking Voice

In the beginning was the Word, and the Word was with God, and the Word was God. (John 1:1)

An intelligent plain man, untaught in the truths of Christianity, coming upon this text, would likely conclude that John meant to teach that it is the nature of God to speak, to communicate His thoughts to others. And he would be right. A word is a medium by which thoughts are expressed, and the application of the term to the eternal Son leads us to believe that self-expression is inherent in the Godhead, that God is forever seeking to speak Himself out to His creation. The whole Bible supports the idea. God is speaking. Not God spoke, but *God is speaking.* He is by His nature continuously articulate. He fills the world with His speaking voice.

One of the great realities with which we have to deal is the voice of God in His world. The briefest and only satisfying cosmogony is this: *For he spoke, and it was done.* The *why* of natural law is the living voice of God immanent in His creation. And this word of God which brought all worlds into being cannot be understood to mean the Bible, for it is not a written or printed word at all, but the expression of the will of God spoken into the structure of all things. This word of God is the breath of God filling the world with living potentiality. The voice of God is the most powerful force in nature; indeed, it is the only force in nature, for all energy is here only because the power-filled Word is being spoken.

The Bible is the written word of God, and because it is written, it is confined and limited by the necessities of ink and paper. The voice of God, however, is alive and free as the sovereign God is free. *The words that I have spoken to you are spirit and are life.* The life is in the speaking words. God's word in the Bible can have power only because it corresponds to God's word in the universe. It is the present voice which makes the written Word all-powerful. Otherwise, it would lie locked in slumber within the covers of a book.

The voice of God is the most powerful force in nature; indeed, it is the only force in nature.

We take a low and primitive view of things when we conceive of God at the creation coming into physical contact with things, shaping and fitting and building like a carpenter. The Bible teaches otherwise: *By the word of the LORD the heavens were made, and by the breath of His mouth all their host. . . . For He spoke, and it was done; He commanded, and it stood fast.* And *By faith we understand that the worlds were prepared by the word of God.* Again, we must remember that God is referring here not to His written Word, but to His speaking voice. His world-filling voice is meant, that voice which predates the Bible by uncounted centuries, that voice which has not been silent since the dawn of creation, but is sounding still throughout the far reaches of the universe.

The Word of God is quick and powerful. In the beginning He spoke to nothing, and it became *something.* Chaos heard it and became order; darkness heard it and became light. *And God said . . . and it was so.* These twin phrases, as cause and effect, occur throughout the Genesis story of the creation. The *said* accounts for the *so.* The *so* is the *said* put into the continuous present.

That God is here and that He is speaking – these truths are the backing for all other Bible truths; without them there could be no revelation at all. God did not write a book and send it by messenger to be read at a distance by unaided minds. He spoke a book and lives in His spoken words, constantly speaking His words and causing the power of them to persist across the years. God breathed on clay and it became a man; He breathes on men and they become clay. "Return ye children of men" was the word spoken at the fall, by which God decreed the death of every man, and no added word has He needed to speak. The sad procession of mankind across the face of the earth from birth to the grave is proof that His original word was enough.

We have not given sufficient attention to that deep

utterance in the gospel of John: *There was the true Light which, coming into the world, enlightens every man.* Shift the punctuation around as we will and the truth is still there: The Word of God affects the hearts of all men as light in the soul. In the hearts of all men the light shines, the Word sounds, and there is no escaping them. Something like this would of necessity be so if God is alive and in His world. And John says that it is so. Even those individuals who have never heard of the Bible have still been preached to with sufficient clarity to remove every excuse from their hearts forever, *which show the work of the law written in their hearts, their conscience bearing witness and their thoughts alternately accusing or else defending them. For since the creation of the world His invisible attributes, His eternal power and divine nature, have been clearly seen, being understood through what has been made, so that they are without excuse.*

This universal voice of God was by the ancient Hebrews often called wisdom, and was said to be everywhere sounding and searching throughout the earth, seeking some response from the sons of men. The eighth chapter of the book of Proverbs begins: *Does not wisdom call, And understanding lift up her voice?* The writer then pictures wisdom as a beautiful woman standing *on top of the heights beside the way, where the paths meet.* She sounds her voice from every quarter so that no one may miss hearing it. *To you, O men, I call, and my voice is to the sons of men.* Then she pleads for the simple and the foolish to give ear to her words. It is a spiritual response for which this wisdom of God is pleading, a response which she has always sought and is but rarely able to secure. The tragedy is that our eternal welfare depends upon our hearing, and we have trained our ears not to hear.

This universal voice has ever sounded, and it has often troubled men even when they did not understand the source of their fears. Could it be that this voice distilling like a living mist upon the hearts of men has been the undiscovered cause of the troubled conscience and the longing for immortality confessed by millions since the dawn of recorded history? We need not fear to face up to this. The speaking voice is a fact. How men have reacted to it is for any observer to note.

When God spoke out of heaven to our Lord, self-centered men who heard it explained it by natural causes: They said, *it thundered*. This habit of explaining the voice by appeals to natural law is at the very root of modern science. In the living, breathing cosmos, there is a mysterious something, too wonderful, too awful for any mind to understand. The believing man does not claim to understand. He falls to his knees and whispers, "God." The man of earth kneels also, but not to worship. He kneels to examine, to search, to find the cause and the how of things. Just now we happen to be living in a secular age. Our thought habits are those of the scientist, not those of the worshipper. We are more likely to explain than to adore. "It thundered," we exclaim, and go our earthly way. But still the voice sounds and searches. The order and life of the world depend upon that voice, but men are mostly too busy or too stubborn to give attention.

Every one of us has had experiences which we have not been able to explain: a sudden sense of loneliness, or a feeling of wonder or awe in the face of the universal vastness. Or we have had a fleeting visitation of light like an illumination from some other sun, giving us in a quick flash an assurance that we are from another world, that our origins are divine. What we saw there, or felt, or heard, may have been contrary to all that we had been taught in the schools and at wide variance with all our former beliefs and opinions. We were forced to suspend our acquired doubts while, for a moment, the clouds were rolled back and we saw and heard for ourselves. Explain such things as we will, I think we have not been fair to the facts until we allow at least the possibility that such experiences may arise from the presence of God in the world and His persistent effort to communicate with mankind. Let us not dismiss such a hypothesis too flippantly.

It is my own belief (and here I shall not feel bad if no one follows me) that every good and beautiful thing which man has produced in the world has been the result of his faulty and sin-blocked response to the creative voice sounding over the earth. The moral philosophers who dreamed their high dreams of virtue, the religious thinkers who speculated about God and immortality, the poets and artists who created out of common stuff pure and lasting beauty – how can we explain them? It is not enough to say simply, "It was genius." What then is genius? Could it be that a genius is a man haunted by the speaking voice, laboring and striving like one possessed to achieve ends which he only vaguely understands? That the great man may have missed God in his labors, that he may have even spoken or written against God, does not destroy the idea I am advancing. God's redemptive revelation in the Holy Scriptures is necessary to saving faith and peace with God. Faith in a risen Savior is necessary if the vague stirrings toward immortality are to bring us to restful and satisfying communion with God. To me, this is a plausible explanation of all that is best out of Christ. But you can be a good Christian and not accept my thesis.

The voice of God is a friendly voice. No one need fear to listen to it unless he has already made up his mind to resist it. The blood of Jesus has covered not only the human race but all creation as well. *Having made peace through the blood of His cross, whether things on earth or things in heaven.* We may safely preach a friendly heaven. The heavens as well as the earth are filled with the good will of Him that dwelt in the bush. The perfect blood of atonement secures this forever.

Religion has accepted the monstrous heresy that noise, size, activity, and bluster make a man dear to God.

Whoever will listen will hear the speaking heaven. This is definitely not the hour when men take kindly to an exhortation to *listen*, for listening is not today a part of popular religion. We are at the opposite end of the pole from there. Religion has accepted the monstrous heresy that noise, size, activity, and bluster make a man dear to God. But we may take heart. To a people caught in the tempest of the last great conflict, God says, *Be still, and know that I am God*, and still He says it, as if He means to tell us that our strength and safety lie not in noise but in silence.

It is important that we get still to wait on God. And it is best that we get alone, preferably with our Bible outspread before us. Then if we will, we may draw near to God and begin to hear Him speak to us in our hearts. I think that for the average person the progression will be something like this: First a sound as of a presence walking in the garden. Then a voice, more intelligible, but still far from clear. Then the happy moment when the Spirit begins to illuminate the Scriptures, and that which had been only a sound, or at best a voice, now becomes an intelligible word, warm and intimate and clear as the word of a dear friend. Then will come life and light, and best of all, the ability to see and rest in and embrace Jesus Christ as Savior and Lord and all.

The Bible will never be a living book to us until we are convinced that God is articulate in His universe. To jump from a dead, impersonal world to a dogmatic Bible is too much for most people. They may admit that they *should* accept the Bible as the Word of God, and they may try to think of it as such, but they find it impossible to believe that the words there on the page are actually for them. A man may *say*, "These words are addressed to me," and yet in his heart not feel and know that they are. He is the victim of a divided psychology. He tries to think of God as mute everywhere else and vocal only in a book.

I believe that much of our religious unbelief is due to a wrong conception of and a wrong feeling for the

Scriptures of truth. A silent God suddenly began to speak in a book, and when the book was finished, He lapsed back into silence again forever. Now we read the book as the record of what God said when He was for a brief time in a speaking mood. With notions like that in our heads, how can we believe? The facts are that God is not silent, has never been silent. It is the nature of God to speak. The second person of the Holy Trinity is called the *Word*. The Bible is the inevitable outcome of God's continuous speech. It is the infallible declaration of His mind for us, put into our familiar human words.

I think a new world will arise out of the religious mists when we approach our Bible with the idea that it is not only a book which was once spoken, but it is also a book which is *now speaking*. The prophets habitually said, *Thus saith the LORD*. They meant their hearers to understand that God's speaking is in the continuous present. We may use the past tense properly to indicate that at a certain time a certain word of God was spoken, but a word of God once spoken continues to be spoken, as a child once born continues to be alive, or a world once created continues to exist. And those are but imperfect illustrations, for children die and worlds burn out, but the Word of our God endureth forever.

If you would follow on to know the Lord, come at once to the open Bible, expecting it to speak to you. Do not come with the notion that it is a *thing* which you may push around at your convenience. It is more than a thing; it is a voice, a word, the very Word of the living God.

Lord, teach me to listen. The times are noisy and my ears are weary with the thousand raucous sounds which continuously assault them. Give me the spirit of the boy Samuel when he said to thee, "Speak, for thy servant heareth." Let me hear thee speaking in my heart. Let me get used to the sound of thy voice, that its tones may be familiar when the sounds of earth die away and the only sound will be the music of thy speaking voice. In Jesus' name, Amen.

Chapter 7

The Gaze of the Soul

Fixing our eyes on Jesus, the author and perfecter of faith. (Hebrews 12:2)

Let us think of our intelligent, plain man mentioned in chapter 6 coming for the first time to the reading of the Scriptures. He approaches the Bible without any previous knowledge of what it contains. He is wholly without prejudice; he has nothing to prove and nothing to defend.

Such a man will not have read long before his mind begins to observe certain truths standing out from the page. They are the spiritual principles behind the record of God's dealings with men, and woven into the writings of holy men as they were *inspired by the Holy Spirit*. As he reads on, he might want to number these truths as they become clear to him and make a brief summary under each number. These summaries will be the tenets of his biblical creed. Further reading will not affect these points except to enlarge and strengthen them. Our man is finding out what the Bible actually teaches.

High up on the list of things which the Bible teaches will be the doctrine of *faith*. The place of weighty importance which the Bible gives to faith will be too plain for him to miss. He will very likely conclude: Faith is of utmost importance. Without faith, it is impossible to please God. Faith will get me anything, take me anywhere in the kingdom of God; but without faith there can be no approach to God, no forgiveness, no deliverance, no salvation, no communion, no spiritual life at all.

By the time our friend has reached the eleventh chapter of Hebrews, the eloquent expression of high praise which is there pronounced upon faith will not seem strange to him. He will have read Paul's powerful defense of faith in his Roman and Galatian epistles. Later, if he goes on to study church history, he will understand the amazing power in the teachings of the Reformers as they showed the central place of faith in the Christian religion.

Faith is of utmost importance.

Now if faith is so vitally important, if it is an indispensable *must* in our pursuit of God, it is perfectly natural that we should be deeply concerned over whether or not we possess this most precious gift. And our minds being what they are, it is inevitable that sooner or later we should get around to inquiring after the nature of faith. The question, What *is* faith? would lie close to the question, Do I *have* faith? and would demand an answer if it were anywhere to be found.

Almost all who preach or write on the subject of faith have much the same things to say concerning it. They tell us that it is believing a promise, that it is taking God at His word, that it is reckoning the Bible to be true and stepping out upon it. The rest of the book or sermon is usually taken up with stories of people who have had their prayers answered as a result of their faith. These answers are mostly direct gifts of a practical and temporal nature such as health, money, physical protection, or success in business. Or if the teacher is of a philosophic turn of mind, he may take another course and lose us in a tumult of metaphysics or snow us under with psychological jargon as he defines and re-defines, paring the slender hair of faith thinner and thinner until it disappears in gossamer shavings at last. When he is finished, we get up disappointed and go out by the same door wherein we went. Surely there must be something better than this.

In the Scriptures there is practically no effort made to define faith. Outside of a brief fifteen-word definition in Hebrews 11:1, I know of no biblical definition, and even there, faith is defined functionally, not philosophically; that is, it is a statement of what faith is *in operation*, not what it is *in essence*. It assumes the presence of faith and shows what it results in, rather than what it is. We will be wise to go just that far and attempt to go no further. We are told from whence it comes and by what means: *Faith . . . is the gift of God*, and *Faith comes from hearing, and hearing by the word of Christ.* This much is clear, and, to paraphrase Thomas à Kempis, "I had rather exercise faith than know the definition thereof."

From here on, when the words "faith is" or their equivalent occur in this chapter, I ask that they be understood to refer to what faith is in operation as exercised by a believing man. Right here we drop the notion of definition and think about faith as it may be experienced in action. The complexion of our thoughts will be practical, not theoretical.

In a dramatic story in the book of Numbers, faith is seen in action. Israel became discouraged and spoke against God, and the Lord sent fiery serpents among them *and they bit the people, so that many people of Israel died.* Then Moses sought the Lord for them and He heard and gave them a remedy against the bite of the serpents. He commanded Moses to make a serpent of brass and put it upon a pole in sight of all the people, *it shall come about, that everyone who is bitten, when he looks at it, he will live.* Moses obeyed, *and it came about, that if a serpent bit any man, when he looked to the bronze serpent, he lived* (Numbers 21:4-9).

In the New Testament, this important bit of history is interpreted for us by no less an authority than our Lord Jesus Christ Himself. He is explaining to His hearers how they may be saved. He tells them that it is by believing. Then to make it clear, He refers to this incident in the book of Numbers. *As Moses lifted up the serpent in the wilderness, even so must the Son of Man be lifted up; so that whoever believes will in Him have eternal life* (John 3:14-15).

Our plain man in reading this would make an important discovery. He would notice that "look" and "believe" were synonymous terms. "Looking" on the Old Testament serpent is identical with "believing" on the New Testament Christ; that is, the *looking* and the *believing* are the same thing. And he would understand that while Israel looked with their external eyes, believing is done with the heart. I think he would conclude that *faith is the gaze of a soul upon a saving God.*

When he had seen this, he would remember passages he had read before, and their meaning would come flooding over him. *They looked to Him and were radiant, and their faces will never be ashamed* (Psalm 34:5). *To You I lift up my eyes, O You who are enthroned in the heavens! Behold, as the eyes of servants look to the hand of their master, as the eyes of a maid to the hand of her mistress, so our eyes look to the LORD our God, until He is gracious to us* (Psalm 123:1-2). Here the man seeking mercy looks straight at the God of mercy and never takes his eyes away from Him until mercy is granted. And our Lord Himself looked always at God. *Looking up toward heaven, He blessed the food, and breaking the loaves He gave them to the disciples* (Matthew 14:19). Indeed, Jesus taught that He wrought His works by always keeping His inward eyes upon His Father. His power lay in His continuous look at God (John 5:19-21).

In full accord with the few texts we have quoted is the whole tenor of the inspired Word. It is summed up for us in the Hebrew epistle when we are instructed to run life's race *fixing our eyes on Jesus, the author and perfecter of faith.* From all this we learn that faith is not a once-done act, but a continuous gaze of the heart at the triune God.

Believing, then, is directing the heart's attention to Jesus. It is lifting the mind to *behold the Lamb of God*, and never ceasing that beholding for the rest of our lives. At first, this may be difficult, but it becomes easier as we look steadily at His wondrous person, quietly and without strain. Distractions may hinder, but once the heart is committed to Him, after each brief excursion away from Him, the attention will return again and rest upon Him like a wandering bird coming back to its window.

I would emphasize this one committal, this one great volitional act which establishes the heart's intention to gaze forever upon Jesus. God takes this intention for our choice and makes what allowances He must for the thousand distractions which beset us in this evil world. He knows that we have set the direction of our hearts toward Jesus, and we can know it too, and comfort ourselves with the knowledge that a habit of soul is forming which will become after a while a sort of spiritual reflex requiring no more conscious effort on our part.

Faith is occupied with the object upon which it rests and pays no attention to itself at all.

Faith is the least self-regarding of the virtues. It is by its very nature scarcely conscious of its own existence. Like the eye which sees everything in front of it and never sees itself, faith is occupied with the object upon which it rests and pays no attention to itself at all. While we are looking at God, we do not see ourselves

– blessed riddance. The man who has struggled to purify himself and has had nothing but repeated failures will experience real relief when he stops tinkering with his soul and looks away to the perfect One. While he looks at Christ, the very things he has so long been trying to do will be getting done within him. It will be God working in him to will and to do.

Faith is not in itself a meritorious act; the merit is in the One toward whom it is directed. Faith is a redirecting of our sight, a getting out of the focus of our own vision and getting God into focus. Sin has twisted our vision inward and made it self-regarding. Unbelief has put self where God should be, and is perilously close to the sin of Lucifer who said, *I will raise my throne above the stars of God*. Faith looks *out* instead of *in* and the whole life falls into line.

All this may seem too simple. But we have no apology to make. To those who would seek to climb into heaven after help or descend into hell, God says, *The word is near you, . . . that is, the word of faith*. The word induces us to lift up our eyes unto the Lord and the blessed work of faith begins.

When we lift our inward eyes to gaze upon God, we are sure to meet friendly eyes gazing back at us, for it is written that the eyes of the Lord run to and fro throughout all the earth. The sweet language of experience is "Thou God seest me." When the eyes of the soul looking out meet the eyes of God looking in, heaven has begun right here on this earth.

"When all my endeavour is turned toward Thee because all Thy endeavour is turned toward me; when

I look unto Thee alone with all my attention, nor ever turn aside the eyes of my mind, because Thou dost enfold me with Thy constant regard; when I direct my love toward Thee alone because Thou, who art Love's self hast turned Thee toward me alone. And what, Lord, is my life, save that embrace wherein Thy delightsome sweetness doth so lovingly enfold me?" So wrote Nicholas of Cusa four hundred years ago.[2]

I should like to say more about this old man of God. He is not much known today anywhere among Christian believers, and among current Fundamentalists he is known not at all. I feel that we could gain much from a little acquaintance with men of his spiritual flavor and the school of Christian thought which they represent. Christian literature, to be accepted and approved by the evangelical leaders of our times, must follow very closely the same train of thought, a kind of "party line" from which it is scarcely safe to depart. A half-century of this in America has made us smug and content. We imitate each other with slavish devotion and our most strenuous efforts are put forth to try to say the same thing that everyone around us is saying – and yet to find an excuse for saying it, some little safe variation on the approved theme or, if no more, at least a new illustration.

Nicholas was a true follower of Christ, a lover of the Lord, radiant and shining in his devotion to the person of Jesus. His theology was orthodox, but fragrant

2 Nicholas of Cusa, *The Vision of God*, E. P. Dutton & Co., Inc., New York, 1928. This and the following quotations used by kind permission of the publishers.

and sweet as everything about Jesus might properly be expected to be. His conception of eternal life, for instance, is beautiful in itself, and, if I mistake not, is nearer in spirit to John 17:3 than that which is current among us today. Life eternal, says Nicholas, is "nought other than that blessed regard wherewith Thou never ceasest to behold me, yea, even the secret places of my soul. With Thee, to behold is to give life; 'tis unceasingly to impart sweetest love of Thee; 'tis to inflame me to love of Thee by love's imparting, and to feed me by inflaming, and by feeding to kindle my yearning, and by kindling to make me drink of the dew of gladness, and by drinking to infuse in me a fountain of life, and by infusing to make it increase and endure."

Now, if faith is the gaze of the heart at God, and if this gaze is but the raising of the inward eyes to meet the all-seeing eyes of God, then it follows that it is one of the easiest things possible to do. It would be like God to make the most vital thing easy and place it within the range of possibility for the weakest and poorest of us.

Several conclusions may fairly be drawn from all this. The simplicity of it, for instance, is one. Since believing is looking, it can be done without special equipment or religious paraphernalia. God has seen to it that the one life-and-death essential can never be subject to the caprice of accident. Equipment can break down or get lost, water can leak away, records can be destroyed by fire, the minister can be delayed, or the church burn down. All these are external to the soul and are subject to accident or mechanical failure; but *looking* is of the heart and can be done successfully by any man standing up or kneeling down or lying in his last agony a thousand miles from any church.

Since believing is looking, it can be done *any time.* No season is superior to another season for this sweetest of all acts. God never made salvation dependent upon new moons or holy days or Sabbaths. A man is not nearer to Christ on Easter Sunday than he is, say, on Saturday, August 3, or Monday, October 4. As long as Christ sits on the mediatorial throne, every day is a good day and all days are days of salvation.

Neither does *place* matter in this blessed work of believing God. Lift your heart and let it rest upon Jesus and you are instantly in a sanctuary, though it be a bed or a factory or a kitchen. You can see God from anywhere if your mind is set to love and obey Him.

Now, someone may ask, "Is not this of which you speak for special people such as monks or ministers who have by the nature of their calling more time to devote to quiet meditation? I am a busy worker and have little time to spend alone." I am happy to say that the life I describe is for every one of God's children, regardless of calling. It is, in fact, happily practiced every day by many hardworking people and is beyond the reach of none.

Many have found the secret of which I speak and, without giving much thought to what is going on within them, constantly practice this habit of inwardly gazing upon God. They know that something inside their hearts sees God. Even when they are compelled to withdraw their conscious attention in order to engage in earthly affairs, there is within them a secret communion always going on. Let their attention but be released for a moment from necessary business and it flies at once to God again. This has been the testimony of many Christians, so many that even as I state it thus, I have a feeling that I am quoting, though from whom or from how many I cannot possibly know.

I do not want to leave the impression that the ordinary means of grace have no value. They most assuredly have. Private prayer should be practiced by every Christian. Long periods of Bible meditation will purify our gaze and direct it. Church attendance will enlarge our outlook and increase our love for others. Service and work and activity – all are good and should be engaged in by every Christian. But the foundation of all these things, giving meaning to them, will be the inward habit of beholding God. A new set of eyes (so to speak) will develop within us, enabling us to be looking at God while our outward eyes are seeing the scenes of this passing world.

The foundation of all these things, giving meaning to them, will be the inward habit of beholding God.

Someone may fear that we are magnifying private religion out of all proportion, that the "us" of the New Testament is being displaced by a selfish "I." Has it ever occurred to you that one hundred pianos all tuned to the same fork are automatically tuned to each other? They are of one accord by being tuned, not to each other, but to another standard to which each one must individually bow. So one hundred worshippers meeting together, each one looking away to Christ, are in heart nearer to each other than they could possibly be were they to become "unity- conscious" and turn their eyes away from God to strive for closer fellowship. Social religion is perfected when private religion is purified. The body becomes stronger as its members become healthier. The whole church of God gains when the members that compose it begin to seek a better and a higher life.

All of the foregoing presupposes true repentance and a full commitment of the life to God. It is hardly necessary to mention this, for only individuals who have made such a commitment will have read this far.

When the habit of inwardly gazing Godward becomes fixed within us, we shall be ushered onto a new level of spiritual life more in keeping with the promises of God and the mood of the New Testament. The triune God will be our dwelling place even while our feet walk the low road of simple duty here among men. We will have found life's *summum bonum* (the highest good) indeed. "There is the source of all delights that can be desired; not only can nought better be thought out by men and angels, but nought better can exist in mode of being! For it is the absolute maximum of every rational desire, than which a greater cannot be."

O Lord, I have heard a good word inviting me to look away to thee and be satisfied. My heart longs to respond, but sin has clouded my vision until I see thee but dimly. Be pleased to cleanse me in thine own precious blood, and make me inwardly pure, so that

I may with unveiled eyes gaze upon thee all the days of my earthly pilgrimage. Then shall I be prepared to behold thee in full splendor in the day when thou shalt appear to be glorified in thy saints and admired in all them that believe. In Jesus' name, Amen.

CHAPTER 8

Restoring the Creator-Creature Relation

Be exalted above the heavens, O God; let Your glory be above all the earth. (Psalm 57:5)

It is a truism to say that order in nature depends upon right relationships; to achieve harmony each thing must be in its proper position relative to each other thing. In human life it is not otherwise.

I have hinted before in these chapters that the cause of all our human miseries is a radical moral dislocation, an upset in our relationship to God and to each other. For whatever else the fall may have been, it was most certainly a sharp change in man's relationship to his Creator. He adopted toward God an altered attitude, and by so doing destroyed the proper Creator-creature relationship in which, unknown to him, his true happiness lay. Essentially salvation is the restoration of a right relationship between man and his Creator, a bringing back to normal of the Creator-creature relationship.

A satisfactory spiritual life will begin with a complete change in relationship between God and the sinner, not a judicial change merely, but a conscious and experienced change affecting the sinner's whole nature. The atonement in Jesus' blood makes such a change judicially possible, and the working of the Holy Spirit makes it emotionally satisfying. The story of the prodigal son perfectly illustrates this latter phase. He had brought a world of trouble upon himself by forsaking the position which he had properly held as son of his father. In reality, his restoration was nothing more than a re-establishing of the father-son relationship which had existed from his birth and had been altered temporarily by his act of sinful rebellion. This story overlooks the legal aspects of redemption, but it makes beautifully clear the experiential aspects of salvation.

In determining relationships, we must begin somewhere. There must be somewhere a fixed center against which everything else is measured, where the law of relativity does not enter and we can say "is" and make no allowances. Such a center is God. When God would make His name known to mankind, He could find no better words than "I AM." When He speaks in the first person He says, "I AM"; when we speak of Him we say, "He is"; when we speak to Him we say, "Thou art." Everyone and everything else measures from that fixed point. *I AM THAT I AM*, says God. *I, the LORD, do not change.*

As the sailor locates his position on the sea by "shooting" the sun, so we may get our moral bearings by looking at God. We must begin with God. We are right when and only when we stand in a right position relative to God, and we are wrong so far and so long as we stand in any other position.

Much of our difficulty as seeking Christians stems from our unwillingness to take God as He is and adjust our lives accordingly. We insist upon trying to modify Him and to bring Him nearer to our own image. The flesh whimpers against the rigor of God's inexorable sentence and begs like Agag for a little mercy, a little indulgence of its carnal ways. It is no use. We can get a right start only by accepting God as He is and learning to love Him for what He is. As we go on to know Him better, we shall find it a source of unspeakable joy that God is just what He is. Some of the most rapturous moments we know will be those we spend in reverent admiration of the Godhead. In those holy moments, the very thought of change in Him will be too painful to endure.

We can get a right start only by accepting God as He is and learning to love Him for what He is.

So let us begin with God. Behind all, above all, before all is God. He is first in sequential order, above in rank and station, exalted in dignity and honor. As the self-existent One, He gave being to all things, and all things exist out of Him and for Him. *Worthy are You, our Lord and our God, to receive glory and honor and power; for You created all things, and because of Your will they existed, and were created.*

Every man, woman, and child belongs to God and exists by His pleasure. God being who and what He is, and we being who and what we are, the only thinkable relationship between us is one of full lordship on His part and complete submission on ours. We owe Him every honor that it is in our power to give Him. Our everlasting grief lies in giving Him anything less.

The pursuit of God will embrace the labor of bringing our total personality into conformity to His, and this not judicially, but actually. I do not here refer to the act of justification by faith in Christ. I speak of a voluntary exalting of God to His proper station over us and a willing surrender of our whole being to the place of worshipful submission, which the Creator-creature circumstance makes proper.

The moment we make up our minds that we are going on with this determination to exalt God over all, we step out of the world's parade. We shall find ourselves out of adjustment to the ways of the world, and increasingly so, as we make progress in the holy way. We shall acquire a new viewpoint; a new and different psychology will be formed within us; a new power will begin to surprise us by its upsurgings and its outgoings.

Our break with the world will be the direct outcome of our changed relationship to God. For the world of fallen men does not honor God. Millions call themselves by His name, it is true, and pay some token respect to Him, but a simple test will show how little He is really honored among them. Let the average man be put to the proof on the question of who is *above*, and his true position will be exposed. Let him be forced into making a choice between God and money, between God and men, between God and personal ambition, God and self, God and human love, and God will take second place every time. Those other things will be exalted above. However the man may protest, the proof is in the choices he makes day after day throughout his life.

Be exalted is the language of victorious spiritual experience. It is a little key to unlock the door to great treasures of grace. It is central in the life of a godly man. Let the seeking man reach a place where life and lips join to say continually, *Be exalted*, and a thousand minor problems will be solved at once. His Christian life ceases to be the complicated thing it had been before and becomes the very essence of simplicity. By the exercise of his will, he has set his course, and on that course he will stay as if guided by an automatic pilot. If blown off course for a moment by some adverse wind, he will surely return again as by a secret bent of the soul. The hidden motions of the Spirit are working in his favor, and "the stars in their courses" fight for him. He has met his life problem at its center, and everything else must follow along.

Let no one imagine that he will lose anything of human dignity by this voluntary sellout of his all to his God. He does not by this degrade himself as a man; rather, he finds his right place of high honor as one made in the image of his Creator. His deep disgrace lay in his moral derangement, his unnatural usurpation of the place of God. His honor will be proved by restoring again that stolen throne. In exalting God over all, he finds his own highest honor upheld.

Anyone who might feel reluctant to surrender his will to the will of another should remember Jesus' words: *Everyone who commits sin is the slave of sin.*

We must of necessity be a servant to someone, either to God or to sin. The sinner prides himself on his independence, completely overlooking the fact that he is the weak slave of the sins that rule his members. The man who surrenders to Christ exchanges a cruel slave driver for a kind and gentle Master whose yoke is easy and whose burden is light.

Made as we were in the image of God, we scarcely find it strange to take again our God as our all. God was our original habitat, and our hearts cannot but feel at home when they enter again that ancient and beautiful abode.

I hope it is clear that there is logic behind God's claim to pre-eminence. That place is His by every right in earth or heaven. When we take to ourselves the place that is His, the whole course of our lives gets out of joint. Nothing will or can restore order until our hearts make the great decision: God shall be exalted above.

Those who honor Me I will honor, said God once to a priest of Israel, and that ancient law of the kingdom stands today unchanged by the passing of time or the changes of dispensation. The whole Bible and every page of history proclaim the perpetuation of that law. *If anyone serves Me, the Father will honor him*, said our Lord Jesus, tying in the old with the new and revealing the essential unity of His ways with men.

Sometimes the best way to see a thing is to look at its opposite. Eli and his sons are placed in the priesthood with the stipulation that they honor God in their lives and ministrations. This they fail to do, and God sends Samuel to announce the consequences. Unknown to

Eli, this law of reciprocal honor has been all the while secretly working, and now the time has come for judgment to fall. Hophni and Phinehas, the degenerate priests, fall in battle; the wife of Phinehas dies in childbirth; Israel flees before her enemies; the ark of God is captured by the Philistines; and the old man Eli falls backward and dies of a broken neck. Thus, stark and utter tragedy followed upon Eli's failure to honor God.

Now set over against this almost any Bible character who honestly tried to glorify God in his earthly walk. See how God winked at weaknesses and overlooked failures as He poured upon His servants grace and blessing untold. Let it be Abraham, Jacob, David, Daniel, Elijah, or whom you will; honor followed honor as harvest follows the seed. The man of God set his heart to exalt God above all; God accepted his intention as fact and acted accordingly. Not perfection, but holy intention made the difference.

Not perfection, but holy intention made the difference.

In our Lord Jesus Christ this law was seen in simple perfection. In His lowly manhood He humbled Himself and gladly gave all glory to His Father in heaven. He sought not His own honor, but the honor of God who sent Him. *If I glorify myself,* He said on one occasion, *my glory is nothing; it is my Father who glorifies me.* So far had the proud Pharisees departed from this law that they could not understand one who honored God at his own expense. *I honor my Father,* said Jesus to them, *and you dishonor Me.*

Another saying of Jesus, and a most disturbing

one, was put in the form of a question: *How can you believe, when you receive glory from one another and you do not seek the glory that is from the one and only God?* If I understand this correctly, Christ taught here the alarming doctrine that the desire for honor among men made belief impossible. Is this sin at the root of religious unbelief? Could it be that those "intellectual difficulties" which men blame for their inability to believe are but smoke screens to conceal the real cause that lies behind them? Was it this greedy desire for honor from man that made men into Pharisees and Pharisees into those who killed God? Is this the secret behind religious self-righteousness and empty worship? I believe it may be. The whole course of the life is upset by failure to put God where He belongs. We exalt ourselves instead of God, and the curse follows.

In our desire after God, let us keep always in mind that God also has desire, and His desire is toward the sons of men, and more particularly toward those sons of men who will make the once-for-all decision to exalt Him over all. Such as these are precious to God above all treasures of earth or sea. In them, God finds a theater where He can display His exceeding kindness toward us in Christ Jesus. With them God can walk unhindered, and toward them He can act like the God He is.

In speaking thus, I have one fear: It is that I may convince the mind, before God can win the heart. For this God-above-all position is one not easy to take. The mind may approve it while not having the consent of the will to put it into effect. While the imagination races ahead to honor God, the will may lag behind and the man may never guess how divided his heart is. The whole man must make the decision before the heart can know any real satisfaction. God wants us all, and He will not rest until He gets us all. No part of the man will do.

Let us pray over this in detail, throwing ourselves at God's feet and meaning everything we say. No one who prays thus in sincerity need wait long for tokens of divine acceptance. God will unveil His glory before His servant's eyes, and He will place all His treasures at the disposal of such a one, for He knows that His honor is safe in such consecrated hands.

> *O God, be thou exalted over my possessions. Nothing of earth's treasures shall seem dear unto me if only thou art glorified in my life. Be thou exalted over my friendships. I am determined that thou shalt be above all, though I must stand deserted and alone in the midst of the earth. Be thou exalted above my comforts. Though it mean the loss of bodily comforts and the carrying of heavy crosses, I shall keep my vow made this day before thee. Be thou exalted over my reputation. Make me ambitious to please thee even if as a result I must sink into obscurity and my name be forgotten as a dream. Rise, O Lord, into thy proper place of honor, above my ambitions, above my likes and dislikes, above my family, my health, and even my life itself. Let me decrease that thou*

mayest increase; let me sink that thou mayest rise above. Ride forth upon me as thou didst ride into Jerusalem mounted upon the humble little beast, a colt, the foal of an ass, and let me hear the children cry to thee, "Hosanna in the highest." In Jesus' name, Amen.

Chapter 9

Meekness and Rest

Blessed are the gentle, for they shall inherit the earth. (Matthew 5:5)

A fairly accurate description of the human race might be furnished to one unacquainted with it by taking the Beatitudes, turning them wrong side out and saying, "Here is your human race." For the exact opposite of the virtues in the Beatitudes are the very qualities which distinguish human life and conduct.

In the world of men, we find nothing approaching the virtues of which Jesus spoke in the opening words of the famous Sermon on the Mount. Instead of poverty of spirit, we find the rankest kind of pride. Instead of mourners, we find pleasure seekers. Instead of meekness, we find arrogance. Instead of hungering after righteousness, we hear men saying, *I am rich and increased with goods and have need of nothing.* Instead of mercy, we find cruelty. Instead of purity of heart, we find corrupt imaginings. Instead of peacemakers, we find men quarrelsome and resentful. Instead of rejoicing in mistreatment, we find them fighting back with every weapon at their command.

Of this kind of moral stuff civilized society is composed. The atmosphere is charged with it; we breathe it with every breath and drink it with our mother's milk. Culture and education refine these things slightly but leave them basically untouched. A whole world of literature has been created to justify this kind of life as the only normal one. And this is the more to be wondered at, seeing that these are the evils which make life the bitter struggle it is for all of us. All our heartaches and a great many of our physical ills spring directly out of our sins. Pride, arrogance, resentfulness, evil imaginings, malice, and greed are the sources of more human pain than all the diseases that ever afflicted mortal flesh.

Into a world like this, the sound of Jesus' words comes wonderful and strange, a visitation from above. It is well that He spoke, for no one else could have done it as well; and it is good that we listen. His words are the essence of truth. He is not offering an opinion; Jesus never uttered opinions. He never guessed; He knew, and He knows. His words are not as Solomon's were, the sum of sound wisdom or the results of keen observation. He spoke out of the fullness of His Godhead, and His words are very truth itself. He is the only one who could say "blessed" with complete authority, for He is the blessed One come from the world above to confer blessedness upon mankind. And His words were supported by deeds mightier than any performed on this earth by any other man. It is wisdom for us to listen.

As was often so with Jesus, He used this word "gentle" in a brief, crisp sentence, and not until some time later did He go on to explain it. In the same book of Matthew He tells us more about it and applies it to our lives. *Come to Me, all who are weary and heavy-laden, and I will give you rest. Take My yoke upon you and learn from Me, for I am gentle and humble in heart, and you will find rest for your souls. For My yoke is easy and My burden is light.* Here we have two things standing in contrast to each other: a burden and a rest. The burden is not a local one, peculiar to those first hearers, but one which is borne by the whole human race. It consists not of political oppression or poverty or hard work. It is far deeper than that. It is felt by the rich as well as the poor, for it is something from which wealth and idleness can never deliver us.

Rest is simply release from that burden.

The burden borne by mankind is a heavy and a crushing thing. The word Jesus used means a load carried or toil borne to the point of exhaustion. Rest is simply release from that burden. It is not something we do; it is what comes to us when we cease to do. His own meekness, that is the rest.

Let us examine our burden. It is altogether an interior one. It attacks the heart and the mind and reaches the body only from within. First, there is the burden of *pride*. The labor of self-love is a heavy one indeed. Think for yourself whether much of your sorrow has not arisen from someone speaking slightingly of you. As long as you set yourself up as a little god to which you must be loyal, there will be those who will delight to offer an affront to your idol. How then can you hope to have inward peace? The heart's fierce effort to protect itself from every slight, to shield its touchy honor from the bad opinion of friend and enemy, will never let the mind have rest. Continue this fight through the years and the burden will become intolerable. Yet the sons of earth are carrying this burden continually, challenging every word spoken against them, cringing under every criticism, smarting under each fancied slight, tossing sleepless if another is preferred before them.

Such a burden as this is not necessary to bear. Jesus calls us to His rest, and gentleness is His method. The meek man cares not at all who is greater than he, for he has long ago decided that the esteem of the world is not worth the effort. He develops toward himself a kindly sense of humor and learns to say, "Oh, so you have been overlooked? They have placed someone else before you? They have whispered that you are pretty small stuff after all? And now you feel hurt because the world is saying about you the very things you have been saying about yourself? Only yesterday you were telling God that you were nothing, a mere worm of the dust. Where is your consistency? Come on, humble yourself, and cease to care what men think."

The gentle man is not a human mouse afflicted with a sense of his own inferiority. Rather, he may be in his moral life as bold as a lion and as strong as Samson; but he has stopped being fooled about himself. He has accepted God's estimate of his own life. He knows he is as weak and helpless as God has declared him to be, but paradoxically, he knows at the same time that he is in the sight of God of more importance than angels. In himself, nothing; in God, everything. That is his motto. He knows well that the world will never see him as God sees him and he has stopped caring. He rests perfectly content to allow God to set His own values. He will be patient to wait for the day when everything will get its own price tag and real worth will come into its own. Then the righteous shall shine forth in the kingdom of their Father. He is willing to wait for that day.

In the meantime, he will have attained a place of soul rest. As he walks on in gentleness, he will be happy to let God defend him. The old struggle to defend himself is over. He has found the peace which meekness brings.

Then also he will get deliverance from the burden of *pretense*. By this I mean not hypocrisy, but the common human desire to put the best foot forward and hide from the world our real inward poverty. For sin has played many evil tricks upon us, and one has been the infusing into us a false sense of shame. There is hardly a man or woman who dares to be just what he or she is without doctoring up the impression. The fear of being found out gnaws like rodents within their hearts. The man of culture is haunted by the fear that he will someday come upon a man more cultured than himself. The learned man fears to meet a man more learned than he. The rich man sweats under the fear that his clothes or his car or his house will sometime be made to look cheap by comparison with those of another rich man. So-called "society" runs by a motivation not higher than this, and the poorer classes on their level are little better.

Let no one smile this off. These burdens are real, and little by little they kill the victims of this evil and unnatural way of life. And the psychology created by years of this kind of thing makes true gentleness seem as unreal as a dream, as aloof as a star. To all the victims of the gnawing disease, Jesus says, *become like children.* For little children do not compare; they receive direct enjoyment from what they have without relating it to something else or someone else. Only as they get older and sin begins to stir within their hearts do jealousy and envy appear. Then they are unable to enjoy what they have if someone else has something larger or better. At that early age does the galling burden come down upon their tender souls, and it never leaves them until Jesus sets them free.

Another source of burden is *artificiality.* I am sure that most people live in secret fear that someday they will be careless, and by chance, an enemy or friend will be allowed to peep into their poor, empty souls. So they are never relaxed. Bright people are tense and alert in fear that they may be trapped into saying something common or stupid. Traveled people are afraid that they may meet some Marco Polo who is able to describe some remote place where they have never been.

This unnatural condition is part of our sad heritage of sin, but in our day it is aggravated by our whole way of life. Advertising is largely based upon this habit of pretense. "Courses" are offered in this or that field of human learning, frankly appealing to the victim's desire to shine at a party. Books are sold, clothes and cosmetics are peddled, by playing continually upon this desire to appear what we are not. Artificiality is one curse that will drop away the moment we kneel at Jesus' feet and surrender ourselves to His meekness. Then we will not care what people think of us so long as God is pleased. Then *what we are* will be everything; what we appear will take its place far down the scale of interest for us. Apart from sin, we have nothing of which to be ashamed. Only an evil desire to shine makes us want to appear other than we are.

The heart of the world is breaking under this load of pride and pretense.

The heart of the world is breaking under this load of pride and pretense. There is no release from our burden apart from the meekness of Christ. Good, keen reasoning may help slightly, but so strong is this vice that if we push it down one place it will come up somewhere else. To men and women everywhere, Jesus says, *Come to Me, . . . and I will give you rest.* The rest He offers is the rest of meekness, the blessed relief which comes when we accept ourselves for what we are and cease to pretend. It will take some courage at first, but the needed grace will come as we learn that we are sharing this new and easy yoke with the strong Son of God Himself. He calls it *my yoke*, and He walks at one end while we walk at the other.

Lord, make me childlike. Deliver me from the urge to compete with another for place or prestige or position. I would be simple and artless as a little child. Deliver me from pose and pretense. Forgive me for thinking of myself. Help me to forget myself and find my true peace in beholding thee. That thou mayest answer this prayer I humble myself before thee. Lay upon me thy easy yoke of self-forgetfulness that through it I may find rest. In Jesus' name, Amen.

CHAPTER 10

The Sacrament of Living

Whether, then, you eat or drink or whatever you do, do all to the glory of God.
(1 Corinthians 10:31)

One of the greatest hindrances to internal peace which the Christian encounters is the common habit of dividing our lives into two areas, the sacred and the secular. As these areas are conceived to exist apart from each other and to be morally and spiritually incompatible, and as we are compelled by the necessities of living to be always crossing back and forth from the one to the other, our inner lives tend to break up so that we live a divided instead of a unified life.

Our trouble springs from the fact that we who follow Christ inhabit at once two worlds, the spiritual and the natural. As children of Adam, we live our lives on earth subject to the limitations of the flesh and the weaknesses and ills to which human nature is heir. Merely to live among men requires of us years of hard toil and much care and attention to the things of this world. In sharp contrast to this is our life in the Spirit. There we enjoy another and higher kind of life; we are children of God; we possess heavenly status and enjoy intimate fellowship with Christ.

This tends to divide our total life into two departments. We come unconsciously to recognize two sets of actions. The first are performed with a feeling of satisfaction and a firm assurance that they are pleasing to God. These are the sacred acts and they are usually thought to be prayer, Bible reading, hymn singing, church attendance, and such other acts as spring directly from faith. They may be known by the fact that they have no direct relation to this world, and would have no meaning whatsoever except as faith shows us another world, *a house not made with hands, eternal in the heavens.*

Over against these sacred acts are the secular ones. They include all of the ordinary activities of life, which we share with the sons and daughters of Adam: eating, sleeping, working, looking after the needs of the body, and performing our dull and tedious duties here on earth. These we often do reluctantly and with many misgivings, often apologizing to God for what we consider a waste of time and strength. The upshot of this is that we are uneasy most of the time. We go about our common tasks with a feeling of deep frustration, telling ourselves pensively that there's a better day coming when we shall shed this earthly shell and be bothered no more with the affairs of this world.

This is the old sacred-secular antithesis. Most Christians are caught in its trap. They cannot get a satisfactory adjustment between the claims of the two worlds. They try to walk the tightrope between two kingdoms and they find no peace in either. Their strength is reduced, their outlook confused, and their joy taken from them.

I believe this state of affairs to be wholly unnecessary. We have gotten ourselves on the horns of a dilemma, true enough, but the dilemma is not real. It is a creature of misunderstanding. The sacred-secular antithesis has no foundation in the New Testament. Without doubt, a more-perfect understanding of Christian truth will deliver us from it.

The Lord Jesus Christ Himself is our perfect example, and He knew no divided life. In the presence of His Father He lived on earth without strain from babyhood to His death on the cross. God accepted the offering of His total life, and made no distinction between act and act. *I always do those things that please him* was His brief summary of His own life as it related to the Father. As He moved among men, He was poised and restful. What pressure and suffering He endured grew out of His position as the world's sin bearer; they were never the result of moral uncertainty or spiritual maladjustment.

The Lord Jesus Christ Himself is our perfect example, and He knew no divided life.

Paul's exhortation to *do everything for the glory of God* is more than pious idealism. It is an integral part of the sacred revelation and is to be accepted as the very Word of Truth. It opens before us the possibility of making every act of our lives contribute to the glory of God. Lest we should be too timid to include everything, Paul mentions specifically eating and drinking. This humble privilege we share with the beasts that perish. If these lowly animal acts can be so performed as to honor God, then it becomes difficult to conceive of one that cannot.

That monkish hatred of the body, which figures so prominently in the works of certain early devotional writers, is wholly without support in the Word of God. Common modesty is found in the sacred Scriptures, it is true, but never prudery or a false sense of shame. The New Testament accepts as a matter of course that in His incarnation our Lord took upon Himself a real human body, and no effort is made to steer around the downright implications of such a fact. He lived in that body here among men and never once performed a non-sacred act. His presence in human flesh sweeps away forever the evil notion that there is about the human body something innately offensive to the Deity. God created our bodies, and we do not offend Him by placing the responsibility where it belongs. He is not ashamed of the work of His own hands.

Perversion, misuse, and abuse of our human powers should give us cause enough to be ashamed. Bodily acts done in sin and contrary to nature can never honor God. Wherever the human will introduces moral evil, we have no longer our innocent and harmless powers as God made them; we have instead an abused and twisted thing which can never bring glory to its Creator.

Let us, however, assume that perversion and abuse are not present. Let us think of a Christian believer in whose life the twin wonders of repentance and the new birth have been wrought. He is now living according to the will of God, as he understands it from the written Word. Of such a one it may be said that every act of his life is or can be as truly sacred as prayer or baptism or the Lord's Supper. To say this is not to bring all acts down to one dead level; it is rather to lift every act up into a living kingdom and turn the whole life into a sacrament.

If a sacrament is an external expression of an inward grace, then we need not hesitate to accept the above thesis. By one act of consecration of our total selves to God, we can make every subsequent act express that consecration. We need no more be ashamed of our body – the fleshly servant that carries us through life – than Jesus was of the humble beast upon which He rode into Jerusalem. *The Lord has need of him* may well apply to our mortal bodies. If Christ dwells in us, we may carry about the Lord of glory as the little beast did of old and give occasion to the multitudes to cry, *Hosanna in the highest.*

That we *see* this truth is not enough. If we would escape from the toils of the sacred-secular dilemma, the truth must "run in our blood" and condition the complexion of our thoughts. We must practice living to the glory of God, actually and determinedly. By meditation upon this truth, by talking it over with God often in our prayers, by recalling it to our minds frequently as we move about among men, a *sense* of its wondrous meaning will begin to take hold of us. The old painful duality will go down before a restful unity of life. The knowledge that we are all God's, that He has received all and rejected nothing, will unify our inner lives and make everything sacred to us.

This is not quite all. Long-held habits do not die easily. It will take intelligent thought and a great deal of reverent prayer to escape completely from the sacred-secular psychology. For instance, it may be difficult for the average Christian to get hold of the idea that his daily labors can be performed as acts of worship acceptable to God by Jesus Christ. The old antithesis will crop up in the back of his head sometimes to disturb his peace of mind. Nor will that old Serpent the Devil take all this lying down. He will be there in the cab or at the desk or in the field to remind the Christian that he is giving the better part of his day to the things of this world and allotting to his religious duties only a trifling portion of his time. And unless great care is taken, this will create confusion and bring discouragement and heaviness of heart.

It will take intelligent thought and a great deal of reverent prayer to escape completely from the sacred-secular psychology.

We can meet this successfully only by the exercise of an aggressive faith. We must offer all our acts to God and believe that He accepts them. Then hold firmly to that position and keep insisting that every act of every hour of the day and night be included in the transaction. Keep reminding God in our times of private prayer that we mean every act for His glory; then supplement those times by a thousand thought-prayers as we go about the job of living. Let us practice the fine art of making every work a priestly ministration. Let us believe that God is in all our simple deeds and learn to find Him there.

Accompanying the error which we have been discussing is the sacred-secular antithesis as applied to places. It is little short of astonishing that we can read the New Testament and still believe in the inherent sacredness of places as distinguished from other places. This error is so widespread that one feels all alone when he tries to combat it. It has acted as a kind of dye to color the thinking of religious people and has colored the eyes as well, so that it is all but impossible to detect its fallacy. In the face of every New Testament teaching to the contrary, it has been said and sung throughout the centuries and accepted as a part of the Christian message, the which it most surely is not. Only the Quakers, so far as my knowledge goes, have had the perception to see the error and the courage to expose it.

Here are the facts as I see them. For four hundred years, Israel had dwelt in Egypt, surrounded by the crassest idolatry. By the hand of Moses they were brought out at last and started toward the Land of Promise. The very idea of holiness had been lost to them. To correct this, God began at the bottom. He localized Himself in the cloud and fire, and later when the tabernacle had been built, He dwelt in fiery manifestation in the Holy of Holies. By innumerable distinctions God taught Israel the difference between holy and unholy. There were holy days, holy vessels, holy garments. There were washings, sacrifices, offerings of many kinds. By these means, Israel learned that *God is holy.* It was this that

He was teaching them. Not the holiness of things or places, but the holiness of Jehovah was the lesson they must learn.

Then came the great day when Christ appeared. Immediately He began to say, *You have heard that the ancients were told . . . but I say to you.* The Old Testament schooling was over. When Christ died on the cross, the veil of the temple was rent from top to bottom. The Holy of Holies was opened to everyone who would enter in faith. Christ's words were remembered: *An hour is coming when neither in this mountain nor in Jerusalem will you worship the Father. . . . But an hour is coming, and now is, when the true worshipers will worship the Father in spirit and truth; for such people the Father seeks to be His worshipers. God is spirit, and those who worship Him must worship in spirit and truth.*

Shortly after, Paul took up the cry of liberty and declared all meats clean, every day holy, all places sacred, and every act acceptable to God. The sacredness of times and places, a half-light necessary to the education of the race, passed away before the full sun of spiritual worship.

The essential spirituality of worship remained the possession of the church until it was slowly lost with the passing of the years. Then the natural *legality* of the fallen hearts of men began to introduce the old distinctions. The church came to observe again days and seasons and times. Certain places were chosen and marked out as holy in a special sense. Differences were observed between one and another day or place or person. The "sacraments" were first two, then three, then four, until with the triumph of Romanism they were fixed at seven.

In all charity, and with no desire to reflect unkindly upon any Christian, however misled, I would point out that the Roman Catholic Church represents today the sacred-secular heresy carried to its logical conclusion. Its deadliest effect is the complete division it introduces between religion and life. Its teachers attempt to avoid this snare by many footnotes and multitudinous explanations, but the mind's instinct for logic is too strong. In practical living, the division is a fact.

From this bondage, reformers and puritans and mystics have labored to free us. Today, the trend in conservative circles is back toward that bondage again. It is said that a horse, after it has been led out of a burning building, will sometimes by a strange obstinacy break loose from its rescuer and dash back into the building again to perish in the flame. By some such stubborn tendency toward error, Fundamentalism in our day is moving back toward spiritual slavery. The observation of days and times is becoming more and more prominent among us. "Lent" and "Holy Week" and "Good" Friday are words heard more and more frequently upon the lips of gospel Christians. We do not know when we are well off.

In order that I may be understood and not be misunderstood, I would throw into relief the practical implications of the teaching for which I have been arguing, namely, the sacramental quality of everyday living. Over against its positive meanings, I should like to point out a few things it does not mean.

It does not mean, for instance, that everything we do is of equal importance with everything else we do or may do. One act of a good man's life may differ widely from another in importance. Paul's sewing of tents was not equal to his writing of an epistle to the Romans, but both were accepted of God and both were true acts of worship. Certainly it is more important to lead a soul to Christ than to plant a garden, but the planting of the garden *can* be as holy an act as the winning of a soul.

Again, it does not mean that every man is as useful as every other man. Gifts differ in the body of Christ. A Billy Bray is not to be compared with a Martin Luther or a John Wesley for sheer usefulness to the church and to the world; but the service of the less-gifted brother is as pure as that of the more-gifted, and God accepts both with equal pleasure.

The "layman" need never think of his humbler task as being inferior to that of his minister. Let every man abide in the calling wherein he is called, and his work will be as sacred as the work of the ministry. It is not what a man does that determines whether his work is sacred or secular; it is *why* he does it. The motive is everything. Let a man sanctify the Lord God in his heart and he can thereafter do no common act. All he does is good and acceptable to God through Jesus Christ. For such a man, living itself will be sacramental and the whole world a sanctuary. His entire life will be a priestly ministration. As he performs his never-so-simple task, he will hear the voice of the seraphim saying, *Holy, holy, holy, is the LORD of the hosts; the whole earth is full of his glory.*

Lord, I would trust thee completely; I would be altogether thine; I would exalt thee above all. I desire that I may feel no sense of possessing anything outside of thee. I want constantly to be aware of thy overshadowing presence and to hear thy speaking voice. I long to live in restful sincerity of heart. I want to live so fully in the Spirit that all my thoughts may be as sweet incense ascending to thee and every act of my life may be an act of worship. Therefore, I pray in the words of thy great servant of old, "I beseech thee so for to cleanse the intent of mine heart with the unspeakable gift of thy grace, that I may perfectly love thee and worthily praise thee." And all this I confidently believe thou wilt grant me through the merits of Jesus Christ thy Son. In Jesus' name, Amen.

About the Author

Hailing from a tiny farming community in western La Jose, Pennsylvania, A.W. Tozer's conversion was as a teenager in Akron, Ohio. While on his way home from work at a tire company, he overheard a street preacher say "If you don't know how to be saved... just call on God, saying, 'Lord, be merciful to me a sinner.'" Upon returning home, he climbed into the attic and heeded the preacher's advice.

In his first editorial, dated June 3, 1950, he wrote, "It will cost something to walk slow in the parade of the ages, while excited men of time rush about, confusing motion with progress. But it will pay in the long run and the true Christian is not much interested in anything short of that."

Among the more than 40 books that he authored, at least two are regarded as Christian classics: *The Pursuit of God* and *The Knowledge of the Holy.* His books impress on the reader the possibility and necessity for a deeper relationship with God.[3]

3 http://en.wikipedia.org/wiki/A._W._Tozer
Image: "A W Tozer" by Unknown. Licensed under Fair use via Wikipedia

Other Similar Titles